SOLUTIONS
TO ACCOMPANY
HEAT TRANSFER

Adrian Bejan

J.S. Jones Professor of Mechanical Engineering
Duke University

John Wiley & Sons, Inc.
New York Chichester Brisbane Toronto Singapore

ISBN 0-471-57878-9

Printed in the United States of America
Printed and bound by Malloy Lithographing, Inc.

10 9 8 7 6 5 4 3 2 1

CONTENTS

Chapter 1

INTRODUCTION

Problem 1.1. Treating the solid bar of Fig. 1.4 as a closed system in the steady state, we recognize that the second law (1.3) reduces to

$$\frac{q_0}{T_0} - \frac{q_L}{T_L} \leq 0$$

in which q_L is the heat current exiting through the x = L cross-section. Under the same conditions, the first law (1.2) states that

$$q_0 - q_L - 0 = 0$$

or that q_0 is conserved as it flows through the entire conductor, $q_0 = q_L$. Combining the above equations we obtain

$$q_0 (T_L - T_0) \leq 0$$

proving that when q_0 is positive the "exit" end temperature T_L cannot be greater than the "entrance" end temperature T_0. Said another way, the heat current must proceed toward lower temperatures, which is why in the Fourier law (1.20) the thermal conductivity k cannot be negative.

Problem 1.2. First, we distinguish between the heat fluxes listed in eqs. (1.42r) - (1.42z) and the heat currents (watts) associated with the same directions:

$$\begin{aligned} q_r &= q_r'' \, r\Delta\theta\Delta z = -k \frac{\partial T}{\partial r} r\Delta\theta\Delta z \\ q_\theta &= q_\theta'' \, \Delta r\Delta z = -\frac{k}{r} \frac{\partial T}{\partial \theta} \Delta r\Delta z \\ q_z &= q_z'' \, (\Delta r)(r\Delta\theta) = -k \frac{\partial T}{\partial z} r\Delta r\Delta\theta \end{aligned} \quad \text{(a)}$$

With reference to the infinitesimal chunk of volume $r\Delta r\Delta\theta\Delta z$ shown in Fig. 1.8, the first law states that

$$q_r - q_{r+\Delta r} + q_\theta - q_{\theta+\Delta\theta} + q_z - q_{z+\Delta z} +$$

$$+ \dot{q} r\Delta r\Delta\theta\Delta z = \rho c r\Delta r\Delta\theta\Delta z \frac{\partial T}{\partial t} \quad \text{(b)}$$

The first six terms on the left-hand side produce in order

$$q_r - q_{r+\Delta r} = q_r - (q_r + \frac{\partial q_r}{\partial r}\Delta r) = -\frac{\partial q_r}{\partial r}\Delta r$$

$$q_\theta - q_{\theta+\Delta\theta} = q_\theta - (q_\theta + \frac{\partial q_\theta}{\partial \theta}\Delta\theta) = -\frac{\partial q_\theta}{\partial \theta}\Delta\theta \qquad \text{(c)}$$

$$q_z - q_{z+\Delta z} = q_z - (q_z + \frac{\partial q_z}{\partial z}\Delta z) = -\frac{\partial q_z}{\partial z}\Delta z$$

Combining eqs. (a) and (c), substituting the resulting expressions into the energy conservation equation (b), and dividing the latter by $r\Delta r\Delta\theta\Delta z$ yields finally

$$\frac{1}{r}\frac{\partial}{\partial r}(kr\frac{\partial T}{\partial r}) + \frac{1}{r^2}\frac{\partial}{\partial \theta}(k\frac{\partial T}{\partial \theta}) + \frac{\partial}{\partial z}(k\frac{\partial T}{\partial z}) + \dot{q} = \rho c\frac{\partial T}{\partial t} \qquad (1.43)$$

Problem 1.3. In terms of heat currents (watts), the first-law statement for the infinitesimal system of Fig. 1.9 is

$$q_r - q_{r+\Delta r} + q_\phi - q_{\phi+\Delta\phi} + q_\theta - q_{\theta+\Delta\theta} +$$

$$+ \dot{q}(\Delta r)(r\Delta\phi)(r\sin\phi\,\Delta\theta) = \rho c(\Delta r)(r\Delta\phi)(r\sin\phi\,\Delta\theta)\frac{\partial T}{\partial t} \qquad \text{(a)}$$

The heat currents are related to the heat fluxes of eqs. (1.48r) - (1.48θ) via

$$q_r = q_r''(r\Delta\phi)(r\sin\phi\Delta\theta)$$

$$q_\phi = q_\phi''(\Delta r)(r\sin\phi\Delta\theta) \qquad \text{(b)}$$

$$q_\theta = q_\theta''(\Delta r)(r\Delta\phi)$$

The first six terms of eq. (a) become

$$q_r - q_{r+\Delta r} = -\frac{\partial q_r}{\partial r}\Delta r = \frac{\partial}{\partial r}(kr^2\frac{\partial T}{\partial r})\,\Delta r\sin\phi\,\Delta\phi\,\Delta\theta$$

$$q_\phi - q_{\phi+\Delta\phi} = -\frac{\partial q_\phi}{\partial \phi}\Delta\phi = \frac{\partial}{\partial \phi}(k\sin\phi\frac{\partial T}{\partial \phi})\,\Delta r\;\Delta\phi\,\Delta\theta \qquad \text{(c)}$$

$$q_\theta - q_{\theta+\Delta\theta} = -\frac{\partial q_\theta}{\partial \theta}\Delta\theta = \frac{\partial}{\partial \theta}(k\frac{\partial T}{\partial \theta})\,\Delta r\;\Delta\phi\,\Delta\theta\frac{1}{\sin\phi}$$

Substituting eqs. (c) into eq. (a), and dividing by the infinitesimal volume $r^2 \sin\phi \Delta r \Delta\phi \Delta\theta$ yields

$$\frac{1}{r^2}\frac{\partial}{\partial r}\left(k\, r^2 \frac{\partial T}{\partial r}\right) + \frac{1}{r^2 \sin\phi}\frac{\partial}{\partial \phi}\left(k \sin\phi \frac{\partial T}{\partial \phi}\right) + \frac{1}{r^2 \sin^2\phi}\frac{\partial}{\partial \theta}\left(k \frac{\partial T}{\partial \theta}\right) + \dot{q} = \rho c \frac{\partial T}{\partial t} \quad (1.49)$$

Problem 1.4. By drawing a tangent to the T(x) curve at the ground-level point (x = 0), we estimate that

$$\frac{dT}{dx} \cong 0.57 \frac{°C}{m}$$

The heat flux that passes through the earth's surface toward the atmosphere (i.e. in the negative x direction) is

$$q'' = k\left(\frac{dT}{dx}\right)_{x=0}$$

$$= 17.4 \frac{W}{m\,K}\, 0.57 \frac{°C}{m} = 9.92 \frac{W}{m^2}$$

The total heat transfer rate through the area $A = 1\ km^2$ is

$$q = q''A = 9.92 \frac{W}{m^2}\left(10^3 m\right)^2$$

$$= 9.92 \times 10^6 W$$

Problem 1.5. The heat flux q" that leaves the heat exchanger surface must pass through the precipitated layer, therefore

$$q'' = k\frac{\Delta T}{L}$$

In this equation, k, ΔT and L are the thermal conductivity of the accumulated solid, the temperature drop across this layer, and the layer thickness. The temperature drop is therefore

$$\Delta T = q''\frac{L}{k} = 0.1 \frac{W}{10^{-4} m^2} \frac{10^{-4} m}{0.6\ W/m \cdot K}$$

$$= 0.17\ °C$$

Problem 1.6. The coal layer generates the following heating rate per unit ground surface:

$$q'' = \dot{q} H = 50 \frac{W}{m^3} 2m = 100 \frac{W}{m^2}$$

A fraction of this amount (30 W/m^2) escapes into the ground, leaving the rest to be discharged into the atmosphere,

$$q''_{atm} = (100 - 30) \frac{W}{m^2} = 70 \frac{W}{m^2}$$

Recalling the definition of heat transfer coefficient,

$$q''_{atm} = h (T_w - T_\infty)$$

we can now calculate the temperature of the upper surface of the layer of coal,

$$T_w = T_\infty + \frac{q''_{atm}}{h} = 30°C + 70 \frac{W}{m^2} \frac{m^2 K}{15 W}$$

$$= 30°C + 4.7°C = 34.7°C$$

Problem 1.7. Proceeding upward across the ice layer, in the steady state the water-side convective heat flux $h_w(T_w - T_0)$ becomes equal to the conduction heat flux through the ice,

$$h_w (T_w - T_0) = k \frac{T_0 - T_s}{L} \qquad \text{(a)}$$

Similarly, the ice conduction heat flux must equal also the convective heat flux on the air side of the ice layer,

$$k \frac{T_0 - T_s}{L} = h_a(T_s - T_a) \qquad \text{(b)}$$

Eliminating $k(T_0 - T_s)/L$ between eqs. (a) and (b), we can first calculate the temperature of the upper surface of the ice layer,

$$T_s = T_a + \frac{h_w}{h_a}(T_w - T_0)$$

$$= -30°C + \frac{500\ W}{m^2K}\frac{m^2K}{100\ W}(4 - 0)°C$$

$$= -30°C + 20°C = -10°C$$

Substituting this result in eq. (a) we obtain the thickness of the ice layer,

$$L = \frac{k}{h_w}\frac{T_0 - T_s}{T_w - T_0} = \frac{2.25\ W}{m \cdot K}\frac{m^2K}{500\ W}\frac{0 - (-10)}{4 - 0}$$

$$= 0.011\ m = 1.1\ cm$$

Problem 1.8. a) The heat flux that leaves the skin is conducted across the sub-skin layer of thickness δ:

$$q'' = k\frac{T_a - T_w}{\delta} \cong 0.42\frac{W}{m{\cdot}K}\frac{(36.5 - 30)\ K}{0.01m}$$

$$\cong 273\frac{W}{m^2}$$

b) The same heat flux is carried away as convection,

$$q'' = h\left(T_w - T_\infty\right)$$

$$h = \frac{q''}{T_w - T_\infty} \cong 273\frac{W}{m^2}\frac{1}{(30 - 20)\ K}$$

$$\cong 27.3\frac{W}{m^2\ K}$$

Problem 1.9. The speed of the metal sheet is

$$V = \frac{30\,\mathrm{m}}{2 \times 3600\,\mathrm{s}} = 0.00417\,\frac{\mathrm{m}}{\mathrm{s}}$$

Next, we write $\delta = 0.01\,\mathrm{m}$ for the sheet thickness, and W for the width of the oven (and the sheet) in the direction normal to the plane of the figure, and calculate in order

$$q'W = \underbrace{\dot{m}c\left(T_{out} - T_{in}\right)}_{\substack{\text{enthalpy rise of}\\ \text{metal stream}}} + \underbrace{hA_{\text{oven wall}}\left(T_w - T_\infty\right)}_{\substack{\text{heat transfer from}\\ \text{wall to ambient}}}$$

$$q' = \rho c \delta V \left(T_{out} - T_{in}\right) + hp\left(T_w - T_\infty\right)$$

$$= 7817\,\frac{\mathrm{kg}}{\mathrm{m}^3}\,0.46\,\frac{10^3\,\mathrm{J}}{\mathrm{kgK}}\,0.01\,\mathrm{m}\;0.00417\,\frac{\mathrm{m}}{\mathrm{s}}\,(1000 - 25)\,\mathrm{K}$$

$$+\, 20\,\frac{\mathrm{W}}{\mathrm{m{\cdot}K}}\,\underbrace{(30 + 3 + 30 + 3)\,\mathrm{m}}_{\substack{\text{oven perimeter,}\\ \text{visible in the figure}}}\,(50 - 25)\,\mathrm{K}$$

$$= 1.462 \times 10^5\,\frac{\mathrm{W}}{\mathrm{m}} + 3.3 \times 10^4\,\frac{\mathrm{W}}{\mathrm{m}}$$

$$= 1.79 \times 10^5\,\frac{\mathrm{W}}{\mathrm{m}}$$

In conclusion, the heat loss through the oven wall amounts to about 18 percent of the electric power input to the oven.

Problem 1.10. The heat flux out of the oven surface is

$$q'' = h\left(T_w - T_\infty\right)$$
$$= 20\,\frac{W}{m^2K}\,(50 - 25)°C = 500\,\frac{W}{m^2}$$

The same heat flux must penetrate (by conduction) the brick wall,

$$q'' = k\,\frac{T_i - T_w}{t} \qquad \left(T_i = 900°C,\; T_w = 50°C\right)$$

and this means that the wall thickness must be

$$t = k\,\frac{T_i - T_w}{q''}$$
$$= 0.1\,\frac{W}{m\cdot K}\cdot\frac{(900 - 50)\,K}{500\ W/m^2} = 0.17\,m$$

Chapter 2

UNIDIRECTIONAL STEADY CONDUCTION

Problem 2.1. The relations between the cylindrical-shell notation and the thin-wall notation are

$$L = r_o - r_i$$

$$A = 2\pi r_i \ell$$

In the limit $r_i \rightarrow r_o$, or $L/r_i \rightarrow 0$, the logarithm appearing in eq. (2.33) becomes

$$\ell n \left(\frac{r_o}{r_i} \right) = \ell n \left(1 + \frac{L}{r_i} \right) \cong \frac{L}{r_i}$$

The thermal resistance of the cylindrical shell can therefore be rewritten as

$$R_t \cong \frac{L/r_i}{2\pi k \ell} = \frac{L}{(2\pi r_i \ell)k} = \frac{L}{Ak}$$

which is identical to the thin-wall resistance, eq. (2.7).

Problem 2.2. The starting point in the analysis is eq. (1.52), in which $T = T(r)$,

$$\frac{1}{r^2} \frac{d}{dr} \left(r^2 \frac{dT}{dr} \right) = 0$$

or, integrating twice,

$$T = \frac{c_1}{r} + c_2$$

The two constants are determined next from the two temperature boundary conditions,

$$T = T_i \quad \text{at} \quad r = r_i$$

$$T = T_o \quad \text{at} \quad r = r_o$$

and the final T(r) expression is

$$T = T_i - \frac{1 - r_i/r}{1 - r_i/r_o} (T_i - T_o)$$

The heat flux q" at any radial position r inside the shell follows from eq. (1.48r),

$$q'' = \frac{k\, r_i\, r_o}{r^2(r_o - r_i)} (T_i - T_o)$$

The total heat transfer rate q is the area integral of q", where the "area" is the spherical surface of radius r (i.e. $4\pi r^2$),

$$q = (4\pi r^2) q'' = 4\pi k \frac{r_i\, r_o}{r_o - r_i} (T_i - T_o)$$

In the thin-wall limit, the product $r_i r_o$ approaches r_i^2, therefore

$$\lim_{r_i \to r_o} q = \frac{k(4\pi r_i^2)}{r_o - r_i} (T_i - T_o)$$

Since $(4\pi r_i^2)$ is the surface of the now thin wall (A), and $(r_o - r_i)$ its thickness (L), we conclude that the above expression for q is identical to eq. (2.6), i.e. that the thermal resistance formula becomes the same as eq. (2.7).

Problem 2.3. The overall thermal resistance consists of the conduction-type resistance of the spherical shell, plus the outer convective resistance, eq. (2.41),

$$R_t = \frac{1}{4\pi k}\left(\frac{1}{r_i} - \frac{1}{r_o}\right) + \frac{1}{4\pi r_o^2 h}$$

To find the R_t extremum, we solve $dR_t/dr_o = 0$; the critical radius is

$$r_{o,c} = 2\frac{k}{h}$$

When $r_o = r_{o,c}$, the second derivative of R_t with respect to r_o is positive,

$$\frac{d^2 R_t}{dr_o^2} = \frac{1}{4\pi k\, r_{o,c}^3} > 0$$

therefore the corresponding value reached by R_t is a minimum.

Problem 2.4. The starting point is eq. (1.44),

$$\frac{1}{r}\frac{d}{dr}\left(r\frac{dT}{dr}\right) = -\frac{\dot{q}}{k}$$

which must be subjected to two boundary conditions,

$$\frac{dT}{dr} = 0 \quad \text{at} \quad r = 0, \text{ (radial symmetry)}$$

$$-k\frac{dT}{dr} = h(T - T_\infty) \quad \text{at} \quad r = r_o, \qquad \text{(convective heat transfer)}$$

Integrated twice, the conduction equation yields

$$T = -\frac{\dot{q}\, r^2}{4k} + c_1 \ln r + c_2$$

and, after relying on the two boundary conditions in order to pinpoint c_1 and c_2,

$$T(r) = T_\infty + \frac{\dot{q}\, r_o^2}{4k}\left[1 - \left(\frac{r}{r_o}\right)^2\right] + \frac{\dot{q}\, r_o}{2h}$$

The maximum temperature occurs in the middle of the cylindrical cross-section, $T_{max} = T(0)$.

Problem 2.5. The total conduction heat transfer rate through the cylindrical ice shell must equal the convection heat transfer rate received by the inner surface of the ice shell,

$$\frac{2\pi k l}{\ln (r_o/r_i)}(T_0 - T_s) = 2\pi r_i l h\, (T_w - T_0)$$

On the right side of this equation, the group $2\pi r_i l$ is the area in contact with the 4°C-water stream. Rearranged, this equation yields

$$\frac{\ln (r_o/r_i)}{r_o/r_i} = \frac{k}{h\, r_o}\frac{T_0 - T_s}{T_w - T_0}$$

$$= \frac{2.25\ \text{W}}{\text{m}\cdot\text{K}}\ \frac{\text{m}^2\text{K}}{1000\ \text{W}}\ \frac{1}{0.04\ \text{m}}\ \frac{0 - (-10)}{4 - 0}$$

$$= 0.141$$

the solution of which is $r_o/r_i = 1.181$. In conclusion, the inner radius of the ice shell is $r_i = 3.39$ cm, which means that the ice shell is 0.61 cm thick.

Problem 2.6. The conduction heat transfer rate across the spherical shell of solidified paraffin must equal the convection heat transfer rate from the liquid paraffin to the outer surface of the solidified shell,

$$4\pi k \frac{r_i r_o}{r_o - r_i} (T_m - T_s) = 4\pi r_o^2 h \, (T_\infty - T_m)$$

In numerical terms, this reduces to

$$\frac{r_o}{r_i}\left(\frac{r_o}{r_i} - 1\right) = \frac{T_m - T_s}{T_\infty - T_m} \frac{k}{h r_i}$$

$$= \frac{27.5 - 10}{35 - 27.5} \frac{0.36\ \mathrm{W}}{\mathrm{m \cdot K}} \frac{\mathrm{m^2K}}{100\ \mathrm{W}} \frac{1}{0.01\mathrm{m}}$$

$$= 0.84$$

which means that $r_o/r_i = 1.54$, and that the thickness of the solidified shell is 0.54 cm.

Problem 2.7. The conductivities of plywood and fiberglass are $k_{1,3} = 0.11$ W/m·K and, respectively, $k_2 = 0.035$ W/m·K. The total heat transfer rate escaping from the box is

$$q = UA\,(T_h - T_c) \qquad \text{(a)}$$

where A is an "effective" area of insulated surface (the surface pierced by q). With reference to the attached figure, we calculate first the overall heat transfer coefficient:

$$\frac{1}{U} = \frac{1}{h_h} + \frac{L_3}{k_3} + \frac{L_2}{k_2} + \frac{L_1}{k_1} + \frac{1}{h_c}$$

$$= \left(\frac{1}{5} + \frac{0.01}{0.11} + \frac{0.1}{0.035} + \frac{0.01}{0.11} + \frac{1}{15}\right) \frac{\mathrm{m^2K}}{\mathrm{W}}$$

$$= 3.31\ \mathrm{m^2K/W}$$

$$U = 0.303\ \mathrm{W/m^2K} \qquad \text{(b)}$$

In writing the q formula (a) we are treating the entire insulating shell (the wall of the box) as a plane (unfolded) wall of frontal area A. As an engineering estimate, this area can be taken as equal to the lateral area of the parallelepiped formed by the midplane of the insulating wall. This new parallelepiped is larger than the x×y×z box exhibited in the problem statement. Its three dimensions can be estimated by adding twice the wall half-thickness to each of the dimensions of the original parallelepiped:

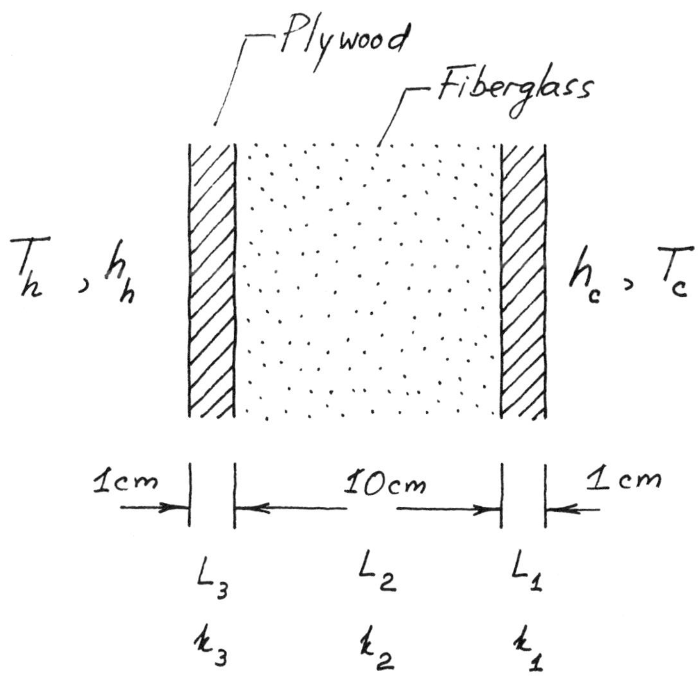

$$X = x + 0.12 \text{ m} = 1.12 \text{ m}$$

$$Y = y + 0.12 \text{ m} = 0.52 \text{ m}$$

$$Z = x + 0.12 \text{ m} = 0.42 \text{ m}$$

The total lateral area of the X×Y×Z box is

$$A = 2\,(XY + YZ + XZ)$$

$$= 2\,(1.12 \times 0.52 + 0.52 \times 0.42 + 1.12 \times 0.42) \text{ m}^2$$

$$= 2.54 \text{ m}^2$$

and the total heat transfer rate becomes, cf. eq. (a)

$$q = 0.303 \frac{\text{W}}{\text{m}^2\text{K}}\, 2.54 \text{ m}^2\, (50 - 10)°\text{C}$$

$$= 30.8 \text{ W}$$

In the steady state, this heat transfer rate must be equal to the electric power dissipated in the resistance placed in the center of the box.

Problem 2.8. According to the unidirectional conduction model shown on the right side of the figure, on each unit cross-sectional area A stands a solid column of cross-sectional area $(1 - \phi)A$, and a fluid space of cross-sectional area ϕA. The total heat transfer rate through the area A is therefore

$$q = k_s \left(1 - \phi\right) A \frac{T_h - T_c}{L} + k_f \phi A \frac{T_h - T_c}{L}$$

which corresponds to the A-averaged heat flux

$$q'' = \frac{q}{A} = \left[\left(1 - \phi\right) k_s + \phi\, k_f\right] \frac{T_h - T_c}{L}$$

This heat flux formula is clearly of the type

$$q'' = k_{eff} \frac{T_h - T_c}{L}$$

in which k_{eff} is the effective thermal conductivity of the composite medium,

$$k_{eff} = (1 - \phi)\, k_s + \phi\, k_f$$

Problem 2.9. a) The heat flux must overcome two resistances in series, one for internal conduction, and the other for air convection:

$$q''_a = \frac{T_a - T_\infty}{\frac{1}{h} + \frac{\delta}{k}}$$

$$= \frac{(36.5 - 20)^\circ\mathrm{C}}{\frac{m^2\,K}{30W} + \frac{0.02\ m}{0.42\ W/m\cdot K}} = 289 \frac{W}{m^2}$$

$$T_w - T_\infty = \frac{q''}{h} = 289 \frac{W}{m^2} \frac{m^2K}{30W}$$

$$= 9.6^\circ\mathrm{C}$$

$$T_w = T_\infty + 9.6^\circ\mathrm{C} = 29.6^\circ\mathrm{C}$$

b) The heat flux is impeded by the resistance to internal conduction added in series to the resistance to convection to the external water:

$$q_b'' = \frac{T_a - T_\infty}{\frac{1}{h} + \frac{\delta}{k}}$$

$$= \frac{(36.5 - 10)°C}{\frac{1}{500}\frac{W}{m^2K} + \frac{0.01\,m}{0.42\,W/m\cdot K}} = 988\,\frac{W}{m^2}$$

$$T_w - T_\infty = \frac{q''}{h} = 988\,\frac{W}{m^2}\,\frac{1}{500}\,\frac{m^2K}{W} \cong 2°C$$

$$T_w \cong T_\infty + 2°C = 12°C$$

c) $$\frac{q_b''}{q_a''} = \frac{988}{289} = 3.42$$

Problem 2.10. Since the thermal conductivity of polystyrene is $k = 0.16$ W/m·K, the critical radius of the present insulation would be

$$r_{o,c} = \frac{k}{h} = \frac{0.16\,W}{m\cdot K}\,\frac{m^2K}{10\,W} = 1.6\text{ cm}$$

This critical radius is smaller than the radius of the bare pipe, therefore it is not unreasonable to expect a "thermal insulation" effect from the installation of a 1 cm-thick layer of polystyrene.

Without the polystyrene layer, the per-unit-length heat transfer rate from the steam pipe to the ambient is

$$q' = 2\pi r_i h\,(T_i - T_\infty)$$

$$= 2\pi\,0.04\text{ m }10\,\frac{W}{m^2K}\,(100 - 15)\text{ K} = 213.6\,\frac{W}{m}$$

With the polystyrene layer, the heat transfer rate is

$$q' = \frac{q}{l} = \frac{T_i - T_\infty}{R_t l}$$

for which $R_t l$ is given by eq. (2.42):

$$R_t l = \frac{\ln (r_o/r_i)}{2\pi k} + \frac{1}{2\pi r_o h}$$

$$= \frac{\ln (5/4)}{2\pi} \frac{m \cdot K}{0.16\ W} + \frac{1}{2\pi\ (0.05)\ m} \frac{m^2 K}{10\ W}$$

$$= (0.222 + 0.318) \frac{m \cdot K}{W} = 0.54 \frac{m \cdot K}{W}$$

In conclusion, the heat transfer rate is 26 percent smaller than in the case where the polystyrene shell is absent,

$$q' = \frac{T_i - T_\infty}{R_t l} = \frac{(100 - 15)\ K}{0.54\ m \cdot K/W} = 157.4\ W$$

Problem 2.11. The mechanical support is a unidirectional conductor with variable thermal conductivity, k(T), and variable cross-sectional area, A(x). Its overall thermal resistance can be calculated using eqs. (2.54) and (2.55),

$$R_t = \frac{1}{k_{avg}} \int_0^{L_1 + L_2} \frac{dx}{A(x)}$$

$$= \frac{1}{k_{avg}} \left(\frac{L_1}{A_1} + \frac{L_2}{A_2} \right) \qquad \text{(i)}$$

The average thermal conductivity depends only on k(T) and the temperature extremes,

$$k_{avg} = \frac{\theta\ (T_{high}) - \theta\ (T_{low})}{T_{high} - T_{low}} \qquad \text{(ii)}$$

in other words, k_{avg} does not vary from design "a" to design "b". Similarly, the geometric quantity shown in parentheses in eq. (i) has the same value in both designs.

In conclusion, both designs promise to have the same R_t value, and, consequently, the same heat leak from T_{high} to T_{low}. Designs "a" and "b" are equally good (or bad) from the point of view of preventing the leakage of heat toward the cold region of the apparatus.

Problem 2.12. In the steady state, the conduction equation (1.50) reduces to

$$\frac{1}{r^2}\frac{d}{dr}\left(r^2\frac{dT}{dr}\right) = -\frac{\dot{q}}{k}$$

or, after two integrations,

$$T = -\frac{\dot{q}\, r^2}{6k} - \frac{c_1}{r} + c_2$$

The relevant boundary conditions are two,

$$\frac{dT}{dr} = 0 \quad \text{at} \quad r = 0, \text{ (radial symmetry)}$$

$$-k\frac{dT}{dr} = h(T - T_\infty) \quad \text{at} \quad r = r_o, \text{ (convective heat transfer)}$$

These conditions are sufficient for pinpointing the constants c_1 and c_2; the final T(r) result is

$$T(r) = T_\infty + \frac{\dot{q}\, r_o^2}{6k}\left[1 - \left(\frac{r}{r_o}\right)^2\right] + \frac{\dot{q}\, r_o}{3h}$$

The center of the sphere is the point where the temperature is maximum, $T_{max} = T(0)$.

Problem 2.13. The problem consists of solving the conduction equation (2.69),

$$\frac{1}{r}\frac{d}{dr}\left(r\frac{dT}{dr}\right) + C_1 + C_2 T = 0 \qquad \text{(i)}$$

subject to the boundary conditions

$$\frac{dT}{dr} = 0 \quad \text{at} \quad r = 0, \text{ (symmetry)} \qquad \text{(ii)}$$

$$-k\frac{dT}{dr} = h(T - T_\infty) \quad \text{at} \quad r = r_o, \text{ (contact with external flow)} \qquad \text{(iii)}$$

In the first step, we select the parabolic temperature profile

$$T = a + b\left(1 - \frac{r^2}{r_o^2}\right) \qquad \text{(iv)}$$

which satisfies already the boundary condition (ii). Forcing the profile (iv) to satisfy the condition (iii), we obtain a profile that depends on only one unknown (b),

$$T = b\left(\frac{2}{Bi} + 1 - \frac{r^2}{r_0^2}\right) + T_\infty \qquad \text{(v)}$$

in which $Bi = hr_0/k$.

In the second step, we substitute the profile (v) into the cross-section integral of the energy equation (i),

$$\left(r\frac{dT}{dr}\right)_{r = r_0} + C_1 \frac{r_0^2}{2} + C_2 \int_0^{r_0} T\, r\, dr = 0 \qquad \text{(vi)}$$

This operation yields an algebraic equation for the unknown amplitude b, the solution of which is

$$b = \frac{2(C_1 + C_2 T_\infty)}{\frac{8}{r_0^2} - C_2\left(\frac{4}{Bi} + 1\right)} \qquad \text{(vii)}$$

In view of eq. (2.70) in the text, the denominator of this expression remains positive and finite if

$$J < \frac{2^{3/2}}{r_0}\left(\frac{k}{\rho_e'}\right)^{1/2}\left(\frac{Bi}{Bi + 4}\right)^{1/2} \qquad \text{(viii)}$$

This allowable-J criterion is more stringent than eq. (2.77) because the new factor $[Bi/(Bi + 4)]^{1/2}$ is less than one, particularly when $Bi << 1$. In conclusion, the effect of the finite Biot number is to reduce further the J domain in which the electrical conductor is thermally stable.

Problem 2.14. a) With reference to the y coordinate shown in the problem statement, we write the equation and boundary conditions for unidirectional steady-state conduction across the coal stockpile,

$$\frac{d^2T}{dy^2} + \frac{\dot{q}}{k} = 0 \qquad \text{(i)}$$

$$T = T_0 \quad \text{at} \quad y = 0 \qquad \text{(ii)}$$

$$-k\frac{dT}{dy} = h(T - T_0) \quad \text{at} \quad y = H \qquad \text{(iii)}$$

Integrating eq. (i) twice in y, we obtain

$$T = -\frac{\dot{q}}{k}\frac{y^2}{2} + c_1 y + c_2 \qquad \text{(iv)}$$

The boundary condition (ii) shows that $c_2 = T_0$. Substituting this value in eq. (iv), and combining eqs. (iv) and (iii), we write, in order,

$$-k\left(-\frac{\dot{q}}{k}H + c_1\right) = h\left(-\frac{\dot{q}}{k}\frac{H^2}{2} + c_1 H + T_0 - T_0\right)$$

$$c_1 = \frac{\dot{q}H}{k}\frac{1 + Bi/2}{1 + Bi}, \quad \text{where} \quad Bi = \frac{hH}{k}$$

The temperature distribution across the coal layer is therefore

$$T - T_0 = \frac{\dot{q}H^2}{k}\left[\frac{1 + Bi/2}{1 + Bi}\frac{y}{H} - \frac{1}{2}\left(\frac{y}{H}\right)^2\right] \qquad \text{(v)}$$

Solving the equation $dT/d(y/H) = 0$, we locate the plane of maximum temperature,

$$\left(\frac{y}{H}\right)_{max.temp.} = \frac{1 + Bi/2}{1 + Bi} \qquad \text{(vi)}$$

This y/H ratio decreases from 1 at Bi = 0, to 1/2 as $Bi \to \infty$. Substituted in eq. (v), it yields the temperature maximum

$$T_{max} - T_0 = \frac{\dot{q}H^2}{2k}\left(\frac{1 + Bi/2}{1 + Bi}\right)^2 \qquad \text{(vii)}$$

b) We know the following numerical data

$$\dot{q} = 50\,\frac{W}{m^3} \qquad h = 2\,\frac{W}{m^2K}$$

$$k = 0.2\,\frac{W}{m \cdot K} \qquad T_0 = 25°C$$

$$H = 2\,m$$

therefore, we can evaluate in order

$$Bi = 2\,\frac{W}{m^2K}\,2\,m\,\frac{m \cdot K}{0.2\,W} = 20$$

$$\left(\frac{y}{H}\right)_{max.temp.} = \frac{1 + 20/2}{1 + 20} = 0.524$$

$$T_{max} - T_0 = \frac{50\ W}{m^3}\ 4m2 \frac{m \cdot K}{(2)\ 0.2\ W}\ (0.524)^2 = 137.2°C$$

$$T_{max} = 25°C + 137.2°C = 162.2°C$$

Problem 2.15. With reference to the right side of the conducting slab ($0 \leq x \leq L/2$), we write the conduction equation and the two boundary conditions:

$$\frac{d^2T}{dx^2} = 0 \qquad (1)$$

$$-k\frac{dT}{dx} = \frac{q''}{2} \quad \text{at} \quad x = 0 \qquad (2)$$

$$T = T_0 \quad \text{at} \quad x = \frac{L}{2} \qquad (3)$$

The general solution to eq. (1) is,

$$T = c_1 x + c_2, \qquad (4)$$

while the boundary conditions (3) and (2) require, in order

$$c_2 = T_0 - c_1 \frac{L}{2} \qquad (5)$$

$$c_1 = -\frac{q''}{2k} \qquad (6)$$

Put together, eqs. (4)-(6) yield the T(x) distribution to the right of the plane heat source,

$$T(x) = T_0 + \frac{q''}{2k}\left(\frac{L}{2} - x\right), \quad x \geq 0 \qquad (7)$$

The temperature distribution on the left side of the plane heat source can be derived in the same way. A more direct approach is to note that the solution for $x < 0$ can be written down by replacing x with -x in eq. (7)

$$T(x) = T_0 + \frac{q''}{2k}\left(\frac{L}{2} + x\right), \quad x \leq 0 \qquad (8)$$

The two halves of the solution, eqs. (7,8), can be condensed as

$$T(x) = T_0 + \frac{q''}{2k}\left(\frac{L}{2} - |x|\right) \qquad (9)$$

to stress the fact that the T(x) distribution is linear and symmetric about the midplane. The temperature of the plane heat source is the center-plane temperature (x = 0):

$$T_c = T_0 + \frac{q''L}{4k} \qquad (10)$$

Problem 2.16. Consider the concentric cylindrical surface of radius r. The radial heat flux that pierces this surface is

$$q'' = -k\frac{dT}{dr} \qquad (1)$$

Energy conservation requires that

$$q' = 2\pi r q'' \qquad (2)$$

By eliminating q" between eqs. (1) and (2) we obtain the equation

$$q' = -k\frac{dT}{dr}2\pi r \qquad (3)$$

which can be integrated from $r = r_o$ (where $T = T_0$) to any r inside the conducting rod:

$$\ln r_o - \ln r = -\frac{2\pi k}{q'}(T_0 - T)$$

Rearranged, the solution reads

$$T(r) = T_0 + \frac{q'}{2\pi k}\ln\frac{r_o}{r}$$

and shows that the line heat source temperature is infinite:

$$T \to \infty \qquad \text{as} \qquad r \to 0$$

Furthermore, the temperature of the conducting medium in the immediate vicinity of the line heat source blows up as $\ln(r_o/r)$ as $r \to 0$.

Problem 2.17. The radial heat flux through the concentric spherical surface of radius r is

$$q'' = -k\frac{dT}{dr}$$

Energy conservation requires that

$$q = 4\pi r^2 q''$$

which yields, sequentially

$$-\frac{dr}{r^2} = \frac{4\pi k}{q}\, dT$$

$$\frac{1}{r} - \frac{1}{r_o} = \frac{4\pi k}{q}(T - T_o)$$

$$T = T_o + \frac{q}{4\pi k}\left(\frac{1}{r} - \frac{1}{r_o}\right)$$

$$= T_o + \frac{q}{4\pi r_o}\left(\frac{r_o}{r} - 1\right)$$

This solution shows that the internal temperature blows up as r_o/r as r approaches 0.

Problem 2.18. a) With reference to Fig. 2.11, we integrate eq. (2.85) from x = 0 to x = L:

$$k\,A_c \frac{d^2T}{dx^2} - hp\,(T - T_\infty) = 0 \qquad (2.85)$$

$$\underbrace{k\,A_c\left(\frac{dT}{dx}\right)_{x=L}}_{-q_{tip}} \; \underbrace{-\,k\,A_c\left(\frac{dT}{dx}\right)_{x=0}}_{q_b} - \int_0^L hp\,(T - T_\infty)\,dx = 0$$

$$q_b = q_{tip} + \int_0^L (T - T_\infty)\,p\,dx$$

In this last expression we make the following substitutions

$$q_{tip} = h\,(T_{tip} - T_\infty)\,A_{tip}$$

$$p\,dx = d\,A_{lateral}$$

$$A_{total} = A_{lateral} + A_{tip}$$

and obtain:

$$q_b = \int_{A_{tip}} h\left(T - T_\infty\right) dA_{tip} + \int_{A_{lateral}} h\left(T - T_\infty\right) dA_{lateral}$$

$$= \int_{A_{total}} h\left(T - T_\infty\right) dA_{total} \qquad (1)$$

b) In summary, the heat current through the base of the fin matches the integral of heat flux over the entire exposed (wetted) area. The same conclusion can be reached more directly by invoking the first law of thermodynamics in relation to the control volume of size $A_c \times L$:

$$q_b = \int_{A_{total}} q''_{\text{out of } A_{total}} \, dA_{total}$$

$$= \int_{A_{total}} h\left(T - T_\infty\right) dA_{total} \qquad (1)$$

This alternative derivation shows that eq. (1) holds for a fin of arbitrary geometry, not just for the one-dimensional fin represented by eq. (2.85) and Fig. 2.11.

Problem 2.19. The geometry of the two-dimensional fin shown in the figure is characterized by

$$A_c = W\delta \qquad\qquad p = 2W$$

Noting further that $dT/dx = \theta_b/L$, the conduction equation for fins with variable cross-sectional area, eq. (2.122), reduces to

$$\frac{d\delta}{dx} = \frac{2h}{k} x$$

Integrating away from the "sharp" tip (i.e. from $x = 0$, where $\delta = 0$), we obtain

$$\delta(x) = \frac{h}{k} x^2$$

The fin thickness at the base is therefore $t = hL^2/k$.

The total heat transfer rate through the base area (tW) is

$$q_b = k\,tW \frac{dT}{dx} = k\,tW \frac{\theta_b}{L}$$

$$= h\,WL\,\theta_b$$

Since the total volume of the parabolic-profile fin is

$$V = \int_0^L W\delta dx = W \frac{hL^3}{3k},$$

the q_b expression can be written in terms of V, not L,

$$q_b = 3^{1/3}\,\theta_b\,(Wh)^{2/3}\,(kV)^{1/3}$$

The maximum heat transfer rate through the plate fin of the same volume (section 2.7.5.) can be expressed similarly, by combining eqs. (2.138) and (2.135):

$$q_{b,max} = 1.257\,\theta_b\,(Wh)^{2/3}\,(kV)^{1/3}$$

Dividing the two results we conclude that the parabolic-profile fin is more effective than the rectangular-profile fin,

$$\frac{q_b}{q_{b,max}} = \frac{3^{1/3}}{1.257} = 1.148$$

Problem 2.20. After noting that x points toward the base of the fin, we write that the fin cross-section varies with x,

$$A_c = \frac{x}{L} Wt \qquad p = 2W \qquad \text{(i)}$$

The appropriate fin conduction equation is eq. (2.122),

$$\frac{d}{dx}\left(k \frac{x}{L} Wt \frac{d\theta}{dx}\right) - 2\,Wh\,\theta = 0 \qquad \text{(ii)}$$

which reduces to

$$x \frac{d^2\theta}{dx^2} + \frac{d\theta}{dx} - a\theta = 0 \qquad \text{(iii)}$$

Next, we substitute the assumed θ series

$$\theta = c_0 + c_1 x + c_2 x^2 + \ldots \qquad \text{(iv)}$$

into each of the terms of eq. (iii). Starting with the last term, we have

$$-a\theta = -ac_0 - ac_1 x - ac_2 x^2 - ac_3 x^3 - \ldots.$$

$$\frac{d\theta}{dx} = c_1 + 2c_2x + 3c_3x^2 + 4c_4x^3 + ...$$

$$x\frac{d^2\theta}{dx^2} = 2c_2x + 6c_3x^2 + 12c_4x^3 + ...$$

Adding these three equations term by term, on the left side we obtain zero [cf. eq. (iii)],

$$0 = (c_1 - ac_0) + (4c_2 - ac_1)x + (9c_3 - ac_2)x^2 + (16c_4 - ac_3)x^3 + ...$$

The right side must vanish for any x, therefore

$$c_1 - ac_0 = 0 \quad \rightarrow \quad c_1 = ac_0$$

$$4c_2 - ac_1 = 0 \quad \rightarrow \quad c_2 = \frac{ac_1}{4} = c_0\frac{a^2}{4}$$

$$9c_3 - ac_2 = 0 \quad \rightarrow \quad c_3 = \frac{ac_2}{9} = c_0\frac{a^3}{36}$$

$$16c_4 - ac_3 = 0 \quad \rightarrow \quad c_4 = \frac{ac_3}{16} = c_0\frac{a^4}{576}$$

The $\theta(x)$ series (iv) becomes

$$\theta = c_0\left(1 + ax + \frac{a^2x^2}{4} + \frac{a^3x^3}{36} + \frac{a^4x^4}{576} + ...\right)$$

$$= c_0\left[\frac{(ax)^0}{(0!)^2} + \frac{ax}{(1!)^2} + \frac{(ax)^2}{(2!)^2} + \frac{(ax)^3}{(3!)^2} + \frac{(ax)^4}{(4!)^2} + ...\right]$$

in other words,

$$\theta = c_0F\,(ax), \quad \text{where} \quad F(ax) = \sum_{n=0}^{\infty} \frac{(ax)^n}{(n!)^2}$$

The remaining constant (c_0) is pinpointed by the base condition,

$$\theta = \theta_b \qquad \text{at} \qquad x = L$$

which yields $c_0 = \theta_b/F(aL)$. In conclusion, the excess temperature distribution along the triangular profile is

$$\theta = \theta_b\frac{F(ax)}{F(aL)}$$

Since $F(0) = 1$, the excess temperature at the tip of the fin is $\theta(x = 0) = \theta_b/F(aL)$.

Problem 2.21. The total heat transfer rate that corresponds to the fin temperature distribution (2.109) is

$$q_{b,(2.109)} = -k A_c \left(\frac{\partial \theta}{\partial x}\right)_{x=0}$$

$$= k A_c m \theta_b \frac{\sinh(mL) + (h/mk)\cosh(mL)}{\cosh(mL) + (h/mk)\sinh(mL)}$$

In this expression, we note two dimensionless groups, (mL) and (h/mk). The latter is proportional to the square root of the Biot number based on fin thickness,

$$\frac{h}{mk} = \left(\frac{h A_c}{kp}\right)^{1/2} = \left(\frac{Bi}{2}\right)^{1/2}$$

where Bi = ht/k, if the fin is two-dimensional with rectangular profile (plate fin).

The approximate formula (2.110) can be written in terms of (mL) and $(Bi/2)^{1/2}$, by using the correction rule $L_c = L + A_c/p$,

$$q_{b,(2.110)} = \theta_b (h A_c hp)^{1/2} \tanh\left[mL + \left(\frac{Bi}{2}\right)^{1/2}\right]$$

Dividing the two q_b expressions listed above, we obtain the ratio "R",

$$R = \frac{q_{b,(2.109)}}{q_{b,(2.110)}} = \frac{\tanh(mL) + (Bi/2)^{1/2}}{\tanh[mL + (Bi/2)^{1/2}]\,[1 + (Bi/2)^{1/2}\tanh(mL)]}$$

Two limits of this result are of interest. First, if the fin is "long" such that (mL) >> 1, then all the "tanh" terms are approximately equal to 1, and R approaches 1. Consequently, in long fins the simpler formula (2.110) is an appropriate substitute for eq. (2.109), regardless of the value of the group $(h A_c/kp)^{1/2}$, or $(Bi/2)^{1/2}$.

In the opposite limit, $(mL) \rightarrow 0$, the ratio R reduces to

$$R = \frac{(Bi/2)^{1/2}}{\tanh(Bi/2)^{1/2}}$$

This ratio approaches 1 [i.e. the formula (2.110) is accurate] when $(Bi/2)^{1/2}$ is small relative to 1,

$$R \cong \frac{(Bi/2)^{1/2}}{(Bi/2)^{1/2} - \frac{1}{3}(Bi/2)^{3/2} + \ldots}$$

Writing (R - 1) for the "error" associated with using eq. (2.110) instead of eq. (2.109), we conclude that

$$\text{error} = R - 1 \cong \frac{1}{1 - \frac{1}{3}\left(\frac{Bi}{2}\right) + \ldots} - 1 \cong \frac{Bi}{6}$$

or, more generally, error = $(1/3)(h\, A_c/kp)$. The corrected-length formula (2.110) is therefore accurate when the Biot number based on thickness is small, i.e. precisely in the Biot number range in which the unidirectional conduction model is valid.

Problem 2.22. The conduction equation for a fin with constant cross-sectional area is

$$k\, A_c \frac{d^2T}{dx^2} - hp\,(T - T_\infty) = 0$$

In the present problem, the fin-fluid temperature difference is constant everywhere along the fin,

$$T(x) - T_\infty(x) = \Delta T, \quad \text{(constant)}$$

The problem to solve is then

$$\frac{d^2T}{dx^2} = m^2 \Delta T, \qquad m^2 = \frac{hp}{k\, A_c}$$

$$T = T_b \qquad \text{at} \qquad x = 0$$

$$-k \frac{dT}{dx} = h\,(T - T_\infty) \qquad \text{at} \qquad x = L$$

in which the last equation accounts for the heat transfer through the tip. The solution is

$$\frac{T_b - T}{\Delta T} = (mL)^2 \left[\frac{x}{L} - \frac{1}{2}\left(\frac{x}{L}\right)^2\right] + \frac{h}{k}\, x$$

The temperature at the tip of the fin is therefore

$$T(L) = T_b - \Delta T \left[\frac{(mL)^2}{2} + \frac{hL}{k}\right]$$

The total heat transfer rate through the base of the fin is

$$q_b = -k A_c \left(\frac{dT}{dx}\right)_{x=0} = k A_c \Delta T \left(m^2 L + \frac{h}{k}\right)$$

Problem 2.23. a) Measuring x clockwise around the ring, we see that the equation for steady unidirectional conduction is the same as eq. (2.89). The general solution to this equation can be written as either eq. (2.93) or eq. (2.101). The latter is more helpful in problems where the conducting medium is finite (as in this case), therefore we write

$$\theta(x) = C_1 \sinh(mx) + C_2 \cosh(mx)$$

where $\theta = T - T_\infty$. The boundary conditions,

$$\underline{x = 0}: \qquad \theta_b = C_2$$

$$\underline{x = L}: \qquad \theta_b = C_1 \sinh(mL) + C_2 \cosh(mL)$$

allows us to determine

$$C_1 = \theta_b \frac{1 - \cosh(mL)}{\sinh(mL)}$$

and to substitute C_1 and C_2 into the $\theta(x)$ expression,

$$\frac{\theta(x)}{\theta_b} = \frac{\sinh(mx) - \cosh(mL)\sinh(mx) + \cosh(mx)\sinh(mL)}{\sinh(mL)}$$

$$= \frac{\sinh(mx) + \sinh[m(L - x)]}{\sinh(mL)}$$

b) The temperature at the point diametrically opposed to x = 0 is obtained by substituting x = L/2 in the above solution:

$$\frac{\theta(L/2)}{\theta_b} = \frac{\sinh(mL/2) + \sinh(mL/2)}{\sinh(mL)}$$

$$= \frac{2\sinh(mL/2)}{2\sinh(mL/2)\cosh(mL/2)}$$

$$= \frac{1}{\cosh(mL/2)}$$

Note that we could have arrived at the same result by treating the first half of the ring ($0 \le x \le L/2$) as a fin with insulated tip at x = L/2, and by replacing L with L/2 in eq. (2.105). The "insulated tip" condition applies at x = L/2 because of symmetry.

c) For calculating the heat input q_b, we rely on this last observation and see that half of q_b proceeds clockwise through the $0 \le x \le L/2$ half-ring (fin with insulated tip), while the other half enters counterclockwise the lower half of the ring. Each $q_b/2$ value is given by eq. (2.106), in which we replace L with the present "tip" position, L/2:

$$\frac{q_b}{2} = \theta_b \, (k \, A_c \, hp)^{1/2} \tanh (mL/2)$$

The total heat transfer input at x = 0 is therefore

$$q_b = 2\theta_b \, (k \, A_c \, hp)^{1/2} \tanh (mL/2).$$

A lengthier derivation of this q_b result would have started with the analytical statement of the splitting of q_b into two conduction currents,

$$q_b = \underbrace{A_c\left(-k \frac{dT}{dx}\right)_{x=0}}_{\substack{\text{current entering} \\ \text{the } 0 \le x \le L/2 \\ \text{half of the ring}}} + \underbrace{A_c\left(k \frac{dT}{dx}\right)_{x=L}}_{\substack{\text{current entering} \\ \text{at } x = L \text{ the other} \\ \text{half of the ring}}}$$

or, because of symmetry,

$$q_b = 2A_c\left(-k \frac{dT}{dx}\right)_{x=0}$$

The gradient $(dT/dx)_{x=0}$ can then be derived from the $\theta(x)$ solution developed in part (a).

Problem 2.24. What distinguishes the hollow-cylinder handle from the one-dimensional fin of Fig. 2.11 is the fact that h is not uniform around the wetted perimeter of the cross-section, $p = \pi D_o + \pi D_i$. Therefore, instead of ph in eq. (2.83), we must write

$$ph = \pi D_o h_o + \pi D_i h_i$$

so that the m parameter of the fin is

$$m = \left(\frac{hp}{k \, A_c}\right)^{1/2} = \left[\frac{\pi \, (D_o h_o + D_i h_i)}{k \frac{\pi}{4} (D_o^2 - D_i^2)}\right]^{1/2}$$

$$= \left[\frac{4 \, (2 \times 10 + 1.5 \times 2) \text{ W cm}}{\text{m}^2\text{K}} \; \frac{\text{m}\cdot\text{K}}{1.03 \text{ W } (4\text{-}2.25) \text{ cm}^2}\right]^{1/2}$$

$$= 71.44 \text{ m}^{-1}$$

According to the long-fin criterion (2.96)

$$mL = 71.44\ m^{-1}\ (0.07)\ m = 5 > 1$$

the hollow-cylinder handle is a "long" fin. This means that we can rely on eq. (2.95) for the purpose of calculating the distance to the base x where T = 30°C:

$$\exp(-mx) = \frac{T - T_\infty}{T_b - T_\infty} = \frac{30 - 15}{50 - 15} = 0.43$$

$$-mx = \ln(0.43) = -0.85$$

$$x = \frac{0.85}{71.44\ m^{-1}} = 1.2\ cm$$

This short distance explains why the handle does not feel hot even when the pot contains boiling water.

The instantaneous heat transfer rate through the base of the handle is, cf. eq. (2.97):

$$q_b = (T_b - T_\infty)\left[k A_c \pi (D_o h_o + D_i h_i)\right]^{1/2}$$

$$= (T_b - T_\infty)\ k A_c m$$

$$= (50 - 15)\ K\ 1.03 \frac{W}{m\cdot K} \frac{\pi}{4} (4 - 2.25)\ cm^2 \frac{71.44}{m}$$

$$= 0.35\ W$$

Problem 2.25. Let T_b be the temperature of the interface between solder and fin material. Then we can write the expression for the fin heat transfer rate in two ways,

$$q_b = k_s \frac{bW}{t}(T_w - T_b) \tag{1}$$

$$q_b = hp\ L_c \left(T_b - T_\infty\right)\eta \tag{2}$$

We can determine T_b by eliminating q_b between eqs. (1) and (2),

$$T_b = \frac{T_w + BT_\infty}{1 + B} \tag{3}$$

where B is shorthand for the dimensionless group

$$B = \frac{hp\, L_c\, t\, \eta}{k_s\, bW} \tag{4}$$

Taken together, eqs. (1) and (3) deliver the wanted expression for q_b:

$$q_b = k_s \frac{bW}{t} \left(T_w - T_\infty\right) \frac{B}{1 + B} \tag{5}$$

The B number can be evaluated numerically by recognizing the following relations

$$p = 2b + 2W = 2\,(b + W) \tag{6}$$

$$L_c = L + \frac{A_c}{p}$$

$$= L + \frac{bW}{2\,(b + W)} \tag{7}$$

therefore

$$B = \frac{ht}{k_s}\left[2\left(\frac{L}{W} + \frac{L}{b}\right) + 1\right]\eta \tag{8}$$

Finally, for the fin efficiency we write

$$\eta = \frac{\tanh\,(m\, L_c)}{m\, L_c} \tag{9}$$

for which L_c is given by eq. (7), and

$$m = \left(\frac{hp}{k\, A_c}\right)^{1/2} = \left[\frac{2h\,(b + W)}{k\, b\, W}\right]^{1/2} \tag{10}$$

The solution consists of eqs. (5) and (7)-(10).

Problem 2.26. a) According to eq. (2.97), the heat transfer rate through the base of one hair strand is

$$q_b = \theta_b\, (k\, A_c\, hp)^{1/2} = \theta_b \left(k \frac{\pi}{4} D^2\, h\, \pi\, D\right)^{1/2}$$

$$= \theta_b\, \frac{\pi}{2} \left(k\, h\, D^3\right)^{1/2}$$

$$= 10\, K\, \frac{\pi}{2} \left(0.37\, \frac{W}{m\,K}\, 100\, \frac{W}{m^2 K}\, 27^3\, 10^{-18}\, m^3\right)^{1/2}$$

$$= 1.34 \times 10^{-5} W$$

Since on the total area $A = (10\ cm)^2$ the total number of hair strands is

$$N = 9000\ \frac{strands}{cm^2}\ 100\ cm^2 = 9 \times 10^5\ strands$$

the total heat transfer rate through the hair is

$$q_{hair} = q_b N = 1.34 \times 10^{-5}\ W\ 9 \times 10^5 =$$

$$= 12.06\ W$$

b) The skin area that is not covered by the roots of the hair strands is

$$A_{skin} = A - N\ \frac{\pi\ D^2}{4}$$

$$= 100\ cm^2 - 9 \times 10^5\ \frac{\pi}{4}\ 27^2\ 10^{-12}\ 10^4\ cm^2$$

$$= 94.8\ cm^2$$

in other words, the roots occupy roughly 5 percent of the skin area. The direct heat transfer rate between the skin and the surrounding air is

$$q_{skin} = h\ A_{skin}\ \theta_b$$

$$= 10\ \frac{W}{m^2\ K}\ 0.00948\ m^2\ 10\ K = 0.95\ W$$

This heat transfer rate is approximately 7 percent of the total heat transfer rate (hair + skin) of 13 W.

c) The order of magnitude of the distance of conduction penetration along the hair strand is the inverse of the m group defined in eq. (2.92):

$$\text{conduction length} \sim \frac{1}{m} = \left(\frac{k\ A_c}{hp}\right)^{1/2} = \left(\frac{k\ D}{4h}\right)^{1/2}$$

$$= 158 \times 10^{-6} m$$

Note that this length is roughly six times the strand diameter $D = 27 \times 10^{-6}$m. This length is representative of the distance of conduction penetration because if we substitute $x = 1/m$ in the fin temperature distribution

$$\theta = \theta_b\ \exp\ (-mx) \qquad (2.95)$$

the argument of the exponential becomes equal to -1, or $\theta/\theta_b \cong 0.37$. Said another way, at a distance of order 1/m the fin temperature already approaches the temperature of the surrounding fluid.

d) The unidirectional conduction model is valid because the Biot number is smaller than 1, cf. eq. (2.130):

$$\left(\frac{hD}{k}\right)^{1/2} = \left(100 \frac{W}{m^2 K} 27 \times 10^{-6} m \frac{m K}{0.37 W}\right)^{1/2}$$
$$= 0.085 << 1$$

Problem 2.27. Treating the two semiinfinite sides of the junction as long fins, we write eq. (2.97) for each side

$$q_1 = (T_0 - T_\infty)\,(k_1 A_c\, h_1 p)^{1/2} \qquad (1)$$

$$q_2 = (T_0 - T_\infty)\,(k_2 A_c\, h_2 p)^{1/2} \qquad (2)$$

Adding side by side, and noting that $q_1 + q_2 = q$, we obtain the relative temperature of the junction:

$$T_0 - T_\infty = \frac{q}{(A_c p)^{1/2}\left[(k_1 h_1)^{1/2} + (k_2 h_2)^{1/2}\right]} \qquad (3)$$

Finally, eqs. (1) and (3) deliver the fraction of q that is conducted away by cable no. 1,

$$\frac{q_1}{q} = \frac{(T_0 - T_\infty)(k_1 A_c\, h_1 p)^{1/2}}{(T_0 - T_\infty)(A_c p)^{1/2}\left[(k_1 h_1)^{1/2} + (k_2 h_2)^{1/2}\right]}$$
$$= \frac{R_{1,2}}{R_{1,2} + 1}$$

where $R_{1,2}$ is the "partition parameter"

$$R_{1,2} = \left(\frac{k_1 h_1}{k_2 h_2}\right)^{1/2}$$

Similarly, from eqs. (2) and (3), or from $q_2 = q - q_1$, we deduce that

$$\frac{q_2}{q} = \frac{1}{R_{1,2} + 1}$$

Problem 2.28. The equation for unidirectional steady conduction through the cable is

$$\frac{d^2\theta}{dx^2} - m^2\theta + \frac{\dot{q}}{k} = 0 \qquad \text{(a)}$$

where $m^2 = hp/k\,A_c$ and

$$\frac{\dot{q}}{k} = C_1 + C_2\theta$$

The general solution to equation (a) is

$$\theta(x) = K_1 e^{rx} + K_2 e^{-rx} + \frac{C_1}{m^2 - C_2}$$

where r is shorthand for $(m^2 - C_2)^{1/2}$. The constants K_1 and K_2 are determined from the boundary conditions

$$\theta = \theta_b \quad \text{at} \quad x = 0$$

$$\frac{d\theta}{dx} \to 0 \quad \text{as} \quad x \to \infty$$

and the $\theta(x)$ solution becomes

$$\theta(x) = \theta_b e^{-rx} + \frac{C_1}{m^2 - C_2}(1 - e^{-rx})$$

The temperature sufficiently far from the x = 0 wall is

$$\theta(\infty) = \frac{C_1}{m^2 - C_2}$$

This value is finite (i.e. thermal runaway is avoided) as long as $m^2 > C_2$, or

$$\frac{hp}{k\,A_c} > C_2$$

Problem 2.29. The optimization procedure consists of maximizing the function

$$q_b = \theta_b (k A_c hp)^{1/2} \tanh (mL) \qquad (2.106)$$

subject to the volume constraint

$$V = \frac{\pi D^2}{4} L, \quad \text{(constant)} \qquad \text{(i)}$$

In the case of a pin fin, $A_c = \pi D^2/4$ and $p = \pi D$. Using these relations, and eliminating L between the two equations listed above we obtain

$$q_b(D) = a D^{3/2} \tanh (b D^{-5/2}) \qquad \text{(ii)}$$

The two constants (a,b) are shorthand for

$$a = \theta_b \left(\frac{kh}{2}\right)^{1/2} \qquad b = \frac{8}{\pi} V \left(\frac{h}{k}\right)^{1/2} \qquad \text{(iii)}$$

The q_b maximum is located next by solving $dq_b/dD = 0$, which is the same as solving

$$\sinh (2b D^{-5/2}) = \frac{10}{3} bD^{-5/2} \qquad \text{(iv)}$$

The solution to this equation is $bD^{-5/2} = 0.9193$, which translates into the following formula for the optimum diameter:

$$D = 1.503\, V^{2/5} \left(\frac{h}{k}\right)^{1/5} \qquad \text{(v)}$$

From the volume constraint we deduce the corresponding length of the pin fin,

$$L = 0.564\, V^{1/5} \left(\frac{k}{h}\right)^{2/5} \qquad \text{(vi)}$$

The corresponding slenderness ratio is obtained by dividing eqs. (v) and (vi), and later eliminating V using again eq. (v):

$$\frac{D}{L} = 2.175 \left(\frac{hD}{k}\right)^{1/2} \qquad \text{(vii)}$$

This result shows that the slenderness ratio is proportional to the square root of the Biot number. The maximum heat transfer rate is obtained finally by substituting the optimum D of eq. (v) into eq. (ii):

$$q_{b,max} = 0.945\ \theta_b\ V^{3/5}\ k^{1/5}\ h^{4/5}$$

$$= 0.513\ \theta_b\ kD\left(\frac{hD}{k}\right)^{1/2}$$

Project 2.1. The insulation volume constraint means that the integral

$$V = \int_0^L t(x)\, W\, dx \tag{1}$$

is constant, where t(x) is the insulation thickness, and W is the width in the direction perpendicular to the $t \times L$ plane. An equivalent statement is that the L-averaged insulation thickness is constant (fixed)

$$t_{avg} = \frac{1}{L}\int_0^L t(x)\, dx = \frac{V}{WL} \tag{2}$$

a) Uniform thickness

When the insulation thickness is constant, that constant is

$$t = t_{avg} \tag{3}$$

and the total rate of heat loss through the wall area $W \times L$ can be calculated in the following sequence:

$$q_a'' = k\frac{T(x) - T_0}{t_{avg}} = k\frac{T_L - T_0}{t_{avg}}\frac{x}{L} \tag{4}$$

$$q_a = \int_0^L W\, q_a''\, dx = \int_0^L W\, k\frac{T_L - T_0}{t_{avg}}\frac{x}{L}\, dx$$

$$= kWL\frac{\Delta T_{max}}{t_{avg}}\underbrace{\int_0^1 m\, dm}_{\frac{1}{2}} \qquad \left(\text{note: } m = \frac{x}{L},\ \text{and } \Delta T_{max} = T_L - T_0\right)$$

$$= \frac{1}{2}kWL\frac{\Delta T_{max}}{t_{avg}} \tag{5}$$

b) Linear thickness

When the insulation thickness increases linearly to match the variation of the temperature difference $T(x) - T_0$, the average thickness constraint (2) requires that

$$t(x) = 2\frac{x}{L}t_{avg} \tag{6}$$

The total rate of heat loss through the wall is

$$\begin{aligned} q_b &= \int_0^L W q_b'' dx = \int_0^L W k \frac{T(x) - T_0}{t(x)} dx \\ &= \int_0^L W k \frac{\Delta T_{max} \frac{x}{L}}{2\frac{x}{L} t_{avg}} dx \\ &= \frac{1}{2} k W L \frac{\Delta T_{max}}{t_{avg}} \end{aligned} \tag{7}$$

This result is identical to eq. (5). It means that when the same amount of insulation material is used, the insulation with linear thickness is as good as the insulation with uniform thickness.

c) Optimal thickness

The insulation thickness that is proportional to $x^{1/2}$ and satisfies the fixed-t_{avg} condition (2) is

$$t_{opt} = \frac{3}{2}\left(\frac{x}{L}\right)^{1/2} t_{avg} \tag{8}$$

The total heat transfer rate through the wall surface $W \times L$ is

$$\begin{aligned} q_c &= \int_0^L W q_c'' dx = \int_0^L Wk \frac{T(x) - T_0}{t_{opt}(x)} dx \\ &= \int_0^L Wk \frac{\Delta T_{max} \frac{x}{L}}{\frac{3}{2}\left(\frac{x}{L}\right)^{1/2} t_{avg}} dx \\ &= \frac{2}{3} k W L \frac{\Delta T_{max}}{t_{avg}} \underbrace{\int_0^1 m^{1/2} dm}_{\frac{2}{3}} \\ &= \frac{4}{9} k W L \frac{\Delta T_{max}}{t_{avg}} \end{aligned} \tag{9}$$

This heat transfer rate is smaller than in the previous two designs, eqs. (5) and (7):

$$\frac{q_a}{q_c} = \frac{q_b}{q_c} = \frac{\frac{1}{2}}{\frac{4}{9}} = \frac{9}{8} = 1.125 \qquad (10)$$

In conclusion, energy savings of 12.5 percent can be realized by employing an insulation with the optimal thickness (8).

d) Proof that eq. (8) is the optimal thickness

The following analysis is simply for the record, and is beyond the scope of this course. I must report it because the present problem and its conclusion are original (new).

The general problem consists of minimizing the total rate of heat transfer through the insulation,

$$q = \int_0^L W k \frac{T(x) - T_0}{t(x)} dx \qquad (11)$$

subject to the integral constraint (2). In other words, we must find the optimal function $t_{opt}(x)$ that satisfies the integral (2) and minimizes the q integral (11). The method of solution is variational calculus (see, for example, pp. 722-723 in A. Bejan, Advanced Engineering Thermodynamics, Wiley, New York, 1988).

The optimal insulation thickness function is found by seeking the extremum of the aggregate integral obtained by combining the integrands of integrals (2) and (11),

$$\Phi = \int_0^L \left[W k \frac{T(x) - T_0}{t(x)} + \lambda \frac{t(x)}{L} \right] dx \qquad (12)$$

in which λ is a Lagrange multiplier. If we call F the combined integrand in the square brackets, then $t_{opt}(x)$ is found by solving the Euler equation

$$\frac{\partial F}{\partial t} = 0 \qquad (13)$$

namely

$$- W k \frac{T(x) - T_0}{t_{opt}^2} + \frac{\lambda}{L} = 0 \qquad (14)$$

The result is

$$t_{opt}(x) = \left(\frac{WLk}{\lambda}\right)^{1/2} [T(x) - T_0]^{1/2} \tag{15}$$

or

$$t_{opt}(x) = K[T(x) - T_0]^{1/2} \tag{16}$$

Finally, the constant factor K is determined by substituting eq. (16) back in the volume constraint (2),

$$K = \frac{t_{avg} L}{\int_0^L (T - T_0)^{1/2} \, dx} \tag{17}$$

and the optimal thickness function becomes

$$t_{opt}(x) = \frac{t_{avg} L}{\int_0^L (T - T_0)^{1/2} \, dx} [T(x) - T_0]^{1/2} \tag{18}$$

In the special case in which the wall temperature varies linearly,

$$T - T_0 = \Delta T_{max} \frac{x}{L} \tag{19}$$

the general result (18) reduces to

$$t_{opt}(x) = \frac{t_{avg} \left(\frac{x}{L}\right)^{1/2}}{\int_0^1 m^{1/2} \, dm} = \frac{3}{2} \left(\frac{x}{L}\right)^{1/2} t_{avg} \tag{20}$$

which is the thickness assumed in eq. (8).

The same conclusions are reached (based on a completely analogous analysis) when the amount of insulation is minimized subject to a fixed rate of heat transfer to the ambient.

Project 2.2. First we verify that the linearly tapered conductance A(x) satisfies the material volume constraint

$$\int_0^L A(x)\, dx = V, \qquad \text{(constant)} \tag{1}$$

We substitute

$$A(x) = \overline{A}\left[1 - a\left(2\frac{x}{L} - 1\right)\right] \tag{2}$$

in eq. (1) and obtain

$$\overline{A}L \underbrace{\int_0^1 [1 - a\,(2m - 1)]\, dm}_{\underbrace{1 - a\left(2\frac{1}{2} - 1\right)}_{1}} = V \tag{3}$$

i.e. the definition of the L-averaged cross section, $\overline{A} = V/L$.

The heat transfer rate from T_L to T_0 is

$$q = kA\frac{dT}{dx}, \qquad \text{(independent of x)} \tag{4}$$

or, after separating the variables and integrating from x = 0 to x = L,

$$q\int_0^L \frac{dx}{A(x)} = \int_{T_0}^{T_L} k(T)\, dT \tag{5}$$

If we use A(x) of eq. (2), and a linear relation between k and T,

$$k(T) = \overline{k}[1 + b\,(2\theta - 1)] \tag{6}$$

with

$$\theta = \frac{T - T_0}{T_L - T_0} \tag{7}$$

eq. (5) yields, in order,

$$q \frac{L}{\bar{A}} \underbrace{\int_0^1 \frac{d\xi}{1 - a\,(2\xi - 1)}}_{\frac{1}{2a} \ln \frac{1+a}{1-a}} = \bar{k}\,(T_L - T_0) \underbrace{\int_0^1 \left[1 + b\,(2\theta - 1)\right] d\theta}_{1} \qquad (8)$$

In conclusion, the heat transfer rate is independent of the thermal conductivity variation with the temperature (b),

$$q = \bar{k}\,\bar{A}\,\frac{T_L - T_0}{L} \cdot \frac{2a}{\ln \frac{1+a}{1-a}} \qquad (9)$$

It depends only on the taper parameter a, through the factor $2a/\ln[(1 + a)/(1 - a)]$,

a	$2a/\ln[(1 + a)/(1 - a)]$
0	1.00
± 0.25	0.98
± 0.5	0.91
± 0.75	0.77
± 0.95	0.52
± 1	0

which shows that the largest thermal conductance corresponds to a = 0, i.e. to the strut with constant (x-independent) cross-sectional area.

Proof that A(x) = constant represents the optimal distribution of conducting material.

The following analysis is simply for the record, as it is beyond the scope of this course. It deals with the general case in which A(x) can be any function of x, i.e. not just the linear distribution assumed in eq. (2). Similarly, k(T) can be any function of T.

For the general case, the right-hand side of eq. (5) is a constant, because T_L, T_0 and k(T) are known. This means that q is maximum when the geometry-related integral

$$I = \int_0^L \frac{dx}{A(x)} \qquad (10)$$

is minimum. This integral must be minimized subject to the volume constraint (1): this operation is equivalent to minimizing the aggregate integral (e.g. A. Bejan, Advanced Engineering Thermodynamics, Wiley, New York, 1988, pp. 722-723)

$$\Phi = \int_0^L \underbrace{\left[\frac{1}{A(x)} + \lambda\, A(x)\right]}_{F} dx \qquad (11)$$

the integrand of which (F) is a linear combination of the integrands of (10) and (1). The optimal function A that minimizes Φ is the solution to the Euler equation

$$\frac{\partial F}{\partial A} = 0 \qquad (12)$$

namely

$$-\frac{1}{A_{opt}^2} + \lambda = 0 \qquad (13)$$

or

$$A_{opt} = \left(\frac{1}{\lambda}\right)^{1/2}, \quad \text{constant} \qquad (14)$$

This constant cross-sectional area is determined by substituting eq. (14) in eq. (1),

$$\int_0^1 A_{opt}\, dx = V \qquad (15)$$

which yields

$$A_{opt} = \frac{V}{L}, \quad \text{constant} \qquad (16)$$

In conclusion, the maximum thermal conductance occurs when A is independent of x, regardless of how k varies with T.

Project 2.3. The local "unit elongation", or strain of the bar in tension is

$$\varepsilon = \frac{F/A(x)}{E(T)} \tag{1}$$

so that each bar slice of length dx stretches (under tension) to the new length $(1 + \varepsilon)dx$. The total elongation of the bar of length L is obtained by integrating eq. (1) from x = 0 to x = L,

$$\Delta L = \int_0^L \frac{F}{A(x)\, E(T)}\, dx \tag{2}$$

Assuming that A increases linearly with x,

$$A(x) = \overline{A}\left[1 + a\left(2\frac{x}{L} - 1\right)\right] \tag{3}$$

and that E decreases linearly as the temperature T increases (in the figure, the bottom end T_0 is the cold end),

$$E(T) = \overline{E}\left[1 - c\left(2\theta - 1\right)\right] \tag{4}$$

the elongation integral (2) can be put in the following dimensionless form

$$\frac{\Delta L}{L}\frac{\overline{A}\,\overline{E}}{F} = \delta = \int_0^1 \frac{d\xi}{\left[1 + a\left(2\xi - 1\right)\right]\left[1 - c\left(2\theta - 1\right)\right]} \tag{5}$$

where

$$\xi = \frac{x}{L} \tag{6}$$

$$\theta = \frac{T - T_0}{T_L - T_0} \tag{7}$$

Note that $\overline{A}$ and $\overline{E}$ are the average values of A(x) and E(T),

$$\overline{A} = \frac{1}{L}\int_0^L A(x)\, dx = \frac{V}{L} \tag{8}$$

$$\overline{E} = \frac{1}{T_L - T_0}\int_{T_0}^{T_L} E(T)\, dT \tag{9}$$

and that by fixing $\bar{A}$ now all the bar shapes A(x) satisfy the material volume constraint V = constant. The taper parameter a and the modulus of elasticity parameter c cover the ranges

$$-1 < a < 1 \tag{10}$$

$$-1 < c < 1 \tag{11}$$

however, of interest in low temperature applications are positive a and c values. Note also that if the cross section is uniform (a = 0) and the modulus of elasticity is constant (c = 0), the elongation integral reduces to $\delta = 1$.

In general, δ depends not only on taper (a) and elasticity (c) but also on the temperature distribution (θ). The latter can be determined by writing the heat current (conserved) from T_L to T_0,

$$q = k(T)\, A(x)\, \frac{dT}{dx} \tag{12}$$

separating the variables, and integrating away from x = 0

$$\tilde{q} \int_0^{\xi} \frac{d\xi}{1 + a\,(2\xi - 1)} = \int_0^{\theta} \left[1 + b\left(2\theta - 1\right)\right] d\theta \tag{13}$$

This became possible after adopting the linear thermal conductivity model

$$k(T) = \bar{k}\left[1 + b\left(2\theta - 1\right)\right] \tag{14}$$

with

$$\bar{k} = \frac{1}{T_L - T_0} \int_{T_0}^{T_L} k(T)\, dT \tag{15}$$

and

$$-1 < b < 1 \tag{16}$$

with special interest in positive b values. The factor $\tilde{q}$ is dimensionless shorthand for heat current,

$$\tilde{q} = \frac{qL}{\bar{k}\,\bar{A}\left(T_L - T_0\right)} \tag{17}$$

Equation (13) yields implicitly the relation $\theta(\xi)$,

$$\tilde{q}\,\frac{1}{2a}\ln\frac{1 + a\left(2\xi - 1\right)}{1 - a} = \theta + b\left(\theta^2 - \theta\right) \tag{18}$$

which, after setting $\xi = 1$ (where $\theta = 1$), shows that $\tilde{q}$ is independent of b,

$$\tilde{q}(a) = \frac{2a}{\ln \frac{1+a}{1-a}} \tag{19}$$

The function $\tilde{q}(a)$ can be plotted to show that $\tilde{q}$ is maximum when the taper is zero ($a = 0$, $\tilde{q}_{max} = 1$), and that a more tapered bar is a better insulator. This was the mission of the preceding project, and is not repeated here.

The integral (5) can be combined with eqs. (18,19), to obtain δ as a function of only (a,b,c). This is done by replacing ξ (and $d\xi$) in terms of θ, based on eqs. (18,19). This step looks menacing early, but in the end the δ integral reduces to the simple form

$$\delta = \frac{1}{2a} \ln \frac{1+a}{1-a} \int_0^1 \frac{1 + b\,(2\theta - 1)}{1 - c\,(2\theta - 1)}\, d\theta \tag{20}$$

which can be integrated,

$$\delta\,(a,b,c) = \frac{1}{2a} \ln \frac{1+a}{1-a} \left(-\frac{b}{c} + \frac{b+c}{2c^2} \ln \frac{1+c}{1-c} \right) \tag{21}$$

Compare eq. (21) with eq. (19), and note that δ is proportional to $1/\tilde{q}$. This means that

$$\delta\,(a,b,c) = \frac{G(b,c)}{\tilde{q}(a)} \tag{22}$$

where G(b,c) stands for the expression in parentheses in eq. (21), and plotted on the following graph. The effect of geometry (a) separates itself from the effect of material choice (b and c), in such a way that

- the smallest elongation occurs when the heat transfer rate is maximum ($a = 0$)
- smaller heat transfer rates (better insulations) mean larger elongations (less rigid bars in tension).

The designer is faced with the challenge of trading small δ for small $\tilde{q}$, depending on the relative cost benefit of reducing δ versus $\tilde{q}$. This challenge is made easier if the structural material is chosen so that G(b,c) has the smallest possible value. The graph shows how the material properties (b,c) influence G, i.e. the product $\tilde{q}\delta$. The ultimate product that must be minimized is not $\tilde{q}\delta$, but its dimensional counterpart

$$q\,\Delta L = G(b,c) \frac{\bar{k}}{\bar{E}} F\,(T_L - T_0)$$

The chosen structural material should have a small G $\bar{k}/\bar{E}$.

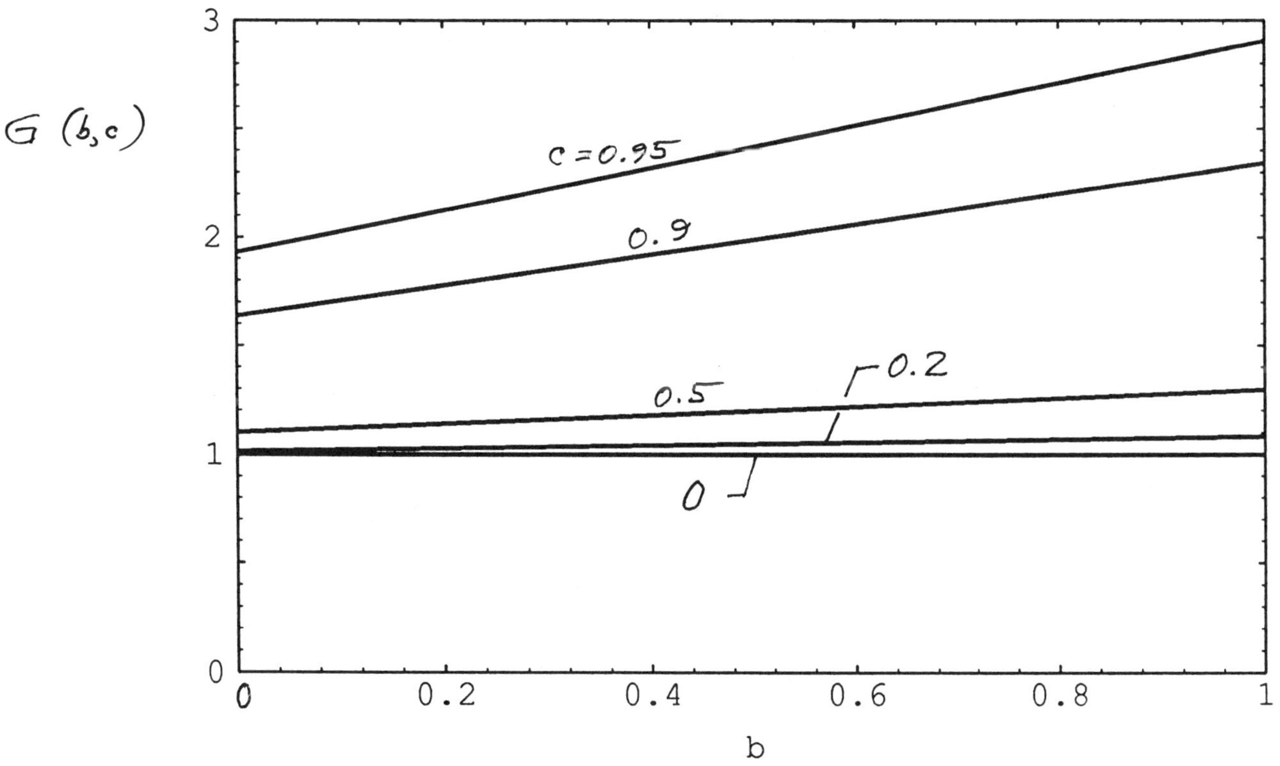

Project 2.4. The equation of flexure of the beam is

$$M = EI\frac{d^2y}{dx^2} \qquad (1)$$

where y(x) is the local deflection (measured downward, i.e. in the direction of F), M(x) is the bending moment at x,

$$M = Fx \qquad (2)$$

I(x) is the local area moment of inertia,

$$I = \frac{B}{12}H^3 \qquad (3)$$

where H = H(x), and B is the width of the beam (in the direction perpendicular to the plane of the figure shown in the text). The tip deflection (y_0) is obtained by integrating eq. (1) twice, in the following sequence

$$\int_x^L \frac{F\,x'}{E\,I(x')}\,dx' = \int_x^L d\left(\frac{dy}{dx}\right)$$

$$= \underbrace{\left(\frac{dy}{dx}\right)_L}_{0} - \frac{dy}{dx} \tag{4}$$

$$\int_0^L \left(\int_x^L \frac{F\,x'}{E\,I(x')}\,dx'\right) dx = \int_0^L (-\,dy)$$

$$= \underbrace{-\,y_L}_{0} + y_0 \tag{5}$$

In conclusion,

$$y_0 = \int_0^L \left(\int_x^L \frac{F\,x'}{E\,I(x')}\,dx'\right) dx \tag{6}$$

in which the beam thickness [see eq. (3)] increases as x^n,

$$H(x) = C\,x^n \tag{7}$$

If we substitute this H(x) expression in the volume constraint V = constant,

$$V = \int_0^L B\,H(x)\,dx = B\,\overline{H}\,L \tag{8}$$

we obtain the proper expression for C in eq. (7), so that in the end eq. (7) becomes

$$H(\xi) = \overline{H}\,(n+1)\,\xi^n \tag{9}$$

with

$$\xi = \frac{x}{L} \quad \text{and} \quad \overline{H} = \frac{V}{BL} \tag{10}$$

We now have all the ingredients for evaluating y_0 as a function of the beam shape (n). We begin with substituting eq. (9) in eq. (3), and eq. (3) in eq. (6), and write eq. (6) so that the double integral will yield a dimensionless number,

$$y_0 = \frac{12\,F\,L^3}{EB\overline{H}^3}\,\frac{1}{(n+1)^3}\,\underbrace{\int_0^1 \left(\int_\xi^1 (\xi')^{1-3n}\,d\xi'\right) d\xi}_{\displaystyle \frac{1}{2-3n}\underbrace{\int_0^1 \left(1-\xi^{2-3n}\right) d\xi}_{\displaystyle \frac{2-3n}{3(1-n)}}} \tag{11}$$

provided $n < 1$. In conclusion, the tip deflection is

$$y_0 = \frac{4F\,L^3}{EB\overline{H}^3}\,\frac{1}{(1-n)\,(1+n)^3} \tag{12}$$

By solving $\partial y_0/\partial n = 0$, we find that y_0 is minimum when

$$n_{opt} = \frac{1}{2} \tag{13}$$

and that the minimum tip deflection is

$$y_{0,min} = \frac{64}{27}\,\frac{F\,L^3}{E\,B\,\overline{H}^3} \tag{14}$$

We want to minimize not only y_0 but also the end-to-end heat transfer rate, which is

$$q = k\,BH\frac{dT}{dx} \tag{15}$$

$$q\int_0^L \frac{dx}{\overline{H}\,(n+1)\,\xi^n} = k\,B\,(T_L - T_0) \tag{16}$$

$$q = k\,\overline{H}\,B\,\frac{T_L - T_0}{L}\left(1 - n^2\right) \tag{17}$$

In the range $0 < n < 1$, the heat transfer rate is the smallest in the limit $n \rightarrow 1$ (triangular profile). This means that the best shape for small y_0 and small q will be such that n is between 1/2 [see eq. (13), for "small y_0" alone] and 1 (the best n for "small q" alone). The actual n value will depend on how important (costly) "small y_0" and "small q" are, and how they are included together in the overall cost formula of the device.

Project 2.5. The equation of flexure of the beam is

$$M = EI \frac{d^2y}{dx^2} \tag{1}$$

where y(x) is the local deflection (measured in the direction of F), M is the bending moment at x,

$$M = Fx \tag{2}$$

I(x) is the local area moment of inertia,

$$I = \frac{B H^3}{12} \tag{3}$$

and B is the width of the beam (the dimension in the direction perpendicular to the plane of the figure shown in the text). The tip deflection (y_0) is obtained by integrating eq. (1) twice, while invoking the boundary conditions

$$\frac{dy}{dx} = 0 \quad \text{at} \quad x = L \tag{4}$$

$$y = 0 \quad \text{at} \quad x = L \tag{5}$$

The analytical details are given in the solution to the preceding project, and the final result is

$$y_0 = \int_0^L \left(\int_x^L \frac{F\,x'\,dx'}{E(T')\,I(x')} \right) dx \tag{6}$$

The temperature T' corresponds to the longitudinal position x' (the dummy variable). The relation between T and x is obtained by integrating from x = 0 to any x the statement

$$q = k(T)\,BH(x)\,\frac{dT}{dx}, \qquad \text{(constant)} \tag{7}$$

in other words, by writing

$$q \int_0^x \frac{dx}{H(x)} = B \int_{T_0}^T k(T)\,dT \tag{8}$$

The blade profile is tapered,

$$H(x) = \overline{H}\left[1 + a\,(2\xi - 1)\right] \qquad (-1 < a < 1) \tag{9}$$

$$\xi = \frac{x}{L} \tag{10}$$

and the thermal conductivity varies linearly with the temperature,

$$k(T) = \bar{k}\left[1 + b\,(2\theta - 1)\right], \qquad (-1 < b < 1) \tag{11}$$

$$\theta = \frac{T - T_0}{T_L - T_0} \tag{12}$$

In these relations, $\overline{H}$ and $\bar{k}$ are the average values

$$\overline{H} = \frac{1}{L}\int_0^L H(x)\,dx \tag{13}$$

$$\bar{k} = \frac{1}{T_L - T_0}\int_{T_0}^{T_L} k(T)\,dT \tag{14}$$

Holding $\overline{H}$, B and L fixed means that the amount of structural material is fixed ($\overline{H}BL$). By substituting eqs. (9) and (11) in eq. (8) we obtain

$$\tilde{q}\int_0^{\xi} \frac{d\xi'}{1 + a\,(2\xi' - 1)} = \int_0^{\theta} \left[1 + b\,(2\theta' - 1)\right] d\theta' \tag{15}$$

where $\tilde{q}$ is the dimensionless total heat current,

$$\tilde{q} = \frac{q}{\bar{k}\,B\,\overline{H}\,(T_L - T_0)/L} \tag{16}$$

When $\xi = 1$ (and $\theta = 1$), eq. (15) yields

$$\tilde{q} = \frac{2a}{\ln \dfrac{1 + a}{1 - a}} \tag{17}$$

For any ξ (and θ) it yields, after using eq. (17),

$$\frac{\ln \dfrac{1 - a}{1 + a\,(2\xi - 1)}}{\ln \dfrac{1 - a}{1 + a}} = (1 - b)\,\theta + b\,\theta^2 \tag{18}$$

which is the relation between T and x, wanted for use in eq. (6). This relation affects the integrand through the modulus of elasticity

$$E(T) = \overline{E}\left[1 - c\,(2\theta - 1)\right], \quad (-1 < c < 1) \tag{19}$$

$$\overline{E} = \frac{1}{T_L - T_0}\int_{T_0}^{T_L} E(T)\, dT \tag{20}$$

The tip deflection integral can be summarized now in the dimensionless form

$$Y = \int_0^1 \left(\int_\xi^1 \frac{\xi'\, d\xi'}{\left[1 - c\,(2\theta' - 1)\right]\left[1 + a\,(2\xi' - 1)\right]^3} \right) d\xi \tag{21}$$

for which $\theta'(\xi')$ is given by eq. (18), and

$$Y = y_0 \frac{\overline{E}\, B\, \overline{H}^3}{12\, F\, L^3} \tag{22}$$

This formulation shows that Y depends on three parameters, one geometric (a), and the other related to the choice of structural material (b,c). The function Y(a,b,c) can be minimized numerically with respect to a: the attached graphs and tables show the optimal taper, $a_{opt}(b,c)$, and the corresponding minimum tip deflection, $Y_{min}(b,c)$, as functions of the material properties b and c. The actual (dimensional) minimum tip deflection depends on $\overline{E}$, b and c,

$$y_{0,min} = Y_{min}(b,c)\, \frac{12\, F\, L^3}{\overline{E}\, B\, \overline{H}^3}$$

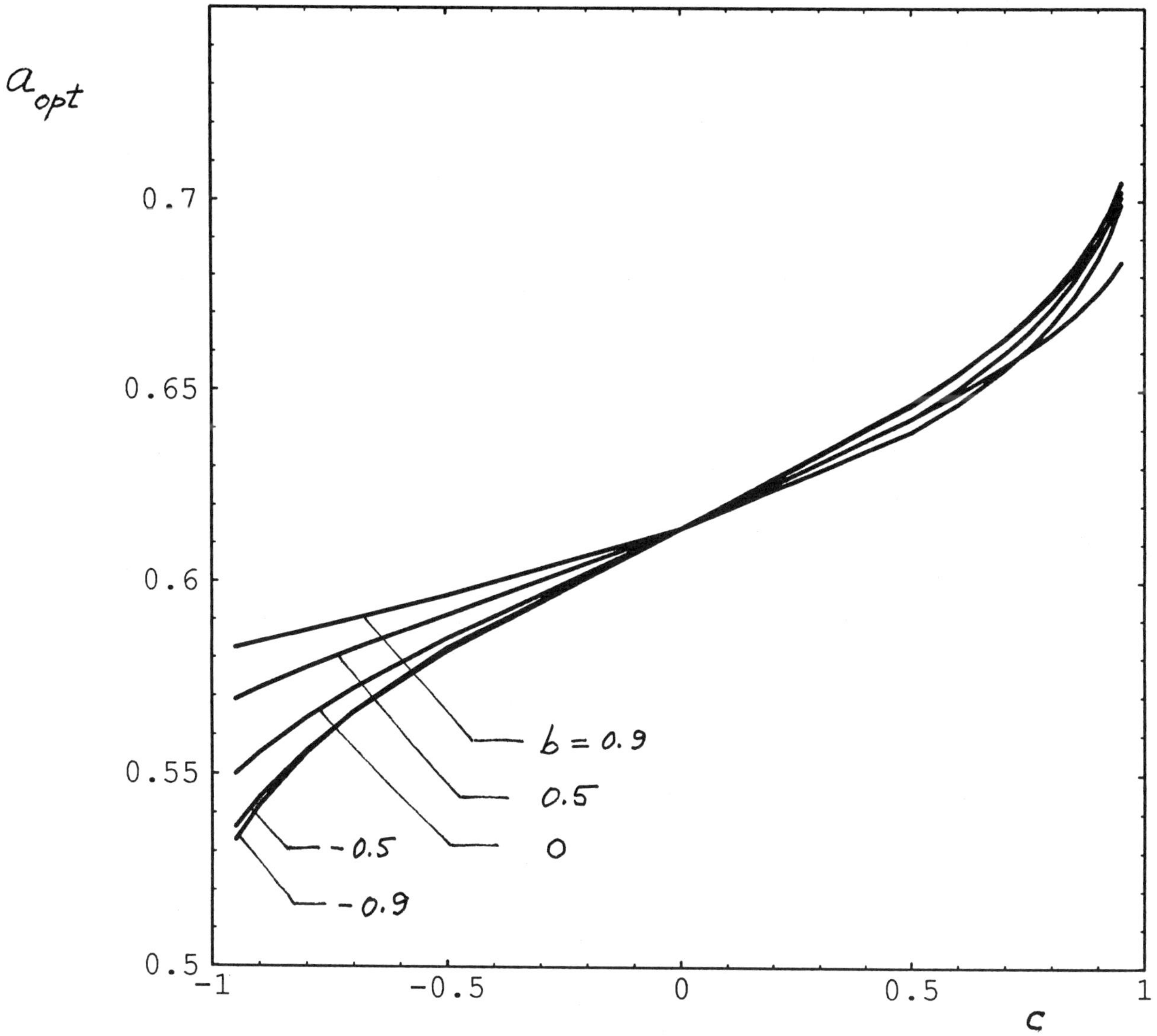

a_{opt}

c \ b	0.9	0.5	0	−0.5	−0.9
-0.95	0.582496	0.569011	0.549986	0.536414	0.533008
-0.9	0.584056	0.571945	0.555350	0.544022	0.541898
-0.8	0.587140	0.577259	0.564267	0.555986	0.555360
-0.7	0.590208	0.582141	0.571864	0.565688	0.565875
-0.5	0.596444	0.591304	0.585042	0.581668	0.582546
0	0.614024	0.614024	0.614024	0.614024	0.614024
0.5	0.639359	0.642745	0.64172	0.646687	0.642816
0.6	0.646687	0.650528	0.654315	0.654550	0.649288
0.7	0.655623	0.659793	0.663767	0.663508	0.656436
0.75	0.660997	0.665263	0.669237	0.668629	0.660405
0.8	0.667274	0.671565	0.675453	0.674358	0.664773
0.85	0.674884	0.679098	0.682771	0.681097	0.669735
0.9	0.684700	0.688665	0.691916	0.689408	0.675687
0.925	0.691035	0.694764	0.697673	0.694611	0.679310
0.95	0.699116	0.702472	0.704885	0.701117	0.683737

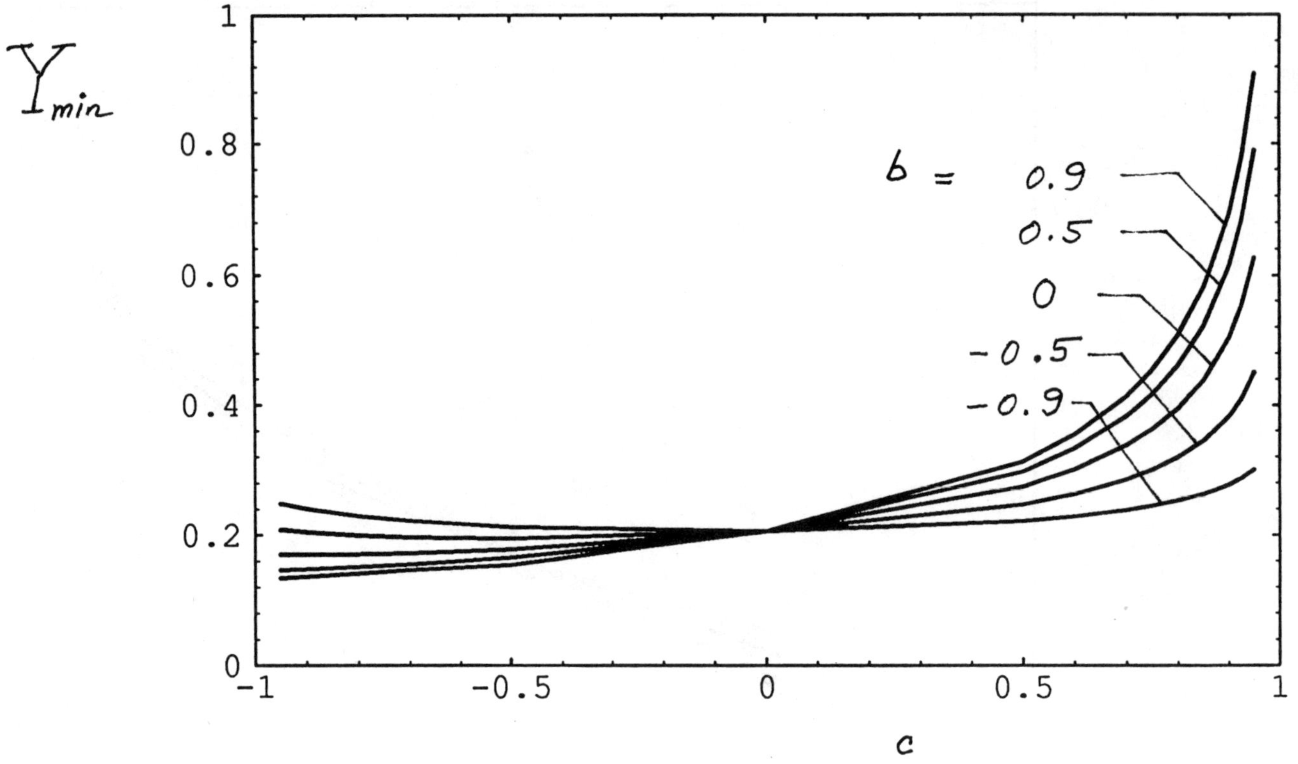

Y_{min}

c \ b	0.9	0.5	0	− 0.5	-0.9
-0.95	0.133789	0.145969	0.170625	0.208013	0.248141
-0.9	0.136010	0.147491	0.170277	0.204206	0.240234
-0.8	0.140792	0.151027	0.170698	0.199253	0.229046
-0.7	0.146068	0.155157	0.172182	0.196398	0.221310
-0.5	0.158383	0.165240	0.177570	0.194255	0.211572
0	0.205394	0.205394	0.205394	0.205394	0.205394
0.5	0.312255	0.297106	0.273981	0.246185	0.220695
0.6	0.354126	0.332660	0.300496	0.262445	0.227930
0.7	0.413554	0.382745	0.337530	0.284991	0.237915
0.75	0.454363	0.416908	0.362572	0.300068	0.244452
0.8	0.507559	0.461191	0.394780	0.319237	0.252537
0.85	0.581240	0.522124	0.438671	0.344955	0.262944
0.9	0.694237	0.614818	0.504626	0.382779	0.277304
0.925	0.780351	0.684966	0.553989	0.410515	0.287142
0.95	0.909114	0.789233	0.626656	0.450582	0.300410

Chapter 3

MULTIDIRECTIONAL STEADY CONDUCTION

Problem 3.1. The one-dimensional-model q_b formula (2.97) can be cast in the language of section 3.1.2. by setting $A_c = HW$ and $p = 2W$, which yields

$$q_{b,(2.97)} = k\theta_b W \left(\frac{2hH}{k}\right)^{1/2}$$

With this and the two-dimensional conduction formula (3.46), we construct the ratio

$$\frac{q_{b,(3.46)}}{q_{b,(2.97)}} = 2\left(\frac{2k}{hH}\right)^{1/2} \sum_{n=1}^{\infty} \frac{\sin^2(a_n)}{a_n + \sin(a_n)\cos(a_n)}$$

The sum can be evaluated by using only the first six characteristic values listed in Table 3.1. The q_b ratio emerges as a function of the Biot number hH/(2k): this function shows that the one-dimensional conduction model is adequate when the Biot number is small. The fact that the q_b ratio is smaller than 1 indicates that the one-dimensional model (2.97) overestimates the value of the total heat current through the plate fin.

$\frac{hH}{2k}$	$\frac{q_{b,(3.46)}}{q_{b,(2.97)}}$
0.01	0.998
0.1	0.986
1	0.902
3	0.797
10	0.619 (1.7%)
30	0.419 (3.9%)
100	0.238 (4.7%)

The percentages listed in parentheses show the relative contribution made to the q_b ratio by the inclusion of the sixth term in the series. We conclude then that the series converges less rapidly as the Biot number increases.

Problem 3.2. In terms of the temperature excess function $\theta = T - T_\infty$, we must solve the equation

$$\frac{\partial^2\theta}{\partial x^2} + \frac{\partial^2\theta}{\partial y^2} = -\frac{\dot{q}}{k} \qquad \text{(a)}$$

subject to the boundary conditions

$$\frac{\partial\theta}{\partial x} = 0 \quad \text{at} \quad x = 0 \qquad \text{(b)}$$

$$\theta = 0 \quad \text{at} \quad x = \frac{L}{2} \qquad \text{(c)}$$

$$\frac{\partial\theta}{\partial y} = 0 \quad \text{at} \quad y = 0 \qquad \text{(d)}$$

$$\theta = 0 \quad \text{at} \quad y = \frac{H}{2} \qquad \text{(e)}$$

Note the use of the symmetry conditions (b) and (d), in place of the boundary conditions at $x = -L/2$ and $y = -H/2$.

The $\theta(x,y)$ solution can be constructed by adding two simpler ones,

$$\theta(x,y) = \theta_1(y) + \theta_2(x,y) \qquad \text{(f)}$$

The complete problem statements for θ_1 and θ_2 are

$$\frac{d^2\theta_1}{dy^2} = -\frac{\dot{q}}{k} \qquad \frac{\partial^2\theta_2}{\partial x^2} + \frac{\partial^2\theta_2}{\partial y^2} = 0 \qquad \text{(g)}$$

$x = 0$: $\quad \frac{\partial\theta_2}{\partial x} = 0 \qquad$ (h)

$x = \frac{L}{2}$: $\quad \theta_2 = -\theta_1(y) \qquad$ (i)

$y = 0$: $\quad \frac{d\theta_1}{dy} = 0 \qquad \frac{d\theta_2}{dy} = 0 \qquad$ (j)

$y = \frac{H}{2}$: $\quad \theta_1 = 0 \qquad \theta_2 = 0 \qquad$ (k) where it should be noted that the governing equations <u>and</u> all the boundary conditions add up to the original problem statement, eqs. (a)-(e). The solution for $\theta_1(y)$ is straightforward,

$$\theta_1(y) = -\frac{\dot{q}}{2k}\left[y^2 - \left(\frac{H}{2}\right)^2\right] \qquad \text{(l)}$$

In the θ_2 problem, we note first the y direction as the direction of homogeneous boundary conditions. Therefore, we assign the characteristic functions (sin, cos) to the y direction:

$$\theta_2(x,y) = K[\cosh(\lambda x) + C_1 \sinh(\lambda x)]\,[\cos(\lambda y) + C_2 \sin(\lambda y)]$$

The constants C_1 and C_2 are both zero because of the θ_2 boundary conditions (h) and (j). Next, the condition (k) pinpoints the characteristic values,

$$\cos\left(\lambda \frac{H}{2}\right) = 0 \qquad \text{, or}$$

$$\lambda_n \frac{H}{2} = (2n+1)\frac{\pi}{2}, \qquad (n = 0, 1, 2 \ldots)$$

and the θ_2 solution reduces to

$$\theta_2 = \sum_{n=0}^{\infty} K_n \cosh(\lambda_n x)\cos(\lambda_n y) \qquad \text{(m)}$$

Finally, the last of the boundary conditions, eq. (i) requires

$$-\theta_1(y) = \sum_{n=0}^{\infty} K_n \cosh\left(\lambda_n \frac{L}{2}\right)\cos(\lambda_n y)$$

where θ_1 is listed in eq. (l). To determine K_n, we multiply both sides of the above equation by $\cos(\lambda_m y)$, and integrate from y = 0 to y = H/2. The end result for K_n, i.e. the end of the analytical solution is

$$K_n = \frac{(-1)^n\, 4H^2\, (-\dot{q}/k)}{(2n+1)^3\, \pi^3 \cosh[(2n+1)\,\pi L/2H]}$$

The maximum θ temperature occurs in the center of the rectangular cross-section,

$$\theta_{max} = \theta(0,0) = \theta_1(0) + \theta_2(0,0) \qquad \text{(n)}$$

where, from eq. (l),

$$\theta_1(0) = \frac{\dot{q}\, H^2}{8k} \tag{o}$$

In the case of a square cross-section, eq. (m) yields

$$\theta_2(0,0) = -\frac{4\,\dot{q}\, H^2}{\pi^3 k} \sum_{n=0}^{\infty} \frac{(-1)^2}{(2n+1)^3 \cosh[(2n+1)\,\pi/2]} \tag{p}$$

The sum listed above converges very rapidly to the value 0.39788 (only the first 2 terms are needed). Putting eqs. (n) - (p) together, we conclude that

$$\theta_{max} = 0.073671\, \frac{\dot{q}\, H^2}{k}$$

The corresponding result based on the integral analysis was

$$\theta_{max} = \frac{3}{32}\, \frac{\dot{q}\, H^2}{k}$$

This estimate is only 27 percent larger than the exact result.

Problem 3.3. The steady-state conduction equation that applies to Fig. 3.9b is eq. (1.46) without the $\partial^2 T/\partial z^2$ term,

$$\frac{1}{r}\frac{\partial}{\partial r}\left(r\frac{\partial T}{\partial r}\right) + \frac{1}{r^2}\frac{\partial^2 T}{\partial \theta^2} = 0 \tag{1.46}$$

in which r is the radial position and θ the angular position measured counterclockwise such that:

$$\theta = 0 \qquad \text{when} \qquad T = T_{hot}$$

$$\theta = \Delta\theta \qquad \text{when} \qquad T = T_{cold}$$

The first question we face is whether T can be a function of both r and θ. The boundary conditions are symmetric about the bisector $\theta = \Delta\theta/2$, therefore, the temperature field must remain unchanged as the entire picture is rotated by 180° around the $\theta = \Delta\theta/2$ axis. This means that the bisector is itself an isotherm, or that the heat flux near the bisector is oriented purely in the θ direction.

The same argument can be applied next to the upper half of the curved domain, $0 < \theta < \Delta\theta/2$, for which the new bisector is $\theta = \Delta\theta/4$. The punchline of the argument is that the line $\theta = \Delta\theta/4$ is also an isotherm. In this way, one can cover the entire field (using smaller segments, and rotating them around their bisectors) to conclude that all the radial lines are isotherms, i.e. that T does not depend on r.

Equation (1.46) reduces then to $d^2T/d\theta^2 = 0$, which can be solved by taking the listed boundary conditions into account,

$$T = T_{hot} - \frac{\Delta T}{\Delta\theta}\,\theta$$

where $\Delta T = T_{hot} - T_{cold}$. This solution shows that the temperature varies linearly in θ, which is also what the graphic solution of Fig. 3.9b shows.

The heat flux can finally be calculated using eq. (1.42θ),

$$q''_\theta = -\frac{k}{r}\frac{\partial T}{\partial\theta} = \frac{k}{r}\frac{\Delta T}{\Delta\theta}$$

According to this result, the heat flux intensifies (the flux lines become denser) as r decreases. This feature is revealed also by the graphic network solution constructed in Fig. 3.9b.

Problem 3.4. The new element in this problem is the boundary condition

$$\theta \to 0 \qquad \text{as} \qquad x \to \infty \tag{a}$$

which replaces eq. (3.6b). Since this condition will be invoked later, it is better to choose instead of eq. (3.11) a linear combination of $\exp(-\lambda x)$ and $\exp(\lambda x)$, i.e.

$$X = C_1\exp(-\lambda x) + C_2\exp(\lambda x)$$

In place of eq. (3.13) we now have the general solution

$$\theta = (C_1e^{-\lambda x} + C_2e^{\lambda x})\,(C_3 \sin \lambda y + C_4 \cos \lambda y)$$

The boundary conditions (a) and (3.6c) require that $C_2 = 0$ and $C_4 = 0$. The $y = H$ condition (3.6d) yields again the characteristic values listed in eq. (3.19) in the text. To summarize, instead of the solution (3.21) we obtain

$$\theta = \sum_{n=0}^{\infty} K_n e^{-\lambda_n x} \sin(\lambda_n y) \tag{b}$$

The final boundary condition, eq. (3.6a), requires

$$\theta_b = \sum_{n=0}^{\infty} K_n \sin(\lambda_n y)$$

Multiplying both sides with $\sin(\lambda_m y)$, and integrating both sides from $y = 0$ to $y = H$, yields

$$K_n = \frac{4\theta_b}{m\pi}, \quad \text{where } m = 2n + 1, \quad (n = 0, 1, 2, ...)$$

The analysis that leads to these K_n values is the same as in eqs. (3.23) - (3.26). Substituting the K_n values into the series solution (b) we obtain finally

$$\frac{\theta}{\theta_b} = \frac{4}{\pi} \sum_{n=0}^{\infty} \frac{1}{2n+1} \exp\left[- (2n + 1)\, \pi \frac{x}{H}\right] \sin\left[(2n + 1)\, \pi \frac{y}{H}\right]$$

Note that this result agrees with the $L/H \to \infty$ limit of eq. (3.28) in the text.

Problem 3.5. The general solution for the temperature distribution inside the semiinfinite rectangular domain is (see also the solution to Problem 3.4.):

$$\theta = (C_1 e^{-\lambda x} + C_2 e^{\lambda x})\,(C_3 \sin \lambda y + C_4 \cos \lambda y)$$

After invoking the $x \to \infty$, $y = 0$ and $y = H$ conditions, we are left with

$$\theta = \sum_{n=0}^{\infty} K_n \exp(- \lambda_n x) \sin(\lambda_n y) \qquad \text{(i)}$$

where $\lambda_n = n\pi/H$. The $x = 0$ boundary condition becomes, in order,

$$by = \sum_{n=0}^{\infty} K_n \sin(\lambda_n y)$$

$$\int_0^H by \sin(\lambda_n y)\, dy = K_n \int_0^H \sin^2(\lambda_n y)\, dy$$

$$n\pi \frac{b}{\lambda_n^2} (- 1)^{n+1} = K_n \frac{H}{2}$$

therefore,

$$K_n = \frac{2bH}{\pi} \frac{(- 1)^{n+1}}{n} \qquad \text{(ii)}$$

Combining (i) and (ii) we obtain the solution

$$\theta = \frac{2}{\pi} bH \sum_{n=0}^{\infty} \frac{(-1)^{n+1}}{n} \exp\left(-n\pi \frac{x}{H}\right) \sin\left(n\pi \frac{y}{H}\right)$$

in which bH is the highest temperature along the x = 0 wall.

Problem 3.6. The superposition scheme outlined in the problem statement is

$$\theta(x,y) = \theta_1(y) + \theta_2(x,y)$$

The θ_1 problem represents one-dimensional conduction (in the y direction) in a plane wall of thickness H:

$$\frac{d^2\theta_1}{dy^2} = 0, \qquad \theta_1(0) = 0, \qquad \theta_1(H) = \theta_a$$

The θ_1 distribution is therefore linear,

$$\theta_1 = \theta_a \frac{y}{H}$$

The θ_2 problem represents two-dimensional conduction in a semiinfinite plate with isothermal top and bottom faces, and linear temperature distribution over the left end (see the θ_1 solution):

$$\frac{\partial^2\theta_2}{dx^2} + \frac{\partial^2\theta_2}{dy^2} = 0$$

$$\theta_2 = -\theta_1(y) \qquad \text{at} \qquad x = 0$$

$$\theta_2 \rightarrow 0 \qquad \text{as} \qquad x \rightarrow \infty$$

$$\theta_2 = 0 \qquad \text{at} \qquad y = 0$$

$$\theta_2 = 0 \qquad \text{at} \qquad y = H$$

The solution to this problem is listed in the statement of Problem 3.5., in which the gradient b is now equal to $-\theta_a/H$:

$$\theta_2 = -\frac{\theta_a}{H} H \frac{2}{\pi} \sum_{n=0}^{\infty} \frac{(-1)^{n+1}}{n} \exp\left(-n\pi \frac{x}{H}\right) \sin\left(n\pi \frac{y}{H}\right)$$

In conclusion, adding θ_1 and θ_2, we obtain

$$\frac{\theta}{\theta_a} = \frac{y}{H} - \frac{2}{\pi} \sum_{n=0}^{\infty} \frac{(-1)^{n+1}}{n} \exp\left(-n\pi \frac{x}{H}\right) \sin\left(n\pi \frac{y}{H}\right).$$

Problem 3.7. Consider the temperature distribution θ(x,y) only in the "end" region, that is, in the region of height H and unknown length δ. This distribution is clearly two-dimensional in this region (θ is a function of both x and y), therefore θ must satisfy the steady two-dimensional conduction equation

$$\frac{\partial^2 \theta}{\partial x^2} + \frac{\partial^2 \theta}{\partial y^2} = 0 \qquad (1)$$

The orders of magnitude of the two terms on the left side of eq. (1) are

$$\frac{\partial^2 \theta}{\partial x^2} = \frac{\partial}{\partial x}\left(\frac{\partial \theta}{\partial x}\right) \sim \frac{\left(\frac{\partial \theta}{\partial x}\right)_\delta - \left(\frac{\partial \theta}{\partial x}\right)_0}{\delta - 0} \sim \frac{0 - \frac{\theta_a}{\delta}}{\delta - 0}$$

$$\sim -\frac{\theta_a}{\delta^2} \qquad (2)$$

$$\frac{\partial^2 \theta}{\partial y^2} = \frac{\partial}{\partial y}\left(\frac{\partial \theta}{\partial y}\right) \sim \frac{\left(\frac{\partial \theta}{\partial y}\right)_H - \left(\frac{\partial \theta}{\partial y}\right)_0}{H - 0} \sim \frac{\frac{\theta_a}{H} - 0}{H - 0}$$

$$\sim \frac{\theta_a}{H^2} \qquad (3)$$

The conduction equation (1) is the same as writing

$$-\frac{\partial^2 \theta}{\partial x^2} = \frac{\partial^2 \theta}{\partial y^2} \qquad (1')$$

therefore, substituting the scales (2) and (3),

$$\frac{\theta_a}{\delta^2} \sim \frac{\theta_a}{H^2} \qquad (4)$$

we conclude that δ ~ H. In other words, the length of the end region (δ) scales with (i.e. is of the same order of magnitude as) the short dimension of the conducting medium (H). This conclusion agrees with the exact drawing: note the general area in which the bends (knees) of the isotherms are located.

Problem 3.8. The temperature distribution inside the triangular cross-section is the same as in the bottom-left half of the square cross-section shown in the figure. Note that since the square boundary conditions are symmetric about the diagonal, the diagonal is also an adiabatic line just like the hypotenuse of the given triangle.

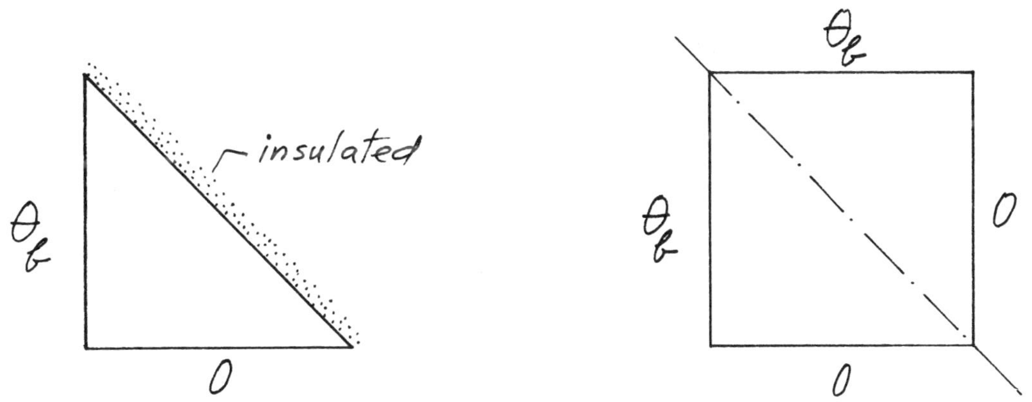

The solution for the temperature distribution inside the square cross-section can be determined by using the superposition rule seen already in Fig. 3.4 in the text,

$$\theta(x,y) = \theta_1(x,y) + \theta_2(x,y) \tag{a}$$

This time, however, the θ_2 problem has the $\theta = \theta_b$ condition specified on the top surface (y = L).

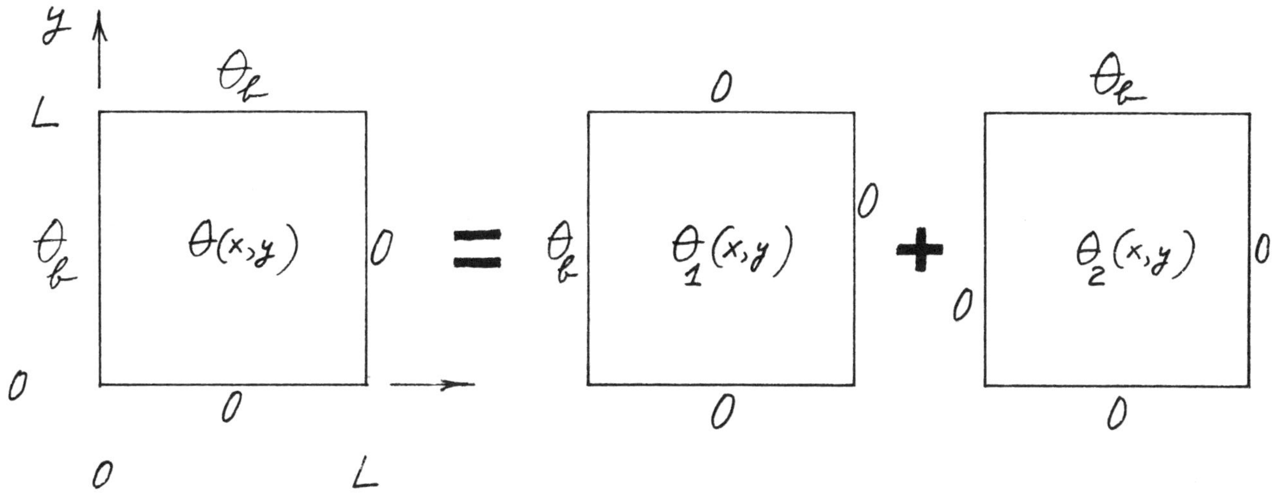

The solution to the first of the simpler problems was developed in the text [see Fig. 3.1 and eq. (3.28)]; therefore, since in the present case H = L, we conclude that

$$\theta_1(x,y) = \frac{4\theta_b}{\pi} \sum_{n=0}^{\infty} \frac{\sinh\left[(2n+1)\pi\left(1-\frac{x}{L}\right)\right]}{\sinh\left[(2n+1)\pi\right]} \cdot \frac{\sin\left[(2n+1)\pi\frac{y}{L}\right]}{2n+1} \qquad \text{(b)}$$

The $\theta_2(x,y)$ field is obtained by rotating the $\theta_1(x,y)$ field clockwise by 90 degrees. The figure below shows that this rotation amounts to replacing y with x, and x with (L - y), in the $\theta_1(x,y)$ expression (b).

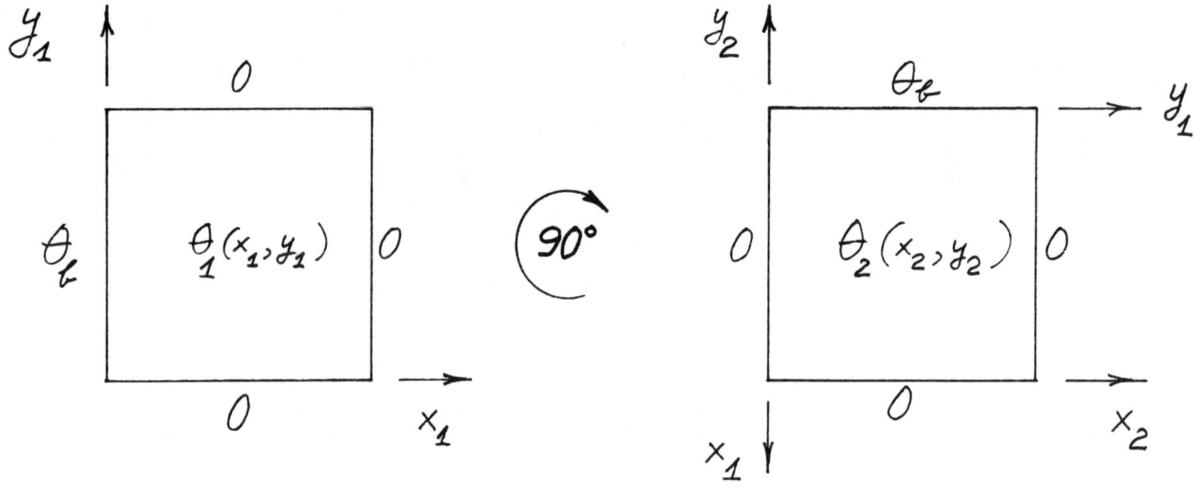

Therefore, the θ_2 solution is

$$\theta_2(x,y) = \frac{4\theta_b}{\pi} \sum_{n=0}^{\infty} \frac{\sinh\left[(2n+1)\pi\frac{y}{L}\right]}{\sinh\left[(2n+1)\pi\right]} \frac{\sin\left[(2n+1)\pi\frac{x}{L}\right]}{2n+1} \qquad \text{(c)}$$

Adding (b) and (c), we construct the solution for the square and the right-angle triangle specified in the problem statement.

Problem 3.9. If we define the excess temperature function $\theta = T - T_c$, two boundary conditions become homogeneous. This state of affairs is shown on the left side of the attached figure.

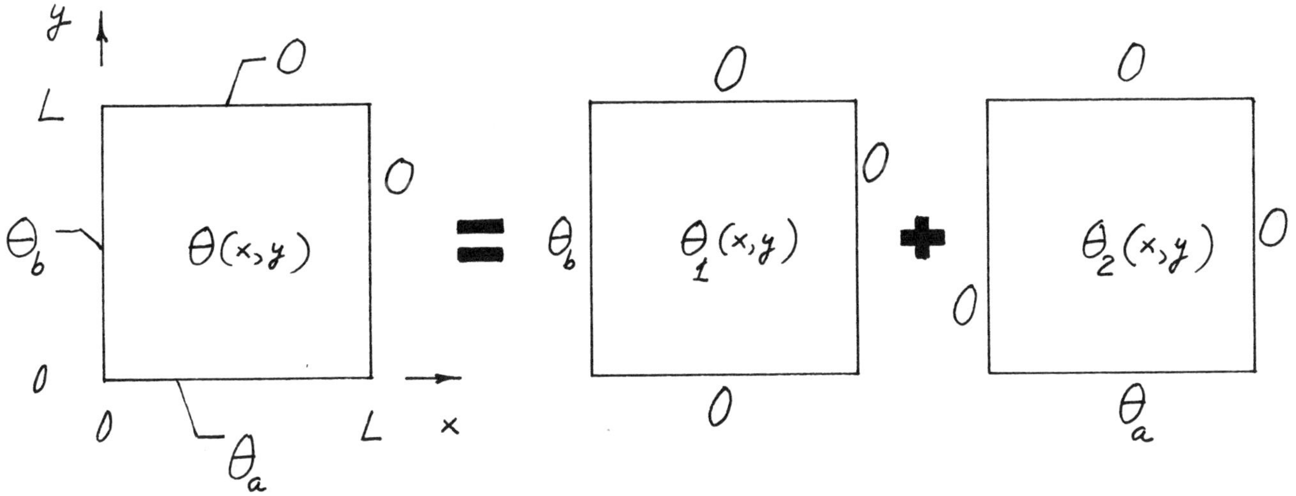

One additional homogeneous boundary condition (on the bottom side, or the left side) is needed in order to produce a direction of homogeneous boundary conditions. This second step is accomplished by the superposition operation shown on the right side of the figure

$$\theta(x,y) = \theta_1(x,y) + \theta_2(x,y)$$

The problem of determining θ_1 has been solved based on Fig. 3.1 in the text. The solution can be read off eq. (3.28), in which we set L = H because the present domain is square:

$$\theta_1 = \frac{4\theta_b}{\pi} \sum_{n=0}^{\infty} \frac{\sinh\left[(2n+1)\pi(1-x/L)\right]}{\sinh\left[(2n+1)\pi\right]} \frac{\sin\left[(2n+1)\pi y/L\right]}{2n+1}$$

Next, we note that the $\theta_2(x,y)$ problem is the same as the $\theta_1(x,y)$ problem rotated counterclockwise by 90 degrees. Therefore, the θ_2 solution can be obtained by replacing x with y, y with x, and θ_b with θ_a in the θ_1 solution:

$$\theta_2 = \frac{4\theta_a}{\pi} \sum_{n=0}^{\infty} \frac{\sinh\left[(2n+1)\pi(1-y/L)\right]}{\sinh\left[(2n+1)\pi\right]} \frac{\sin\left[(2n+1)\pi x/L\right]}{2n+1}$$

Finally, the temperature distribution in the original square domain is obtained by adding the θ_1 and θ_2 solutions listed above.

Problem 3.10. Expressed per unit length in the direction perpendicular to the plane of Fig. 3.1, the heat transfer rate through the "tip" (the x = L surface) is

$$q'_{tip} = \int_0^H -k \left(\frac{\partial \theta}{\partial x}\right)_{x=L} dy$$

The tip temperature gradient follows from the $\theta(x,y)$ solution displayed in eq. (3.28):

$$\left(\frac{\partial \theta}{\partial x}\right)_{x=L} = -\frac{4\theta_b}{H} \sum_{n=0}^{\infty} \frac{\sin[(2n+1)\pi y/H]}{\sinh[(2n+1)\pi L/H]}$$

Combining these two results, we obtain

$$q'_{tip} = \sum_{n=0}^{\infty} \frac{4k\theta_b}{H} \frac{1}{\sinh[(2n+1)\pi L/H]} \frac{H}{\pi(2n+1)} \underbrace{\int_0^{(2n+1)\pi} \sin\alpha \, d\alpha}_{2}$$

$$= \frac{8}{\pi} k\theta_b \sum_{n=0}^{\infty} \frac{1}{(2n+1)\sinh[(2n+1)\pi L/H]}$$

In conclusion, q'_{tip} is a function of the slenderness ratio L/H, and decreases sharply as L/H increases. In the case of L/H = 2, the infinite series is approximated very well by the first term,

$$\sum_{n=0}^{\infty} (\ ) = \frac{1}{1 \sinh 2\pi} + \frac{1}{3 \sinh 6\pi} + \ldots$$

$$= (3.74)\, 10^{-3} + (4.34)\, 10^{-9} + \ldots$$

and the tip heat transfer rate reduces to

$$q'_{tip} = \frac{8}{\pi}(0.00374)\, k\theta_b = 0.00952\, k\theta_b$$

Problem 3.11. a) Stated in terms of the excess temperature function $\theta = T - T_\infty$, both boundary conditions in the y direction are homogeneous. Therefore we assign sine and cosine functions in the y direction, and write the general solution as in eq. (3.13):

$$\theta(x,y) = \left(C_1' \sinh \lambda x + C_2' \cosh \lambda x\right)\left(C_3 \sin \lambda y + C_4 \cos \lambda y\right)$$

However, since the domain is semiinfinite in x, an even more convenient expression is the analogous form in which the hyperbolic functions are replaced by exponentials:

$$\theta(x,y) = \left(C_1 e^{\lambda x} + C_2 e^{-\lambda x}\right)\left(C_3 \sin \lambda y + C_4 \cos \lambda y\right)$$

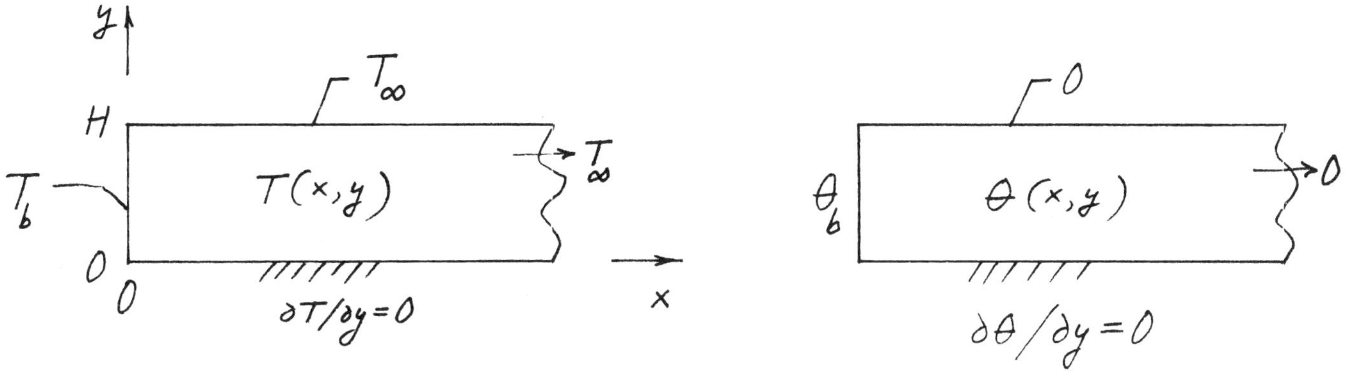

It is easy to see that the boundary condition "$\theta \to 0$ as $x \to \infty$" requires that $C_1 = 0$. Furthermore, the insulated bottom condition $\partial\theta/\partial y = 0$ at $y = 0$ requires $C_3 = 0$. Finally, the isothermal top condition $\theta = 0$ at $y = H$ requires

$$\cos \lambda H = 0, \qquad \text{or} \qquad \lambda_n H = (2n+1)\frac{\pi}{2}$$

$$n = 0,1,2,...$$

In conclusion, the $\theta(x,y)$ solution reduces to

$$\theta(x,y) = \sum_{n=0}^{\infty} K_n e^{-\lambda_n x} \cos \lambda_n y$$

for which the K_n coefficients are pinpointed by the last of the boundary conditions, $\theta = \theta_b$ at $x = 0$:

$$\theta_b = \sum_{n=0}^{\infty} K_n \cos \lambda_n y$$

Multiplying both sides by $\cos \lambda_m y$, and integrating from y = 0 to y = H we obtain, in order,

$$\int_0^H \theta_b \cos \lambda_n y \, dx = K_n \int_0^H \cos^2 \lambda_n y \, dy$$

$$\frac{\theta_b}{\lambda_n} \left| \sin \beta \right|_0^{(2n+1)\frac{\pi}{2}} = K_n \frac{1}{\lambda_n} \left| \frac{\beta}{2} + \frac{1}{4} \sin 2\beta \right|_0^{(2n+1)\frac{\pi}{2}}$$

$$\frac{\theta_b}{\lambda_n} (-1)^n = \frac{K_n}{\lambda_n} \frac{\pi}{4} (2n+1)$$

$$K_n = \frac{4}{\pi} \theta_b \frac{(-1)^n}{2n+1}$$

The solution for $\theta(x,y)$, or $T - T_\infty$ reads now

$$\theta = T - T_\infty = \frac{4}{\pi} \theta_b \sum_{n=0}^{\infty} \frac{(-1)^n}{2n+1} \exp\left[-(2n+1) \frac{\pi}{2} \frac{x}{H}\right] \cos\left[(2n+1) \frac{\pi}{2} \frac{y}{H}\right]$$

b) The problem solved in part (a) is the same as the $h = \infty$ (or $Bi = \infty$) case of the more general problem solved in the text based on Fig. 3.3. Note also that what in Fig. 3.3 was called H/2 is now called H. The solution to the general problem is given in eqs. (3.44)-(3.45), for which the a_n eigenvalues are listed in Table 3.1. The bottom row of that table shows that when $h \to \infty$

$$a_n = (2n+1) \frac{\pi}{2}, \qquad n = 0,1,2, \ldots$$

in which the n index begins with the value n = 0 (i.e. unlike the n index used in Table 3.1). In the same limit, the K_n coefficients of eq. (3.45) become

$$K_n = 2\theta_b \frac{(-1)^n}{(2n+1)\pi/2} = \frac{4}{\pi} \theta_b \frac{(-1)^n}{2n+1}$$

leading to eq. (3.44) and the eventual solution

$$\theta = \sum_{n=0}^{\infty} \frac{4}{\pi} \theta_b \frac{(-1)^n}{2n+1} e^{-\lambda_n x} \cos(\lambda_n y)$$

where $\lambda_n = a_n/H = (2n+1)\pi / (2H)$. This $\theta(x,y)$ solution is identical to the one developed in part (a).

Problem 3.12. Writing $\Delta T = 4°C$ for the temperature difference between the pipe wall and the freezing front, we note that the heat transfer rate from the pipe is equal to

$$q' = \frac{q}{L} = \frac{S}{L} k \, \Delta T$$

The shape factor per unit pipe length S/L is provided by the seventh entry in Table 3.3,

$$\frac{S}{L} = \frac{2\pi}{\cosh^{-1}(2H/D)} = 1.98, \text{ if } H = 3m$$

$$= 2.27, \text{ if } H = 2m$$

In the evaluation of the inverse hyperbolic cosine we relied on one formula out of Appendix E:

$$\cosh^{-1} u = \ln\left[u + \left(u^2 - 1\right)^{1/2}\right]$$

Before proceeding with the calculation of q', it is worth noting that in the present configuration the H/D ratio is large enough so that we could have used the simpler (approximate) formula listed also in Table 3.3:

$$\frac{S}{L} \cong \frac{2\pi}{\ln(4\,H/D)} = 1.98, \text{ if } H = 3m$$

$$= 2.27, \text{ if } H = 2m$$

Substituting these S/L values in eq. (a), and using $K = 1 W/m{\cdot}K$ and $\Delta T = 4°C$, we obtain

$$q' = \frac{S}{L} k \, \Delta T = 7.92 \frac{W}{m}, \text{ if } H = 3m$$

$$= 9.06 \frac{W}{m}, \text{ if } H = 2m$$

In conclusion, the heat transfer rate from the water pipe increases by 15 percent as the freezing front moves downward by 1m.

Problem 3.13. According to Table 3.3, the heat transfer rate from a sphere of diameter $D = 2\, r_o$ and temperature T_0 imbedded in an infinite stationary medium T_∞ is

$$q = S\, k_\infty \left(T_0 - T_\infty\right) = 2\pi D\, k_\infty \left(T_0 - T_\infty\right)$$

where k_∞ is the conductivity of the surrounding medium. The heat flux through the spherical surface is

$$q'' = \frac{q}{4\pi r_o^2} = \frac{2\pi\, 2\, r_o\, k_\infty \left(T_0 - T_\infty\right)}{4\pi r_o^2}$$

$$= \frac{k_\infty}{r_o}\left(T_0 - T_\infty\right)$$

which compared with $q'' = h\,(T_0 - T_\infty)$ states that the "heat transfer coefficient" at the spherical surface is

$$h = \frac{k_\infty}{r_o}$$

Consider now the overall thermal resistance felt by a sphere of radius r_i, surrounded by a spherical shell of conductivity k and outer radius r_o, imbedded in an infinite stationary medium of conductivity k_∞. According to eq. (2.41), we write

$$R_t = \frac{1}{4\pi k}\left(\frac{1}{r_i} - \frac{1}{r_o}\right) + \frac{1}{4\pi r_o^2 h}$$

In view of the conclusion that $h = k_\infty / r_o$, the dimensionless version of this result is

$$\left(4\pi\, r_i\, k_\infty\right) R_t = \frac{k_\infty}{k} + \frac{r_i}{r_o}\left(1 - \frac{k_\infty}{k}\right)$$

where the factor $(4\pi r_i k_\infty)$ is a constant. As the spherical shell becomes thicker (i.e. as r_o/r_i increases), the thermal resistance increases monotonically if $k_\infty > k$. In the opposite case, $k_\infty < k$, the thermal resistance decreases.

Said another way, the installation of a spherical shell has an "insulation" effect only if the shell material is less conductive than the outer medium. If the shell material is more conductive, the presence of the shell has a heat transfer augmentation effect.

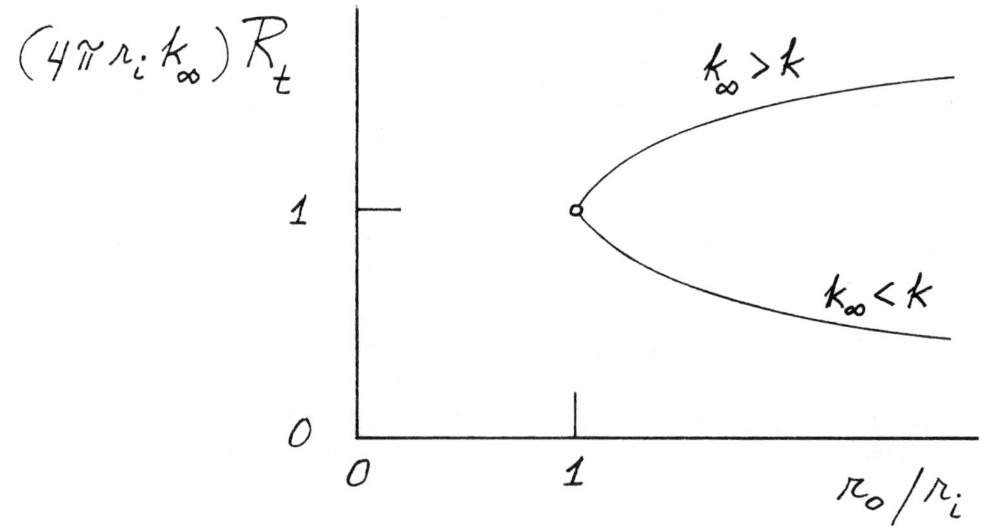

Problem 3.14. The heat transfer rate per unit of pipe length is given by the overall thermal resistance formula

$$q' = \frac{q}{L} = \frac{T_s - T_\infty}{R_t L} \tag{a}$$

If, as an approximation, we assume that heat is conducted radially through the annular layer of insulation, we may use eq. (2.42) to evaluate the overall thermal resistance.

$$R_t = \frac{\ln (D_o/D_i)}{2\pi k_{ins} L} + \frac{1}{\pi\, D_o L h} \tag{b}$$

in which h is the apparent heat transfer coefficient felt by the outer surface of the insulation layer. The coefficient h can be evaluated based on the definition of shape factor

$$S = \frac{q}{k_\infty \Delta T} = \frac{h\, \pi D_o L\, \Delta T}{k_\infty\, \Delta T} \tag{c}$$

which yields:

$$h = \frac{S\, k_\infty}{\pi D_o L} \tag{D}$$

From Table 3.3 we learn that

$$S \cong \frac{2\pi L}{\ln\left(4\, \frac{H}{D_o}\right)} \qquad \left(\text{note that } \frac{H}{D} > \frac{3}{2}\right) \tag{e}$$

and, combining eqs. (d) and (e) we obtain

$$h = \frac{k_\infty}{\pi D_o L}\, \frac{2\pi L}{\ln (4H/D_o)} = \frac{2k_\infty}{D_o \ln (4H/D_o)} =$$

$$= \frac{2 \times 1\text{W}}{\text{m·K}}\, \frac{1}{0.4\text{m} \ln\left(4\, \frac{2}{0.4}\right)} = 1.67\, \frac{\text{W}}{\text{m}^2\text{K}} \tag{f}$$

Now we can use eq. (b)

$$R_t L = \frac{\ln (D_o/D_i)}{2\pi k_{ins}} + \frac{1}{\pi D_o h}$$

$$= \frac{\ln (0.4/0.26)}{2\pi\ 0.2\text{W/m·K}} + \frac{1}{\pi\ 0.4\text{m}\ 1.67\ \text{W/m}^2\text{K}}$$

$$= 0.82\, \frac{\text{m·K}}{\text{W}}$$

and finally calculate the per-unit-length heat transfer rate from the steam pipe,

$$q' = \frac{T_s - T_\infty}{R_t L} = \frac{120°C - 10°C}{0.82\ m\cdot K/W} \cong 134\ \frac{W}{m}$$

Problem 3.15. The total heat transfer rate leaking through the insulation is

$$q = Sk\,(T_h - T_c)$$

where S is the overall conduction shape factor (cf. Table 3.3),

$$S = 2\,S_{plane\ xy} + 2\,S_{plane\ yz} + 2\,S_{plane\ xz} +$$

$$4\,S_{edge\ x} + 4\,S_{edge\ y} + 4\,S_{edge\ z} + 8\,S_{corner}$$

$$= 2\left(\frac{1 \times 0.5}{0.1} + \frac{0.5 \times 0.3}{0.1} + \frac{1 \times 0.3}{0.1}\right) m +$$

$$4 \times 0.54\,(1 + 0.5 + 0.3)\ m + 8 \times 0.15 \times 0.1\ m$$

$$= (19 + 3.89 + 0.12) = 23.01\ m$$

The total heat transfer rate is therefore

$$q = 23.01\ m\ 0.035\ \frac{W}{m\cdot K}\ 10°C \cong 8\ W$$

Problem 3.16. The control volume associated with the boundary node (i,j) has the height Δx and width $\Delta x/2$. Proceeding clockwise around the control volume, the heat currents that enter it from the north, east, south and west are

$$q_N \cong kW\,\frac{\Delta x}{2}\,\frac{T_{i,j+1} - T_{i,j}}{\Delta x}$$

$$q_W \cong kW\,\Delta x\,\frac{T_{i-1,j} - T_{i,j}}{\Delta x} \qquad\qquad q_E \cong q''W\,\Delta x$$

$$q_S \cong kW\,\frac{\Delta x}{2}\,\frac{T_{i,j-1} - T_{i,j}}{\Delta x}$$

In these expressions, W is the length of the control volume in the direction normal to the plane of the figure. The steady-state first law of thermodynamics requires,

$$q_N + q_E + q_S + q_W = 0$$

and, after using the finite difference expressions for the heat currents, we obtain

$$T_{i,j} \cong \frac{1}{4}(T_{i,j-1} + T_{i,j+1}) + \frac{1}{2}T_{i-1,j} + \frac{q''\Delta x}{2k}$$

The adiabatic-boundary limit of this result is the case represented by q" = 0,

$$T_{i,j} \cong \frac{1}{4}(T_{i,j-1} + T_{i,j+1} + 2\,T_{i-1,j})$$

This expression is identical to the h = 0 limit of the "surface node" formula listed in Table 3.4.

Problem 3.17. Recognizing again the symmetry about the $0 - T_3 - T_2 - \Delta T$ diagonal, half of the total heat current q leaves the square through the left-side surface,

$$\frac{q}{2} = q_e + q_f + q_g + q_h \qquad (1)$$

The four minicurrents $q_e,\ldots,q_h$ are associated with the four control volumes that touch the left boundary, therefore,

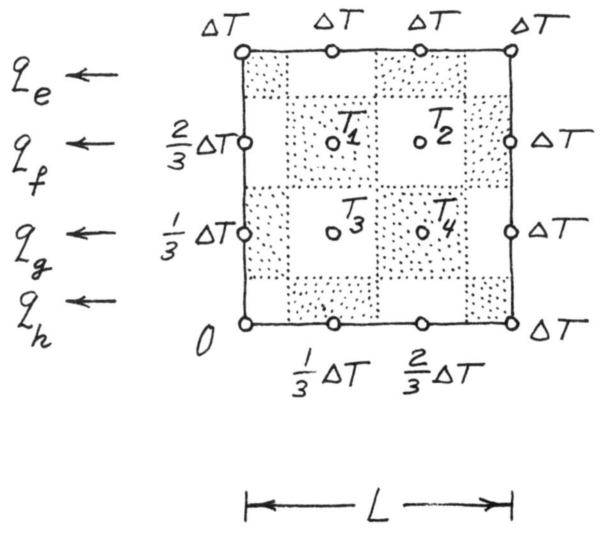

$$q_e \cong kW\frac{L}{6}\frac{\Delta T - \Delta T}{L/3}$$

$$q_f \cong kW\frac{L}{3}\frac{T_1 - 2\Delta T/3}{L/3}$$

$$q_g \cong kW\frac{L}{3}\frac{T_3 - \Delta T/3}{L/3}$$

$$q_h \cong kW\frac{L}{6}\frac{\Delta T/3 - 0}{L/3}$$

Substituting these estimates into eq. (1), we obtain

$$\frac{q}{2} \cong kW\,\Delta T\,(0.7778 - \frac{2}{3} + 0.5556 - \frac{1}{3} + \frac{1}{6})$$

$$\cong 0.5\ kW\,\Delta T$$

which translates into the same shape factor as in eq. (3.102),

$$S = \frac{q}{k\,\Delta T} \cong W$$

Problem 3.18. Consider the "internal" control volume that contains the ith node. This chunk has the volume $A_c\Delta x$. Proceeding clockwise around this control volume, we recognize the heat currents that enter from the north, east, south and west,

$$q_N \cong h\,\frac{p}{2}\,\Delta x\,(T_\infty - T_i)$$

$$q_W \cong k\,A_c\,\frac{T_{i-1} - T_i}{\Delta x} \qquad\qquad q_E \cong k\,A_c\,\frac{T_{i+1} - T_i}{\Delta x}$$

$$q_S \cong h\,\frac{p}{2}\,\Delta x\,(T_\infty - T_i)$$

In these relations, p is the total wetted perimeter of the A_c cross-section. Substituting these estimates into the steady-state first law requirement $q_N + q_E + q_S + q_W = 0$, we obtain

$$T_{i-1} - \left[2 + (m\Delta x)^2\right] T_i + T_{i+1} + (m\Delta x)^2\,T_\infty = 0 \qquad (1)$$

where m^2 is the fin parameter listed in eq. (2.92),

$$m^2 = \frac{hp}{k\,A_c}$$

The control volume attached to the tip node (n + 1) has the volume $A_c\Delta x/2$. The four heat currents entering the tip control volume are

$$q_N \cong h\,\frac{p}{2}\,\frac{\Delta x}{2}\,(T_\infty - T_{n+1})$$

$$q_W \cong k\,A_c\,\frac{T_n - T_{n+1}}{\Delta x} \qquad\qquad q_E \cong h\,A_c\,(T_\infty - T_{n+1})$$

$$q_S \cong h\,\frac{p}{2}\,\frac{\Delta x}{2}\,(T_\infty - T_{n+1})$$

The energy conservation statement $q_N + q_E + q_S + q_W = 0$ becomes now

$$T_n - \left[\frac{h\Delta x}{k} + 1 + \frac{1}{2}(m\Delta x)^2\right] T_{n+1} + \left[\frac{h\Delta x}{k} + \frac{1}{2}(m\Delta x)^2\right] T_\infty = 0 \qquad (2)$$

The insulated-tip limit of this last analysis is the case of $q_E = 0$, or $h\Delta x/k = 0$ in eq. (2).

Problem 3.19. The $\hat{\theta}$ version of the internal node temperature expression (3.87) is

$$\hat{\theta}_{i,j} \cong \frac{1}{4}(\hat{\theta}_{i-1,j} + \hat{\theta}_{i+1,j} + \hat{\theta}_{i,j-1} + \hat{\theta}_{i,j+1}) + \frac{1}{4N^2}$$

The numerical data that I obtained by following the steps (ii) - (ix) listed in the problem statement are:

ε	N	p	$\hat{\theta}_c$	Error	Cost
10^{-4}	4	12	0.070307	0.0456	108
	8	39	0.072746	0.0125	1911
	12	75	0.073177	0.0067	9075
	16	117	0.073262	0.0055	26325
	20	165	0.073238	0.0059	59565
10^{-5}	4	15	0.070312	0.0456	135
	8	53	0.072779	0.0121	2597
	12	108	0.073262	0.0055	13068
	16	177	0.073428	0.0033	39825
	20	258	0.073498	0.0023	93138

The Error and Cost values are displayed also in the attached figure. Generally speaking, a large N and a small ε mean high accuracy and high cost. The use of a N = 10 grid assures a better than 1 percent accuracy in the present problem. The value of the residue ε has a greater impact on both Error and Cost as the grid becomes finer.

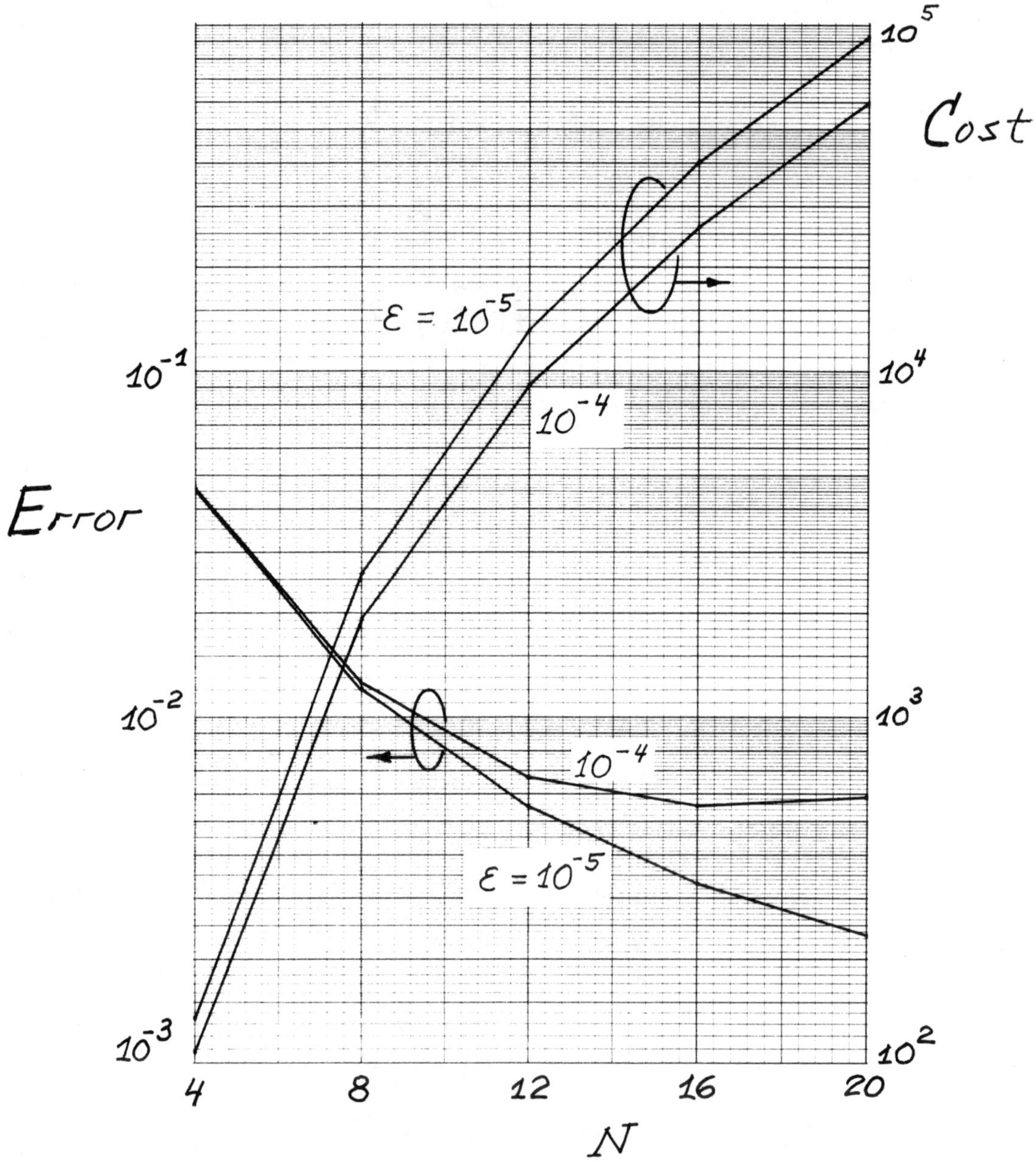

Cost
Error
$\varepsilon = 10^{-5}$
10^{-4}
10^{-4}
$\varepsilon = 10^{-5}$
10^{-1}
10^{-2}
10^{-3}
10^{5}
10^{4}
10^{3}
10^{2}
4
8
12
16
20
N

Problem 3.20. Consider the surface-plate segment of height Δy shown in the figure. The temperature of this infinitesimal segment is the boundary temperature T. In the steady state, the conservation of energy in the segment of height Δy and thickness δ requires (proceeding in the sequence west-south-east-north around the segment),

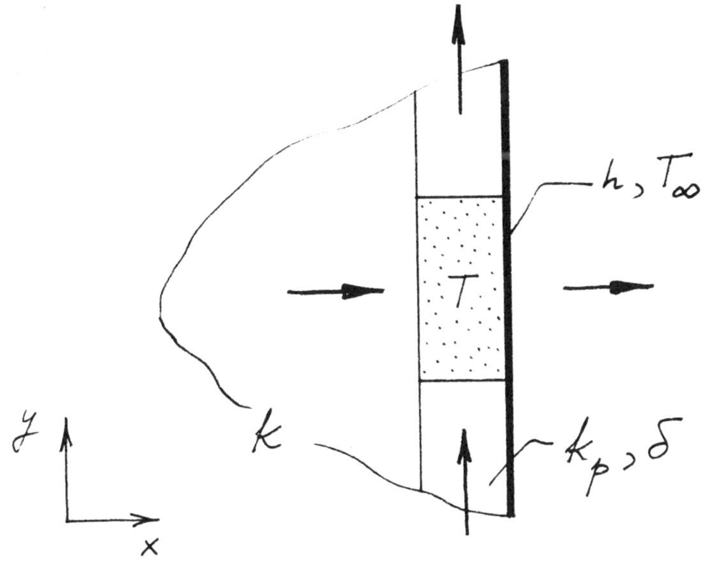

$$- k \, \Delta y \frac{\partial T}{\partial x} - k_p \, \delta \frac{\partial T}{\partial y} - h \, \Delta y \, (T - T_\infty) +$$

$$+ k_p \, \delta \frac{\partial T}{\partial y} + \frac{\partial}{\partial y} (k_p \, \delta \frac{\partial T}{\partial y}) \, \Delta y = 0$$

Dividing this equation by Δy yields the sought boundary condition

$$- k \frac{\partial T}{\partial x} = h \, (T - T_\infty) - k_p \, \delta \frac{\partial^2 T}{\partial y^2}$$

Problem 3.21. a) In the case of a surface node, we focus on the second diagram shown in Table 3.4, and write the heat currents* that enter the $(\Delta x) \cdot \left(\frac{\Delta x}{2}\right)$ control volume:

$$q'_N = k \frac{\Delta x}{2} \frac{T_{i,j+1} - T_{i,j}}{\Delta x}$$

$$q'_W = k \, \Delta x \frac{T_{i-1,j} - T_{i,j}}{\Delta x}, \qquad q'_E = h \, \Delta x \, (T_\infty - T_{i,j})$$

$$q'_S = k \frac{\Delta x}{2} \frac{T_{i,j-1} - T_{i,j}}{\Delta x}$$

The steady-state form of the first law of thermodynamics is

$$q'_N + q'_W + q'_S + q'_E + \dot{q} \, \Delta x \frac{\Delta x}{2} = 0$$

---

* Per unit length in the direction normal to the plane of the diagram shown in Table 3.4.

in which the last term represents the rate of heat generation inside the control volume. In the end, this first law statement reduces to

$$2\,T_{i-1,j} + T_{i,j+1} + T_{i,j-1} - 4\,T_{i,j} -$$

$$- 2\frac{h\Delta x}{k} T_{i,j} + 2\frac{h\Delta x}{k} T_{\infty} + \frac{\dot{q}}{k}(\Delta x)^2 = 0$$

or, in terms of the excess temperature function $\theta = T - T_{\infty}$,

$$2\,\theta_{i-1,j} + \theta_{i,j+1} + \theta_{i,j-1}$$

$$- 2\left(2 + \frac{h\Delta x}{k}\right)\theta_{i,j} + \frac{\dot{q}}{k}(\Delta x)^2 = 0$$

b) For an external corner node, we use the third diagram of Table 3.4 and identify all the heat currents that enter the control volume:

$$q'_N = h\frac{\Delta x}{2}(T_{\infty} - T_{i,j})$$

$$q'_W = k\frac{\Delta x}{2}\frac{T_{i-1,j} - T_{i,j}}{\Delta x}, \qquad q'_E = h\frac{\Delta x}{2}(T_{\infty} - T_{i,j})$$

$$q'_S = k\frac{\Delta x}{2}\frac{T_{i,j-1} - T_{i,j}}{\Delta x}$$

The rate of heat generation inside the control volume is $\dot{q}(\Delta x/2)\cdot(\Delta x/2)$, therefore the steady-state first law statement

$$q'_N + q'_W + q'_S + q'_E + \dot{q}\left(\frac{\Delta x}{2}\right)^2 = 0$$

becomes

$$T_{i-1,j} + T_{i,j-1} - 2\,T_{i,j} + 2\frac{h\Delta x}{k}(T_{\infty} - T_{i,j}) + \frac{\dot{q}}{2k}(\Delta x)^2 = 0$$

or, by setting $\theta = T - T_{\infty}$,

$$\theta_{i-1,j} + \theta_{i,j-1} - 2\left(1 + 2\frac{h\Delta x}{k}\right)\theta_{i,j} + \frac{\dot{q}}{2k}(\Delta x)^2 = 0$$

Problem 3.22. a) We cover the L × L domain with a uniform square mesh as on the right side of Fig. 3.12,

$$\Delta x = \Delta y = \frac{L}{N}$$

where N + 1 is the number of nodes in both directions (i = horizontal, j = vertical).

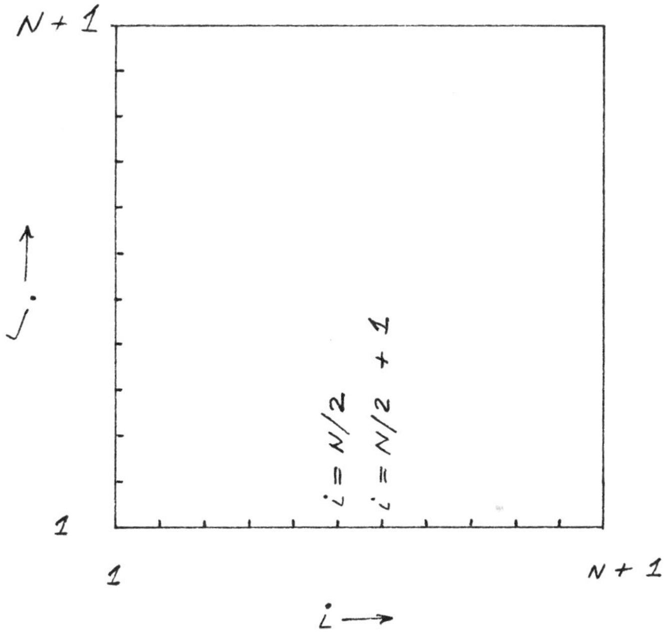

The finite difference equations are deduced from Table 3.4 (where h = 0 for the insulated surfaces):

Internal nodes (i = 2,3,...,N; j = 2,3,...,N):

$$T_{i,j} = \frac{1}{4}\left(T_{i-1,j} + T_{i+1,j} + T_{i,j-1} + T_{i,j+1}\right)$$

Top side (j = N + 1):

$$T_{i,N+1} = 10°C$$

Bottom side (j = 1):

$$T_{i,1} = 0°C, \qquad \left(\text{for } i = 1,2,..., \frac{N}{2} + 1\right)$$

$$2T_{i,2} + T_{i-1,1} + T_{i+1,1} - 4T_{i,1} = 0, \qquad \left(\text{for } i = \frac{N}{2} + 2,...,N\right)$$

$$T_{N,1} + T_{N+1,2} - 2T_{N+1,1} = 0, \qquad (\text{for } i = N + 1)$$

Left side ($i = 1; j = 2,...,N$):

$$2T_{2,j} + T_{1,j-1} + T_{1,j+1} - 4T_{1,j} = 0$$

Right side ($i = N + 1; j = 2,...,N$):

$$2T_{N,j} + T_{N+1,j-1} + T_{N+1,j+1} - 4T_{N+1,j} = 0$$

The convergence and accuracy of the Gauss-Seidel iteration procedure were determined by calculating the total heat transfer rate from T_h to T_c. This quantity was evaluated along the top of the $L \times L$ domain,

$$q'_{top} = k\frac{\Delta x}{2}\frac{T_{1,N+1} - T_{1,N}}{\Delta y} + \sum_{1=2}^{N} k\,\Delta x\frac{T_{i,N+1} - T_{i,N}}{\Delta y} + k\frac{\Delta x}{2}\frac{T_{N+1,N+1} - T_{N+1,N}}{\Delta y}$$

and along the horizontal midplane

$$q'_{mid} = k\frac{\Delta x}{2}\frac{T_{1,\frac{N}{2}+1} - T_{1,\frac{N}{2}}}{\Delta y} + \sum_{1=2}^{N} k\,\Delta x\frac{T_{i,\frac{N}{2}+1} - T_{i,\frac{N}{2}}}{\Delta y} + k\frac{\Delta x}{2}\frac{T_{N+1,\frac{N}{2}+1} - T_{N+1,\frac{N}{2}}}{\Delta y}$$

where $k = 0.1$ W/m·K. The units of q' are W/m.

The calculation was started by assigning 10°C and 0°C to the nodes on the heated and, respectively, cooled boundaries, and 5°C to the remaining nodes (internal, and on insulated surfaces). The acceptable relative error between two successive calculations of each node temperature

$$\max\left[\frac{T_{i,j}^{(new)} - T_{i,j}^{(old)}}{T_{i,j}^{(old)}}\right] \qquad (3.99)$$

$$i = 1,...,N + 1$$

$$j = 1,...,N + 1$$

was fixed at 10^{-4}. This was attained after a number of iterations (p), which is a function of the grid (N). The following table shows how the number of iterations and accuracy increase as N increases.

	grid, N × N			
	6 × 6	10 × 10	14 × 14	20 × 20
p	65	145	239	388
q'_{top}	0.85898	0.845016	0.83748	0.83378
q'_{mid}	0.85851	0.84355	0.84041	0.8399
$\dfrac{q'_{top} - q'_{mid}}{q'_{top}}$	5.4×10^{-4}	1.73×10^{-3}	3.4×10^{-3}	7.3×10^{-3}

The attached drawings show the temperature distribution plotted by using the 20×20 grid.

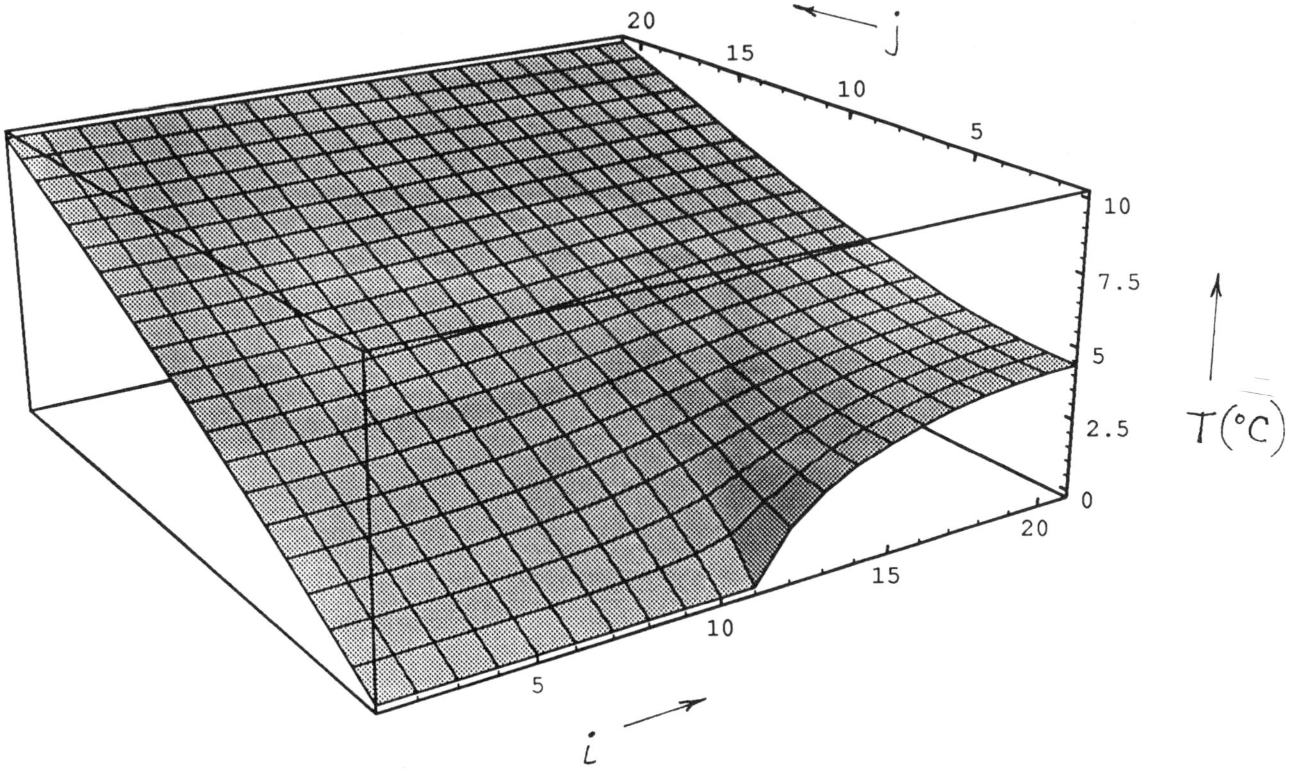

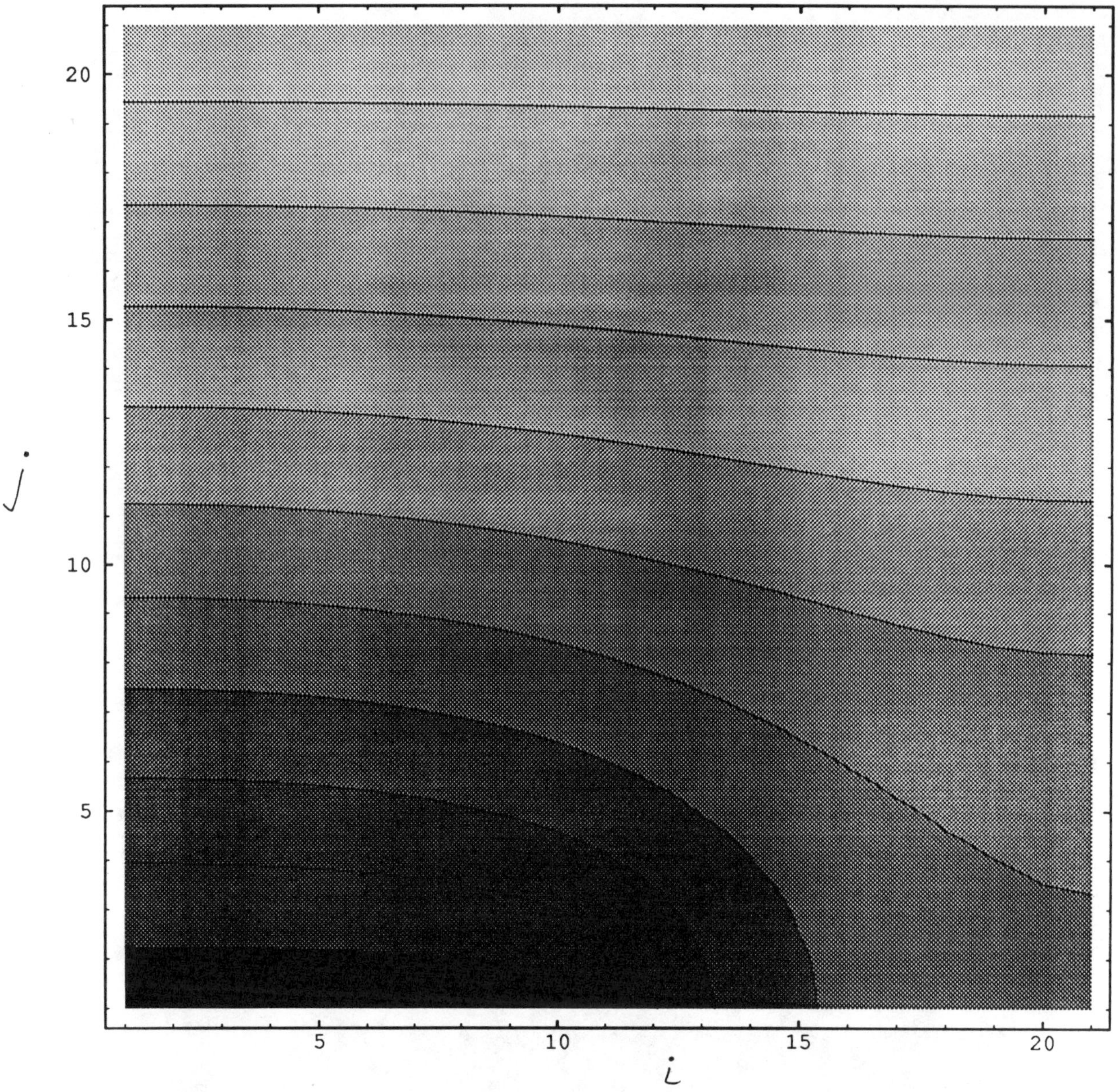

b) The tabulated q'_{top} and q'_{mid} values show that the total heat transfer rate approaches

$$q' \cong 0.83684 \ \frac{W}{m}$$

When the entire base is cooled to 0°C, the total heat transfer rate is

$$q'_0 = kL \frac{T_h - T_c}{L}$$

$$= 0.1 \frac{W}{m \cdot K} (10 - 0)\ K = 1 \frac{W}{m}$$

We conclude that

$$\frac{q'}{q'_0} = \frac{0.83684}{1} = 0.83684$$

or that the strangling of the flow of heat reduces the heat transfer rate by 16.31 percent.

Problem 3.23. a) We cover the L × L area with a uniform square grid,

$$\Delta x = \Delta y = \frac{L}{N}$$

and count the grid lines in terms of i horizontally, and j vertically.

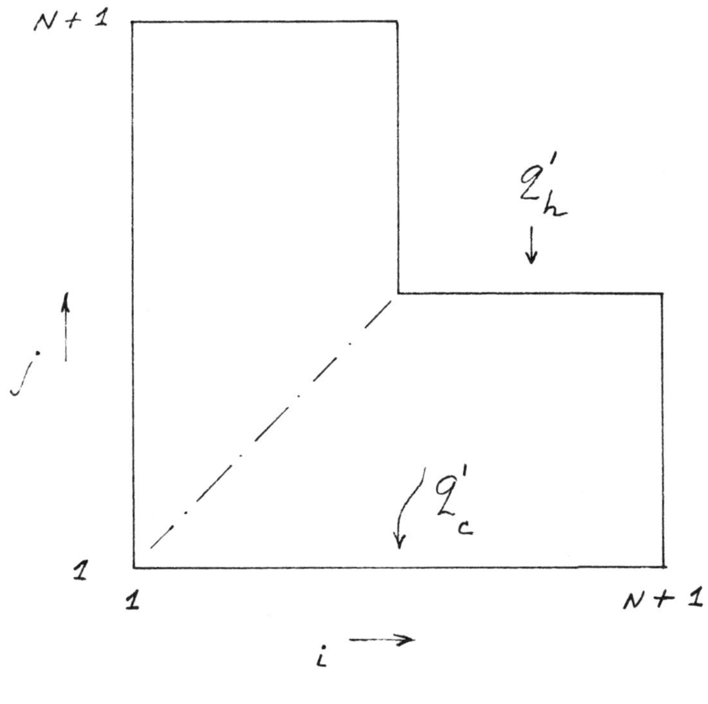

Table 3.4 is the guide for writing the finite difference equations:

Internal nodes:

$$T_{i,j} = \frac{1}{4}\left(T_{i-1,j} + T_{i+1,j} + T_{i,j-1} + T_{i,j+1}\right)$$

Hot boundaries:

$$T_{i,j} = 20°C, \text{ for } \left(i = \frac{N}{2} + 1; j = \frac{N}{2} + 1, ..., N + 1\right)$$
$$\text{and } \left(i = \frac{N}{2} + 1, ..., N + 1; j = \frac{N}{2} + 1\right)$$

Cold boundaries:

$$T_{i,j} = 0°C, \text{ for } (i = 1; j = 1, ..., N + 1)$$
$$\text{and } (i = 1, ..., N + 1; j = 1)$$

Insulated top side ($i = 2, ..., \frac{N}{2}; j = N + 1$):

$$2T_{i,N} + T_{i-1,N+1} + T_{i+1,N+1} - 4T_{i,N+1} = 0$$

Insulated right side ($i = N + 1; j = 2, ..., \frac{N}{2}$):

$$2T_{N,j} + T_{N+1,j-1} + T_{N+1,j+1} - 4T_{N+1,j} = 0$$

The Gauss-Seidel procedure begins with setting $T_{i,j} = (T_h + T_c)/2 = 10°C$ at the nodes situated on the insulated boundaries and in the interior. The convergence and accuracy was monitored by calculating the maximum relative change between two consecutive calculations of each node temperature,

$$\max\left[\frac{T_{i,j}^{(new)} - T_{i,j}^{(old)}}{T_{i,j}^{(old)}}\right] \tag{3.99}$$

Convergence was achieved when this quantity became less than 10^{-5}.

The heat transfer rate through half of the corner-shaped system was calculated in two ways, through a hot side of length L/2,

$$q'_h = k\,\Delta x \frac{T_{\frac{N}{2}+1,\frac{N}{2}+1} - T_{\frac{N}{2}+1,\frac{N}{2}}}{\Delta y} +$$

$$+ \sum_{i=\frac{N}{2}+2}^{N} k\,\Delta x \frac{T_{i,\frac{N}{2}+1} - T_{i,\frac{N}{2}}}{\Delta y}$$

$$+ k\frac{\Delta x}{2} \frac{T_{N+1,\frac{N}{2}+1} - T_{N+1,\frac{N}{2}}}{\Delta y}$$

and through the corresponding cold side of length L,

$$q'_c = k\frac{\Delta x}{2}\frac{T_{1,2} - T_{1,1}}{\Delta y} + \sum_{i=2}^{N} k\,\Delta x \frac{T_{i,2} - T_{i,1}}{\Delta y} + k\frac{\Delta x}{2}\frac{T_{N+1,2} - T_{N+1,1}}{\Delta y}$$

where k = 0.5 W/m·K.

The number of iterations p where convergence is achieved depends on the grid fineness, as shown in the table. The two drawings show the temperature distribution based on the 22 × 22 grid.

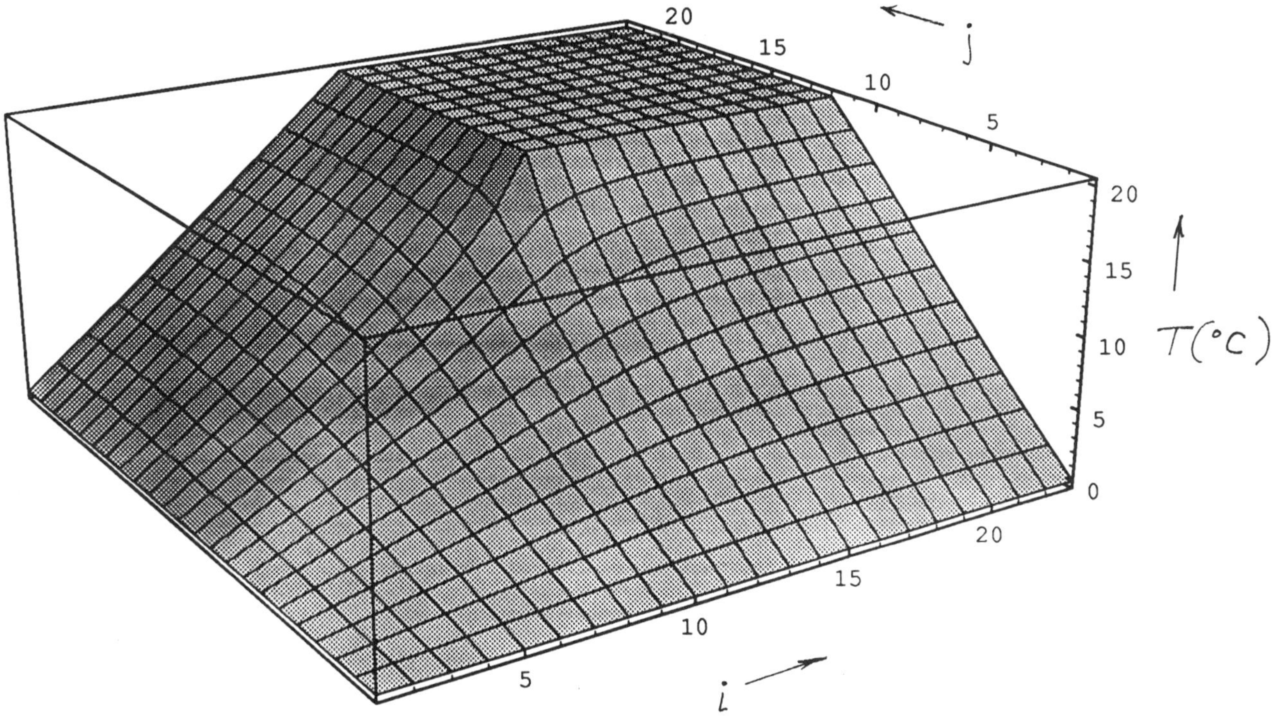

	grid, N × N		
	10 × 10	16 × 16	22 × 22
p	37	113	208
q'_h	13.0021	12.9022	12.8632
q'_c	13.0017	12.903	12.865
$\frac{q'_h - q'_c}{q'_h}$	2.4×10^{-5}	6.5×10^{-5}	1.4×10^{-4}

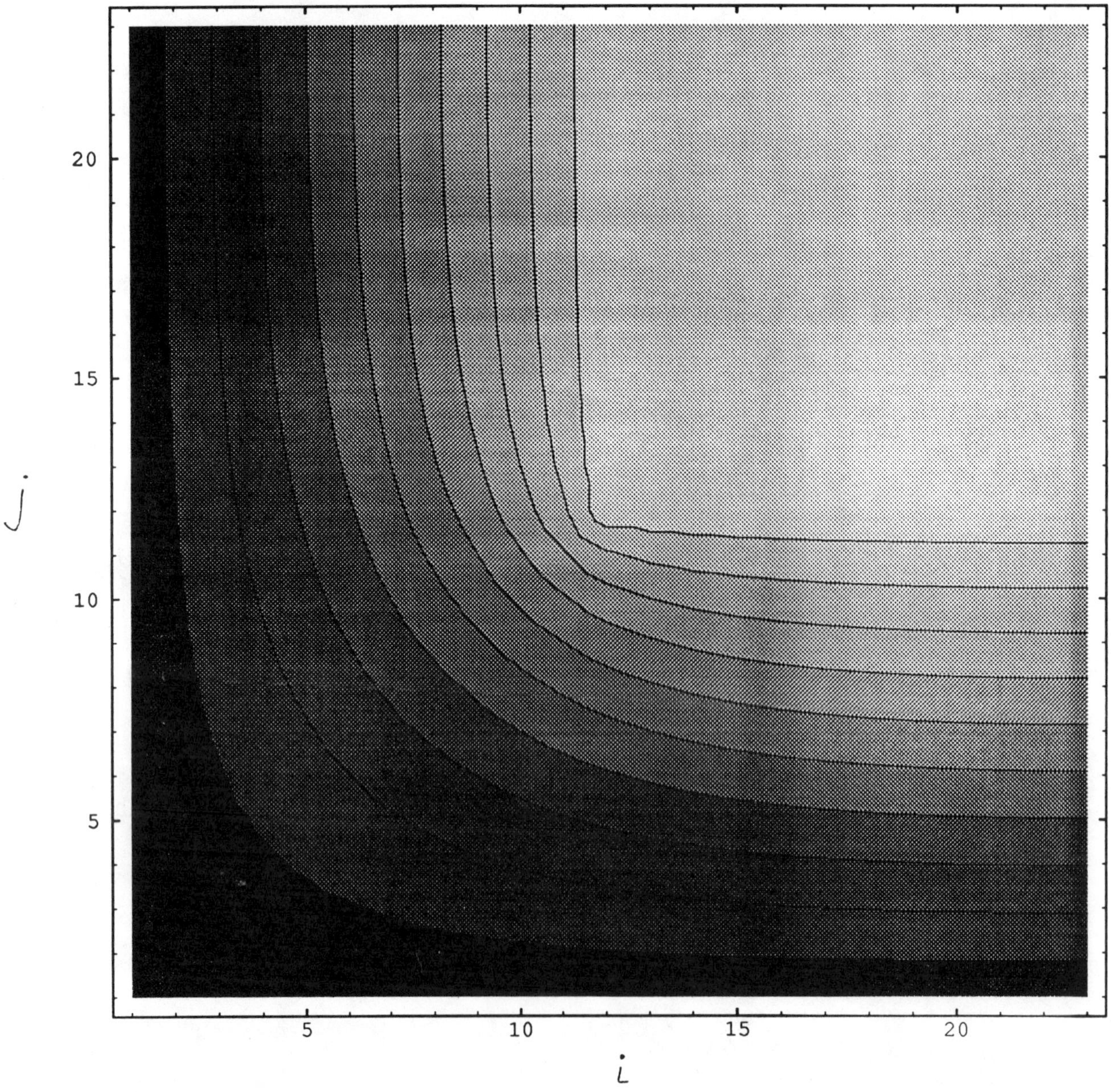

b) The numerical solution performed in part (a) showed that the heat transfer rate q' through half of the corner-shaped region (i.e. q_h', or q_c') is

$$q' = 12.865 \frac{W}{m}$$

The same quantity can be estimated by using the shape factor information listed in Table 3.3:

$$2q = k(T_h - T_c)(S_{edge\ prism} + 2S_{plane\ wall})$$

$$= k(T_h - T_c)\left(0.54W + 2\frac{W\frac{L}{2}}{\frac{L}{2}}\right)$$

Noting that q' = q/W, we obtain

$$q' = k(T_h - T_c)\frac{0.54 + 2}{2}$$

$$= 0.5\frac{W}{m \cdot K}\, 20\, K\, \frac{2.54}{2} = 12.7\frac{W}{m}$$

which is only 1.3 percent smaller than the numerical estimate produced in part (a).

Problem 3.24. a) We cover the $L \times 2L$ with a uniform square grid,

$$\Delta x = \Delta y = \frac{L}{N}$$

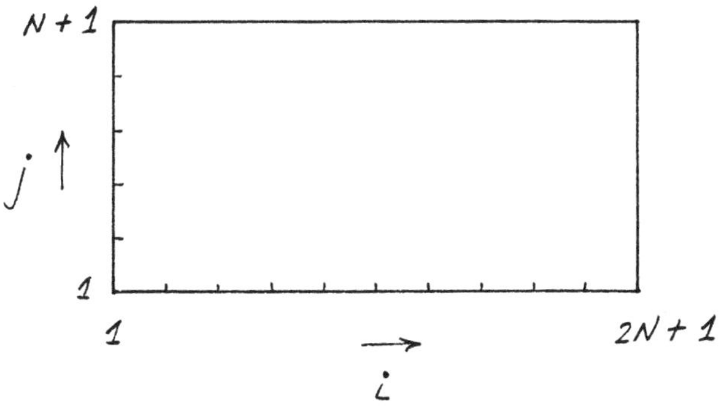

so that i = 1,2,...,2N + 1 horizontally, and j = 1,2,...,N + 1 vertically. Table 3.4 provides the finite-difference equations:

Internal nodes:

$$T_{i,j} = \frac{1}{4}\left(T_{i-1,j} + T_{i+1,j} + T_{i,j-1} + T_{i,j+1}\right)$$

Left side (i = 1, j = 1,2,...,N + 1):

$$T_{1,j} = 40°C$$

Bottom side (j = 1; i = 2,...,2N):

$$T_{i,1} = \left[T_{i,2} + \frac{1}{2}\left(T_{i+1,1} + T_{i-1,1}\right) + Bi \bullet T_\infty\right]\frac{1}{2 + Bi},$$

$$\text{where } Bi = \frac{h\,\Delta x}{k}$$

Bottom-right corner (i = 2N + 1; j = 1):

$$T_{2N+1,1} = \left[\frac{1}{2}\left(T_{2N,1} + T_{2N+1,2}\right) + Bi \cdot T_\infty\right]\frac{1}{1 + Bi}$$

Right side (i = 2N + 1); j = 2,...,N):

$$T_{2N+1,j} = \left[T_{2N,j} + \frac{1}{2}\left(T_{2N+1,j+1} + T_{2N+1,j-1}\right) + Bi \cdot T_\infty\right]\frac{1}{2 + Bi}$$

Top side (i = 2,3,...,2N; j = N + 1):

$$T_{i,N+1} = \left[T_{i,N} + \frac{1}{2}\left(T_{i+1,N+1} + T_{i-1,N+1}\right) + Bi \cdot T_\infty\right]\frac{1}{2 + Bi}$$

Top-right corner (i = 2N + 1; j = N + 1):

$$T_{2N+1,N+1} = \left[\frac{1}{2}\left(T_{2N+1,N} + T_{2N,N+1}\right) + Bi \cdot T_\infty\right]\frac{1}{1 + Bi}$$

The heat transfer rate from T_b to T_∞ was calculated by summing up the contributions made by the three surfaces with convective heat transfer

$$q' = q'_{top} + q'_{bottom} + q'_{right\ side}$$

$$\frac{1}{k} q'_{top} = \frac{\Delta x}{2} \frac{T^*_{1,N} - T^*_{1,N+1}}{\Delta y} +$$

$$+ \sum_{i=2}^{2N} \Delta x \frac{T_{i,N} - T_{i,N+1}}{\Delta y}$$

$$+ \frac{\Delta x}{2} \frac{T_{2N+1,N} - T_{2N+1,N+1}}{\Delta y}$$

$$\text{with } T^*_{1,N} = \frac{3T_{1,N} + T_{2,N}}{4}$$

$$T^*_{1,N+1} = \frac{3T_{1,N+1} + T_{2,N+1}}{4}$$

$$\frac{1}{k} q'_{bottom} = \frac{\Delta x}{2} \frac{T^*_{1,2} - T^*_{1,1}}{\Delta y} +$$

$$+ \sum_{i=2}^{2N} \Delta x \frac{T_{i,2} - T_{i,1}}{\Delta y}$$

$$+ \frac{\Delta x}{2} \frac{T_{2N+1,2} - T_{2N+1,1}}{\Delta y}$$

$$\text{with } T^*_{1,2} = \frac{3T_{1,2} + T_{2,2}}{4}$$

$$T^*_{1,1} = \frac{3T_{1,1} + T_{2,1}}{4}$$

$$\frac{1}{k} q'_{\substack{right\\side}} = \frac{\Delta y}{2} \frac{T_{2N,1} - T_{2N+1,1}}{\Delta x} +$$

$$+ \sum_{j=2}^{N} \Delta y \frac{T_{2N,j} - T_{2N+1,j}}{\Delta x}$$

$$+ \frac{\Delta y}{2} \frac{T_{2N,N+1} - T_{2N+1,N+1}}{\Delta x}$$

The Gauss-Seidel method was used, by starting with $T_{i,j} = (T_b + T_\infty)/2 = 30°C$ at all the interior and convective-surface nodes. The solution was considered converged when the maximum relative change in each node temperature

$$\max\left[\frac{T_{i,j}^{(new)} - T_{i,j}^{(old)}}{T_{i,j}^{(old)}}\right]$$

became less than 10^{-5}. The number of iterations where convergence occurs (p) depends on the grid fineness, as shown in the following table (note: k = 10W/m·K, h = 500 W/m²K, and the units of all q' values are W/m):

	grid, N × 2N			
	6 × 12	10 × 20	14 × 28	16 × 32
p	222	507	854	1043
q'_{bottom}	62.0177	70.098	74.0499	75.3995
q'_{top}	62.0341	70.125	74.0857	75.4401
q'	152.99	165.1	171.44	173.709

b) The following drawings illustrate the temperature distribution calculated based on the N = 14 grid. They show that the temperature varies visibly in the j direction, unlike in the unidirectional conduction model used for treating fins in section 2.7. This should have been expected, because the unidirectional conduction requirement (2.129) is not satisfied,

$$\left(\frac{hL}{k}\right)^{1/2} = \left(\frac{500\text{ W}}{\text{m}^2\text{K}}\, 0.01\text{m}\, \frac{\text{m}\cdot\text{K}}{10\text{ W}}\right)^{1/2} = 0.71$$

This criterion shows that the Biot number hL/k is not much smaller than 1.

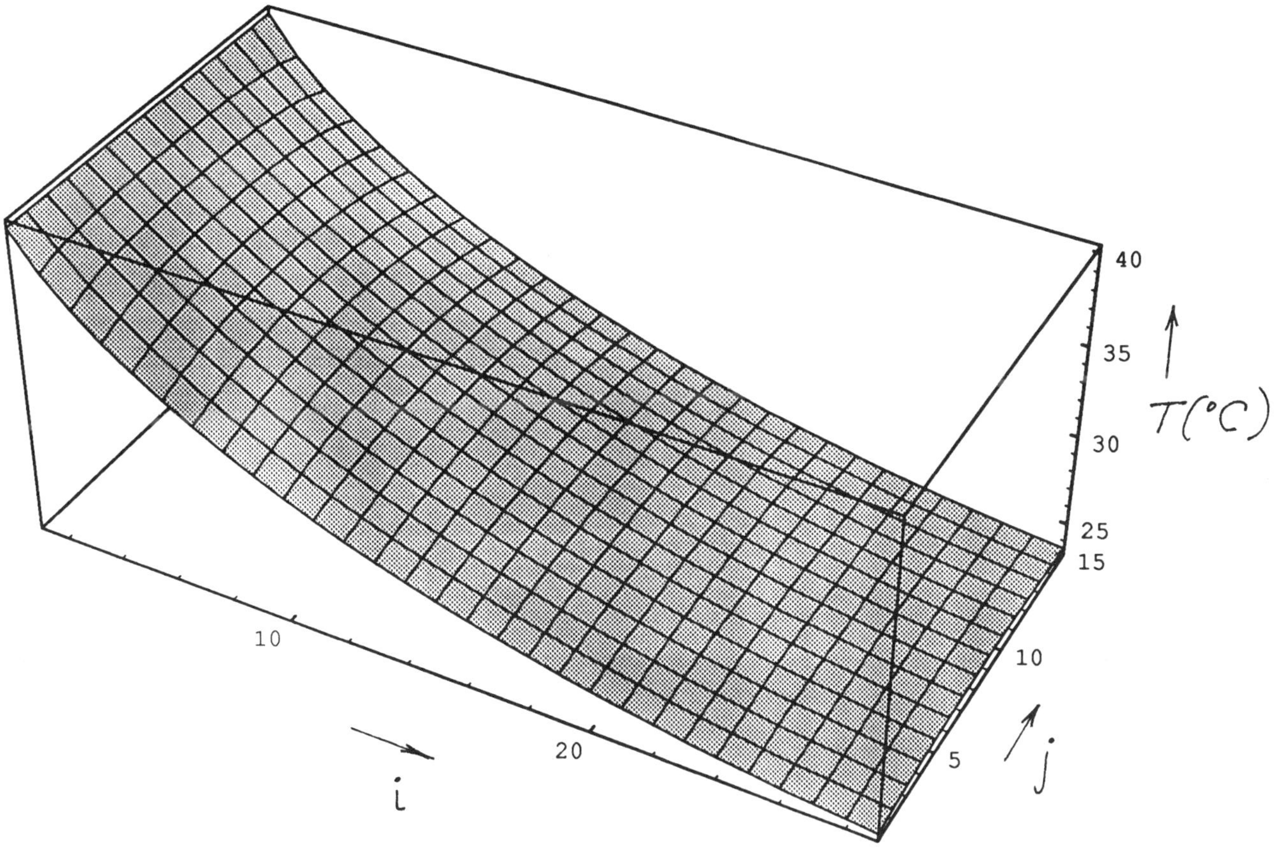
40
35
30
25
T(°C)
15
10
5
j
10
20
i

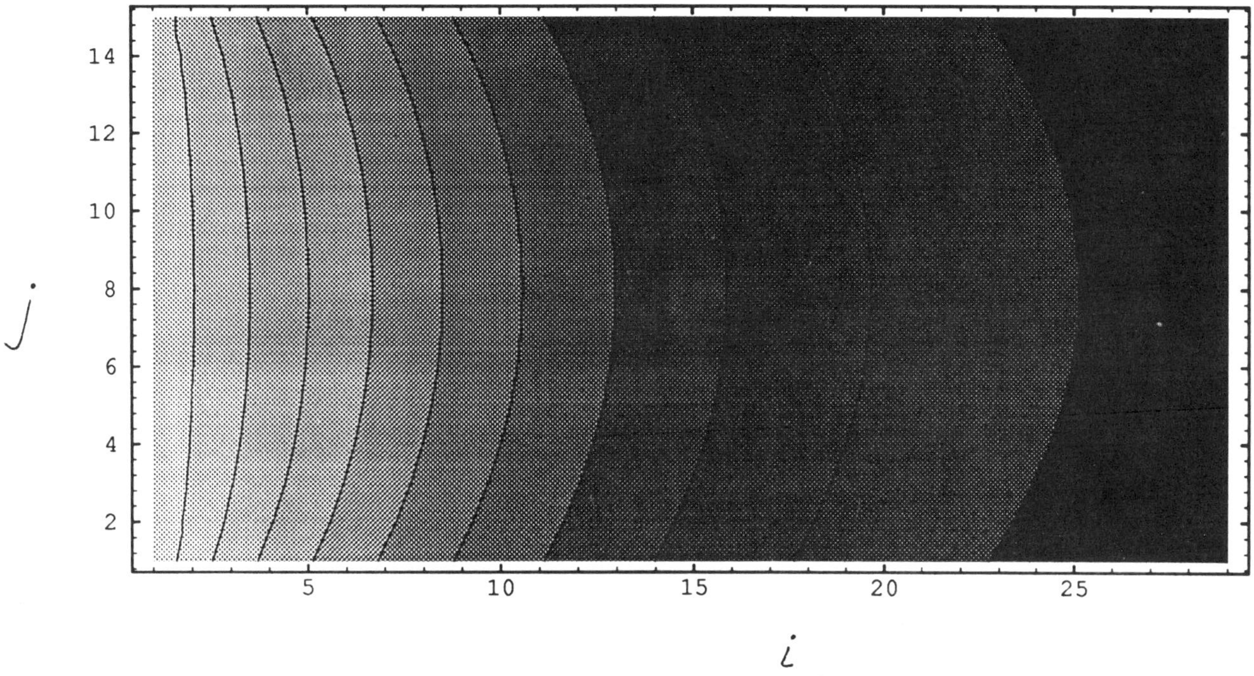
14
12
10
8
6
4
2
j
5
10
15
20
25
i

Problem 3.25. a) We cover the L × 2L rectangle with a uniform square grid of spacing

$$\Delta x = \Delta y = \frac{L/2}{n-1}$$

such that

$$j = 1,2,...,n \qquad \text{vertically}$$

and

$$i = 1,2,...,m \qquad \text{horizontally}$$

and where $m - 1 = 2\,(n - 1)$, to match the geometric aspect ratio of the rectangular domain.

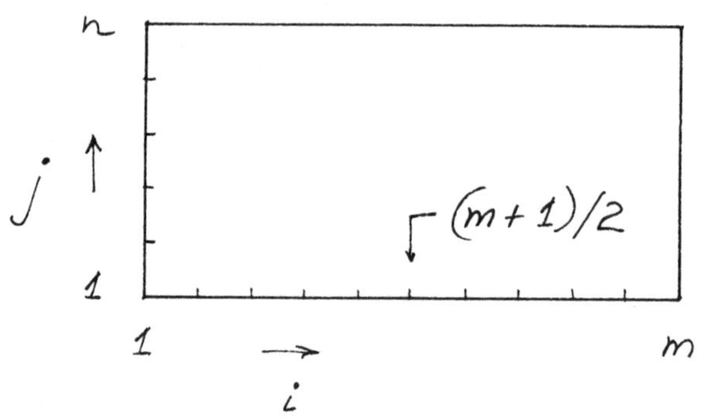

The Gauss-Seidel scheme is constructed based on the finite-difference equations outlined in Table 3.4, in the same way as in the preceding problem. These equations are omitted for brevity. The temperature was fixed at 100°C on the heated portion of the bottom boundary [i = 1,2,...,(m + 1)/2]. The starting temperature of all the other nodes was set at $(T_b + T_\infty)/2 = 60°C$. Convergence was achieved when the maximum relative change experienced by one node temperature

$$\max\left[\frac{T_{i,j}^{(new)} - T_{i,j}^{(old)}}{T_{i,j}^{(new)}}\right] \qquad (3.99)$$

became less than 10^{-5}. The number of iterations needed to achieve convergence, p, is reported in the table. The total heat transfer rate was obtained by integrating the heat flux over all the exposed surfaces,

$$q' = q'_{bottom} + q'_{right} + q'_{top}$$

$$q'_{bottom} = \sum_{i=1+(m+1)/2}^{m-1} h\,\Delta x\,(T_{i,1} - T_\infty) +$$

$$+ h\,\frac{\Delta x}{2}(T_{m,1} - T_\infty)$$

$$q'_{right} = h\,\frac{\Delta x}{2}(T_{m,1} - T_\infty) + \sum_{j=2}^{n-1} h\,\Delta x\,(T_{m,j} - T_\infty) +$$

$$+ h\,\frac{\Delta x}{2}(T_{m,n} - T_\infty)$$

$$q'_{top} = h\,\frac{\Delta x}{2}(T_{1,n} - T_\infty) + \sum_{i=2}^{m-1} h\,\Delta x\,(T_{i,n} - T_\infty) +$$

$$+ h\,\frac{\Delta x}{2}(T_{m,n} - T_\infty)$$

	grid, $m \times n$			
	7×13	9×17	11×21	15×29
p	430	712	1042	1823
q'_{bottom}	7.309	7.471	7.566	7.667
q'_{right}	7.965	7.960	7.953	7.935
q'_{top}	15.951	15.943	15.933	15.908
q'	31.226	31.375	31.453	31.503

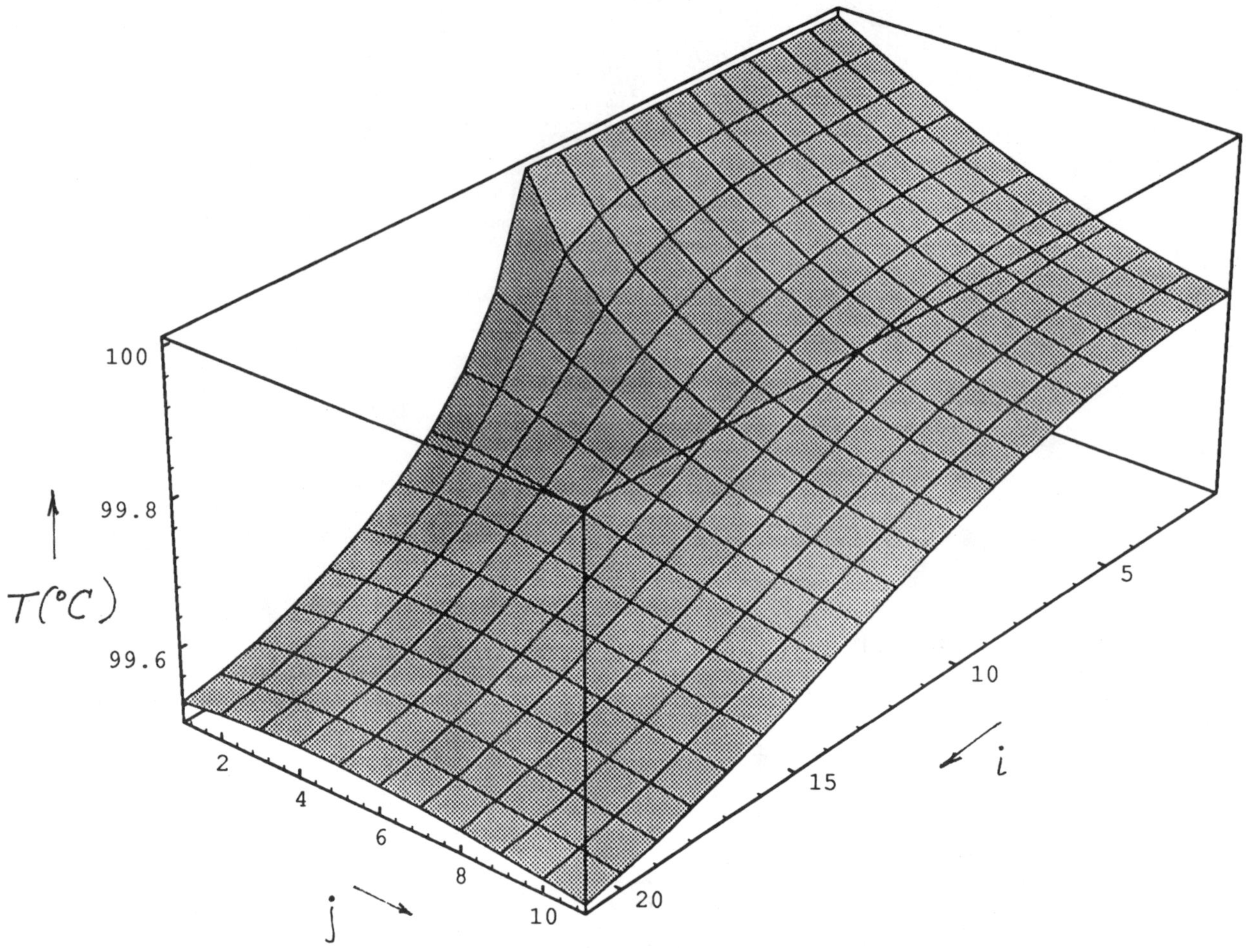

100
99.8
T(°C)
99.6
2
4
6
8
10
j
5
10
15
20
i

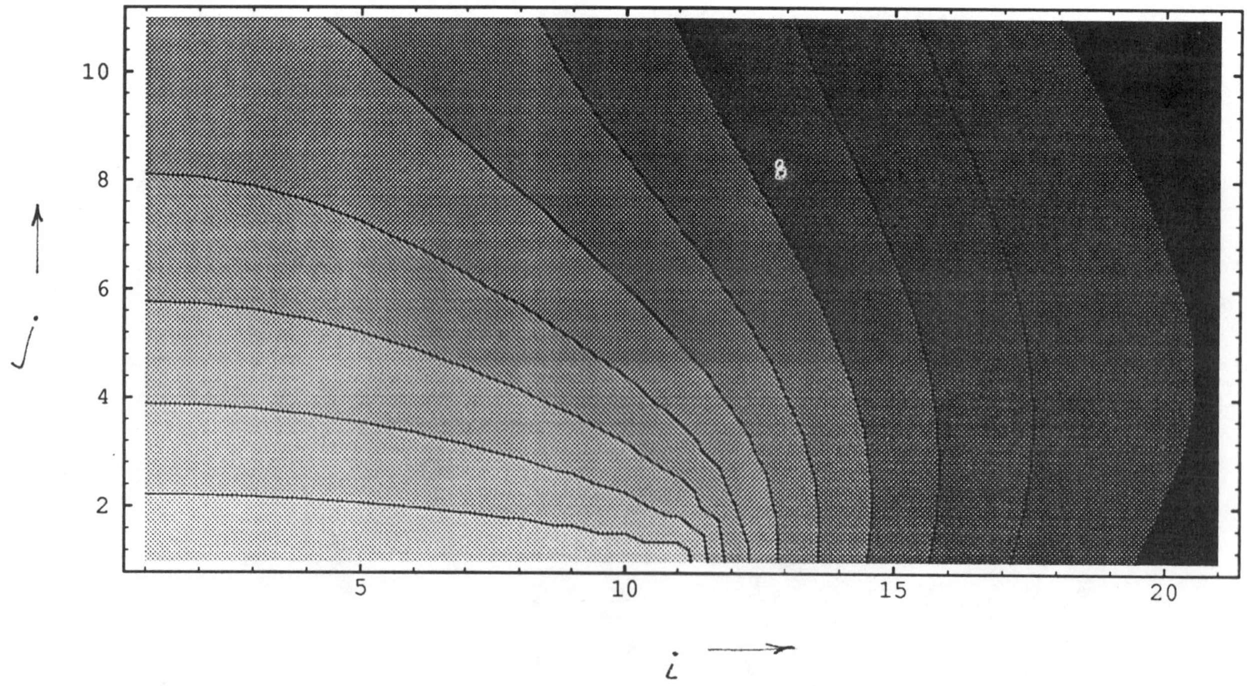

10
8
6
4
2
j
5
10
15
20
i

b) The units of q' (and its components) are W/m. The table shows that the 15 × 29 is adequate for producing a q' estimate that is practically insensitive to changes in Δx. The drawings illustrate the temperature distribution in the flange cross-section. In conclusion, the flange is practically isothermal (at 100°C), and q' could have been calculated much more simply by writing

$$
\begin{aligned}
q' &= h\,(L + L + 2L)\left(T_b - T_\infty\right) \\
&= 2\,\frac{W}{m^2K}\,4 \times 0.05m\,(100 - 20)\,K \\
&= 32\,\frac{W}{m}
\end{aligned}
$$

This estimate is only 1.6 percent larger than the q' value based on the 15 × 29 grid.

c) The nearly isothermal state of the flange cross-section could have been anticipated by calculating the Biot number based on L (flange length scale in both j and i directions),

$$
\begin{aligned}
Bi &= \frac{hL}{k} = 2\,\frac{W}{m^2K}\,\frac{0.05m}{100\ W/m \cdot K} \\
&= 10^{-3} \ll 1
\end{aligned}
$$

and by recalling the "isothermal-across" criterion (2.129), which is satisfied. In conclusion, before setting up the (usually laborious) numerical solution to any problem, it pays to develop a feel for the shape of the answers that might come out. And if those answers appear to be sufficiently simple, well, there may be another method of solution (less costly than finite differences) that deserves to be tried first.

Project 3.1. a) Because of the symmetry about the vertical longitudinal midplane of the coal stockpile, it is sufficient to focus on the right half of the cross-section, in which the x = 0 surface is adiabatic. We cover the (L/2) × H rectangle with a uniform square grid of size

$$\Delta x = \Delta y = \frac{H}{N}$$

where N is the number of steps in the vertical direction. The number of steps in the horizontal direction is, proportionally,

$$M = N \frac{L/2}{H}$$

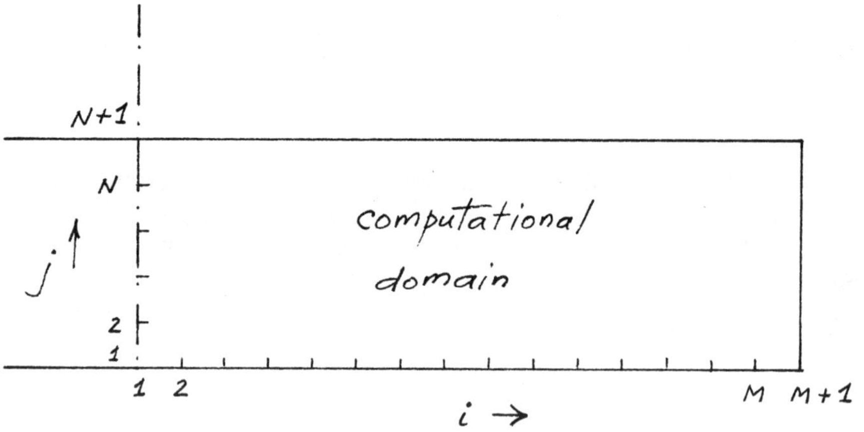

Consider now the finite-difference energy conservation equations written in terms of the dimensionless temperature

$$\tilde{\theta} = \frac{T - T_\infty}{\dot{q} H^2/k}$$

For all the internal nodes, eq. (3.87) reduces to

$$\tilde{\theta}_{i,j} = \frac{1}{4}\left(\tilde{\theta}_{i-1,j} + \tilde{\theta}_{i+1,j} + \tilde{\theta}_{i,j+1} + \tilde{\theta}_{i,j-1}\right) + \frac{1}{4}\left(\frac{\Delta x}{H}\right)^2$$

$$\begin{pmatrix} i = 2,3,\ldots, M \\ j = 2,3,\ldots, N \end{pmatrix}$$

The bottom side (j = 1) is isothermal ($T = T_\infty$), therefore

$$\tilde{\theta}_{i,1} = 0, \qquad (i = 1,2,...,M + 1)$$

The right side (i = M + 1) makes contact with the atmosphere through the heat transfer coefficient h, therefore using the formula listed in the statement to Problem 3.21, we write

$$2\tilde{\theta}_{M,j} + \tilde{\theta}_{M+1,j+1} + \tilde{\theta}_{M+1,j-1} -$$

$$- 2\left(2 + Bi\,\frac{\Delta x}{H}\right)\tilde{\theta}_{M+1,j} + \left(\frac{\Delta x}{H}\right)^2 = 0$$

$$(j = 2,3,...,N)$$

where Bi is the Biot number based on height,

$$Bi = \frac{hH}{k} = \frac{5\ W}{m^2 K}\,\frac{5 m}{0.2\ W/m{\cdot}K} = 125$$

The left side (i = 1) is adiabatic. The equation for this boundary is obtained by setting h = 0 in the second expression of Table 3.4, while keeping track of the roles played by i and j in the present problem:

$$\tilde{\theta}_{1,j} = \frac{1}{4}\left(2\,\tilde{\theta}_{2,j} + \tilde{\theta}_{1,j-1} + \tilde{\theta}_{1,j+1}\right)$$

$$(j = 2,3,...,N)$$

The top side (j = N + 1) makes contact with a flow across h (or Bi):

$$2\,\tilde{\theta}_{i,N} + \tilde{\theta}_{i+1,N+1} + \tilde{\theta}_{i-1,N+1}$$

$$- 2\left(2 + Bi\,\frac{\Delta x}{H}\right)\tilde{\theta}_{i,N+1} + \left(\frac{\Delta x}{H}\right)^2 = 0$$

$$(i = 2,3,...,M)$$

The top-right corner (i = M+1, j = N+1) is exposed to the flow on two sides of the $(\Delta x/2) \times (\Delta x/2)$ control volume:

$$\tilde{\theta}_{M,N+1} + \tilde{\theta}_{M+1,N} - 2\left(1 + 2\,Bi\,\frac{\Delta x}{H}\right)\tilde{\theta}_{M+1,N+1}$$

$$+ \frac{1}{2}\left(\frac{\Delta x}{H}\right)^2 = 0$$

The top-left corner ($i = 1$, $j = N+1$) has one adiabatic side, and one side exposed to the flow,

$$\tilde{\theta}_{1,N} + \tilde{\theta}_{2,N+1} - \left(2 + Bi\,\frac{\Delta x}{H}\right)\tilde{\theta}_{1,N+1} + \frac{1}{2}\left(\frac{\Delta x}{H}\right)^2 = 0$$

The Gauss-Seidel iteration method can be started by assuming $\tilde{\theta}_{i,j} = 0$ at all the nodes, $i = 1,2,...,M+1$, and $j = 1,2,...,N+1$. New $\tilde{\theta}_{i,j}$ values are obtained by sweeping the domain in the i direction while holding j fixed, and then increasing j by 1 and sweeping again in i, etc. In the results described next, the convergence criterion (3.99) was set at

$$\varepsilon = 10^{-4}$$

It was found that the maximum temperature $\tilde{\theta}_{max}$ occurs on the left boundary (in the midplane of the cross-section) at the dimensionless location y_{max}/H. The effect of grid fineness (N) and number of iterations (p) on this key result is summarized in the following table [all the calculations are for $Bi = 125$, and $(L/2)/H = 3$]:

N	p	$\tilde{\theta}_{max}$	y_{max}/H
4	29	0.0832	0.5045
12	181	0.1112	0.5041
16	288	0.1147	0.5041
20	410	0.1169	0.5041
24	545	0.1178	0.5041
28	689	0.1184	0.5041

The altitude of the point of maximum temperature is clearly

$$y_{max} = 0.504\,H = 2.52\text{ m}$$

The maximum temperature is

$$T_{max} = T_\infty + \frac{\dot{q}\,H^2}{k}\,\tilde{\theta}_{max}$$

$$\cong 25°C + \frac{50\text{ W}}{\text{m}^3}\,\frac{25\text{ m}^2}{0.2\text{ W/m}\cdot\text{K}}\,0.118$$

$$\cong 763°C$$

b) This T_{max} estimate can be compared with the theoretical solution listed in Problem 2.14 for the limit L/2 >> H,

$$\tilde{\theta}_{max} = \frac{1}{2}\left(\frac{1 + Bi/2}{1 + Bi}\right)^2 = \frac{1}{2}\left(\frac{1 + 125/2}{1 + 125}\right)^2$$

$$= 0.127$$

which would imply that

$$T_{max} = 25°C + \frac{50\ W}{m^3}\ \frac{25\ m^2}{0.2\ W/m \cdot K}\ 0.127$$

$$\cong 819°C$$

The difference

$$T_{max,theoretical} - T_{max,numerical} = 56°C$$

is positive, because in the numerical model the flow provides an additional cooling effect on the right side of the half-cross-section.

Project 3.2. The conduction equation

$$0 = \frac{\partial}{\partial x}\left[k\,(T)\,\frac{\partial T}{\partial x}\right] + \frac{\partial}{\partial y}\left[k\,(T)\,\frac{\partial T}{\partial y}\right] \tag{1}$$

must be solved subject to the boundary conditions indicated on the figure in the book, and with

$$k\,(T) = k_0 + a\,(T - T_0) \tag{2}$$

We choose the following dimensionless variables,

$$\xi = \frac{x}{H}, \qquad \eta = \frac{y}{H} \tag{3}$$

$$\theta = \frac{T - T_0}{T_L - T_0} \qquad A = \frac{a}{k_0}\,(T_L - T_0) \tag{4}$$

so that eqs. (1) and (2) become

$$0 = \frac{\partial}{\partial \xi}\left[(1 + A\theta)\frac{\partial \theta}{\partial \xi}\right] + \frac{1}{\partial \eta}\left[(1 + A\theta)\frac{\partial \theta}{\partial \eta}\right] \tag{5}$$

$$\frac{k}{k_0} = 1 + A\theta \tag{6}$$

The dimensionless boundary conditions are

$$\theta = 1 \quad \text{at} \quad \xi = 0 \tag{7}$$

$$\theta = 0 \quad \text{at} \quad \xi = \frac{L}{H} \tag{8}$$

$$\frac{\partial \theta}{\partial \eta} = 0 \quad \text{at} \quad \eta = 0 \tag{9}$$

$$\eta = 1 \quad \text{and} \quad 0 < \xi < \frac{L}{2H}$$

$$\eta = \frac{1}{2} \quad \text{and} \quad \frac{L}{2H} < \xi < \frac{L}{H}$$

$$\frac{\partial \theta}{\partial \xi} = 0 \quad \text{at} \quad \xi = \frac{L}{2H} \quad \text{and} \quad \frac{1}{2} < \eta < 1 \tag{10}$$

For the overall thermal resistance we write

$$R_t = \frac{T_0 - T_L}{q'_{0\to L}} \tag{11}$$

where

$$q'_{0\to L} = \int_0^H -k_0 \left(\frac{\partial T}{\partial x}\right)_{x=0} dy \tag{12}$$

The dimensionless thermal resistance that is obtained after using the variables defined in eqs. (3,4) is

$$\tilde{R} = k_0 R_t = \frac{1}{\int_0^1 \left(\frac{\partial \theta}{\partial \xi}\right)_{\xi=0} d\eta} \tag{13}$$

The dimensionless formulation of the problem, eqs. (5)-(10), shows that the overall thermal resistance depends on two parameters, the geometric aspect ratio L/H, and the conductivity slope A,

$$\tilde{R} = \tilde{R}\left(\frac{L}{H}, A\right) \tag{14}$$

In the following numerical solution the grid was square,

$$\Delta\xi = \Delta\eta$$

It was obtained by placing N equidistant nodes along the left side (height H), in other words, by setting

$$\Delta\eta = \frac{1}{N - 1}$$

For example, for the temperature at an internal node (i,j) the discretized version of eq. (5)

$$0 = \frac{\left(1 + A\theta_{i+\frac{1}{2},j}\right)\dfrac{\theta_{i+1,j} - \theta_{i,j}}{\Delta\xi} - \left(1 + A\theta_{i-\frac{1}{2},j}\right)\dfrac{\theta_{i,j} - \theta_{i-1,j}}{\Delta\xi}}{\Delta\xi}$$

$$+ \frac{\left(1 + A\theta_{i,j+\frac{1}{2}}\right)\dfrac{\theta_{i,j+1} - \theta_{i,j}}{\Delta\eta} - \left(1 + A\theta_{i,j-\frac{1}{2}}\right)\dfrac{\theta_{i,j} - \theta_{i,j-1}}{\Delta\eta}}{\Delta\eta}$$

yielded

$$\theta_{i,j} = \left[\left(1 + A\theta_{i+\frac{1}{2},j}\right)\theta_{i+1,j} + \left(1 + A\theta_{i-\frac{1}{2},j}\right)\theta_{i-1,j} + \right.$$

$$\left. + \left(1 + A\theta_{i,j+\frac{1}{2}}\right)\theta_{i,j+1} + \left(1 + A\theta_{i,j-\frac{1}{2}}\right)\theta_{i,j-1}\right] /$$

$$/ \left[4 + A\left(\theta_{i+\frac{1}{2},j} + \theta_{i-\frac{1}{2},j} + \theta_{i,j+\frac{1}{2}} + \theta_{i,j-\frac{1}{2}}\right)\right]$$

with the following notation

$$\theta_{i+\frac{1}{2},j} = \frac{1}{2}\left(\theta_{i+1,j} + \theta_{i,j}\right)$$

$$\theta_{i-\frac{1}{2},j} = \frac{1}{2}\left(\theta_{i-1,j} + \theta_{i,j}\right)$$

$$\theta_{i,j+\frac{1}{2}} = \frac{1}{2}\left(\theta_{i,j+1} + \theta_{i,j}\right)$$

$$\theta_{i,j-\frac{1}{2}} = \frac{1}{2}\left(\theta_{i,j-1} + \theta_{i,j}\right)$$

The convergence criterion (3.99) was

$$\frac{\theta_{i,j}^{(p+1)} - \theta_{i,j}^{(p)}}{\theta_{i,j}^{(p)}} < 10^{-8}$$

for every node in the domain. The accuracy test shown in the attached table shows that the grid with N = 11 (or $\Delta\xi = \Delta\eta = 0.1$) is adequate

L/H	N	$(\tilde{R})^{-1}$
2	5	0.3923
	7	0.3893
	9	0.3879
	11	0.3871
	13	0.3866
	15	0.3862
	21	0.3857

The results for the dimensionless thermal conductance $(\tilde{R})^{-1}$ are listed in the table as a function of L/H and A:

$\frac{L}{H}$	A			
	0	0.25	0.5	0.75
2	0.3097	0.3484	0.3871	0.4258
4	0.1605	0.1806	0.2006	0.2207
10	0.0657	0.0779	0.0821	0.0903

The thermal conductance increases with A, because k increases. At the same time it decreases as the slenderness ratio L/H increases. Finally, an accuracy test conducted for L/H = 10 shows that N = 11 is adequate even when the domain is very slender:

L/H	N	$(\tilde{R})^{-1}$
10	5	0.06580
	7	0.06574
	9	0.06569
	11	0.06566

The next two pages show the beginning and the end of the computation for the case L/H = 2 and A = 0 (constant k). It required 1002 steps. Noteworthy is the global energy conservation check made under the final table of node temperatures: the heat current entering through the x = 0 plane is equal to the current calculated at x = L.

L/H = 2 A = 0

1	1	1	1	1	1	1	1	1	1	1
0.5	0.5	0.5	0.5	0.5	0.5	0.5	0.5	0.5	0.5	0.5
0.5	0.5	0.5	0.5	0.5	0.5	0.5	0.5	0.5	0.5	0.5
0.5	0.5	0.5	0.5	0.5	0.5	0.5	0.5	0.5	0.5	0.5
0.5	0.5	0.5	0.5	0.5	0.5	0.5	0.5	0.5	0.5	0.5
0.5	0.5	0.5	0.5	0.5	0.5	0.5	0.5	0.5	0.5	0.5
0.5	0.5	0.5	0.5	0.5	0.5	0.5	0.5	0.5	0.5	0.5
0.5	0.5	0.5	0.5	0.5	0.5	0.5	0.5	0.5	0.5	0.5
0.5	0.5	0.5	0.5	0.5	0.5	0.5	0.5	0.5	0.5	0.5
0.5	0.5	0.5	0.5	0.5	0.5	0.5	0.5	0.5	0.5	0.5
0.5	0.5	0.5	0.5	0.5	0.5	0.5	0.5	0.5	0.5	0.5
0.5	0.5	0.5	0.5	0.5	0.5					
0.5	0.5	0.5	0.5	0.5	0.5					
0.5	0.5	0.5	0.5	0.5	0.5					
0.5	0.5	0.5	0.5	0.5	0.5					
0.5	0.5	0.5	0.5	0.5	0.5					
0.5	0.5	0.5	0.5	0.5	0.5					
0.5	0.5	0.5	0.5	0.5	0.5					
0.5	0.5	0.5	0.5	0.5	0.5					
0.5	0.5	0.5	0.5	0.5	0.5					
0	0	0	0	0	0					

Final Solution   steps = 1002

1	1	1	1	1	1	1	1	1	1	1
0.966034	0.966174	0.966585	0.967233	0.968062	0.968998	0.969947	0.970814	0.971506	0.971953	0.972107
0.931787	0.932079	0.932933	0.934284	0.936019	0.937981	0.939977	0.941801	0.943259	0.944198	0.944522
0.896956	0.897421	0.898784	0.90095	0.903748	0.90693	0.910181	0.913154	0.91553	0.917059	0.917586
0.861196	0.861865	0.863833	0.866984	0.871094	0.875812	0.880661	0.885104	0.888647	0.890922	0.891704
0.8241	0.825008	0.8277	0.832059	0.837832	0.844561	0.851548	0.857952	0.863034	0.866276	0.867388
0.785187	0.786368	0.789899	0.79572	0.803615	0.813053	0.823018	0.832123	0.839261	0.843762	0.845294
0.743912	0.745376	0.74981	0.757305	0.767855	0.781018	0.795348	0.80826	0.818126	0.824215	0.826265
0.69971	0.701415	0.706661	0.715834	0.729482	0.747817	0.769096	0.787442	0.800768	0.808708	0.811335
0.652099	0.653912	0.659583	0.66989	0.686424	0.711671	0.745776	0.771644	0.788797	0.798514	0.80166
0.600861	0.602553	0.607868	0.617718	0.634651	0.666666	0.730695	0.764561	0.784262	0.794891	0.798276
0.546239	0.547571	0.551618	0.558464	0.567798	0.576656					
0.488952	0.489875	0.492569	0.496721	0.50142	0.504364					
0.429821	0.430406	0.432062	0.434433	0.436795	0.437961					
0.369517	0.369869	0.370838	0.372156	0.373365	0.373892					
0.308511	0.308714	0.309264	0.309987	0.310618	0.310876					
0.247098	0.247212	0.247518	0.247911	0.248243	0.248375					
0.185456	0.185519	0.185685	0.185895	0.18607	0.186138					
0.12369	0.123722	0.123807	0.123915	0.124003	0.124037					
0.0618585	0.0618722	0.0619082	0.0619533	0.0619901	0.0620043					
0	0	0	0	0	0					

Heat current in from the left = 0.309657

Heat current out to the right = 0.309655

Chapter 4

TIME-DEPENDENT CONDUCTION

Problem 4.1. a) According to Fig. 4.5, the lumped capacitance model is valid when the solid is a good enough conductor (large k, large a) so that the Biot number is small,

$$Bi = \frac{hr_o}{k} << 1$$

and the Fourier number is large

$$Fo = \frac{\alpha t}{r_o^2} >> 1$$

These two conditions are met by the present physical configuration:

$$Bi = \frac{20 \text{ W}}{m^2K} \frac{0.005 \text{ m}}{54 \text{ W/m·K}} = 0.00185$$

$$Fo = \frac{0.148 \text{ cm}^2}{s} \frac{2 \times 60s}{0.25 \text{ cm}^2} = 71.04$$

b) We are therefore entitled to use eq. (4.17),

$$\frac{T - T_\infty}{T_1 - T_\infty} = \exp\left[-\frac{hA}{\rho cV}(t - t_c)\right]$$

in which

$T_1 = T_i = 100°C$, [because Bi << 1, cf. eq. (4.21)]

and

$t - t_c \cong t = 120s$, [because Fo >> 1 means

$$\frac{t}{t_c} = \frac{\alpha t}{r_o^2} << 1, \text{ cf. eq. (4.9)}]$$

Furthermore, the numerical coefficient in the exponential temperature decay is

$$\frac{hA}{\rho c V} = \frac{\alpha}{k} h \frac{4\pi r_o^2}{\frac{4\pi}{3} r_o^3} = 3 \frac{h r_o}{k} \frac{\alpha}{r_o^2}$$

$$= 3\ Bi \frac{\alpha}{r_o^2} = 3 \times 0.00185 \frac{0.148\ cm^2/s}{0.25\ cm^2}$$

$$= 0.00329\ s^{-1} = \frac{1}{304.4s}$$

This means that the "time constant" of the temperature decay is approximately 304.4s @ 5 minutes. Finally, the temperature of the steel ball after 2 minutes can be calculated numerically

$$\frac{T - 10°C}{100°C - 10°C} = \exp\left(- \frac{2 \times 60s}{304.4s}\right) = 0.674$$

$$T = 10°C + 0.674 \times 90°C = 70.7°C$$

c) In an argument similar to eqs. (4.18)-(4.19) in the text, we write that the scale of the convective heat flux at the surface must be the same as the scale of the radial conduction heat flux through the ball,

$$h\left(T - T_\infty\right) \sim k \frac{T_c - T}{r_o}$$

In this balance, the unknown is the temperature difference between the center of the sphere (T_c) and the surface (represented by the lumped-capacitance temperature T),

$$T_c - T \sim \frac{h r_o}{k} (T - T_\infty)$$

$$\sim 0.00185\ (70.7 - 10)°C$$

$$\sim 0.1°C$$

In conclusion, the temperature variation through the sphere is of order of only one tenth of a degree Celsius, which is why the lumped-capacitance model was an appropriate model for this problem.

Problem 4.2. The immersed body of heat capacity $(mc)_1$ and the liquid pool of finite heat capacity $(mc)_2$ are two lumped capacitances linked through the same thermal conductance hA. First-law statements like eq. (4.15) can be written for each of these systems

$$(mc)_1 \frac{dT_1}{dt} = hA\,(T_2 - T_1) \tag{1}$$

$$(mc)_2 \frac{dT_2}{dt} = hA\,(T_1 - T_2) \tag{2}$$

in which the two temperature histories $T_1(t)$ and $T_2(t)$ must satisfy the initial conditions

$$T_1(0) = T_{1,0}, \qquad \text{and} \qquad T_2(0) = T_{2,0} \tag{3}$$

This problem can be solved by first eliminating T_2 between eqs. (1) and (2), and obtaining

$$\frac{d^2T_1}{dt^2} + n\frac{dT_1}{dt} = 0 \tag{4}$$

in which

$$n = hA\left[\frac{1}{(mc)_1} + \frac{1}{(mc)_2}\right] \tag{5}$$

The general solution to eq. (4) is

$$T_1 = C_1 + C_2 \exp(-nt) \tag{6}$$

with the corresponding solution for T_2 [obtained after combining eqs. (6) and (1)]:

$$T_2 = C_1 + C_2\left[1 - n\frac{(mc)_1}{hA}\right]\exp(-nt) \tag{7}$$

The constants C_1 and C_2 are obtained next by subjecting the solutions (6,7) to the initial conditions (4); the resulting constants are, in order

$$C_2 = \frac{hA}{n(mc)_1}\left(T_{1,0} - T_{2,0}\right) \tag{8}$$

$$C_1 = T_{1,0} - C_2 = T_{2,0} - C_2\left[1 - \frac{n(mc)_1}{hA}\right] \tag{9}$$

After some rearranging, the temperature expressions (6,7) become

$$T_1 = T_{1,0} - \frac{T_{1,0} - T_{2,0}}{1 + (mc)_1/(mc)_2}(1 - e^{-nt}) \tag{10}$$

$$T_2 = T_{2,0} + \frac{T_{1,0} - T_{2,0}}{1 + (mc)_2/(mc)_1}(1 - e^{-nt}) \tag{11}$$

The validity of these final expressions can be demonstrated by recognizing the first law of thermodynamics for the aggregate system (immersed body, plus liquid pool), which is an "isolated" system:

$$(mc)_1 dT_1 + (mc)_2 dT_2 = 0 \tag{12}$$

$$(mc)_1 T_1 + (mc)_2 T_2 = \underbrace{(mc)_1 T_{1,0} + (mc)_2 T_{2,0}}_{\text{constant}} \tag{13}$$

It is easy to show that if we multiply both sides of eq. (10) by $(mc)_1$, and of eq. (11) by $(mc)_2$, and if we add the resulting equations side by side we obtain eq. (13). In conclusion, the solution (10,11) satisfies the first law of thermodynamics written for the aggregate system.

The other test concerns the $t \to \infty$ limit of the solution (10,11):

$$\lim_{t \to \infty} T_1 = \frac{(mc)_1}{(mc)_1 + (mc)_2} T_{1,0} + \frac{(mc)_2}{(mc)_1 + (mc)_2} T_{2,0} \tag{14}$$

$$\lim_{t \to \infty} T_2 = \frac{(mc)_1}{(mc)_1 + (mc)_2} T_{1,0} + \frac{(mc)_2}{(mc)_1 + (mc)_2} T_{2,0} \tag{15}$$

These limits show that both T_1 and T_2 approach the same value as the time increases. This common value is the <u>equilibrium temperature</u> of the immersed body and the surrounding liquid.

<u>Problem 4.3</u>. a) In order to use the $T_1(t)$ and $T_2(t)$ formulas listed in the preceding problem statement, we assign the index 1 to the milk bottle, and index 2 to the amount of cold water, and calculate sequentially

$$(mc)_1 = 0.125 \text{ kg } 4 \frac{\text{kJ}}{\text{kg K}} = 0.5 \frac{\text{kJ}}{\text{K}}$$

$$(mc)_2 = (\rho V c)_2 = 1 \frac{\text{g}}{\text{cm}^3} 200 \text{ cm}^3 \; 4.19 \frac{\text{kJ}}{\text{kg K}} = 0.84 \frac{\text{kJ}}{\text{K}}$$

$$A = \pi DH + \frac{\pi}{4} D^2 = (\pi \times 4 \times 6 + \frac{\pi}{4} 4^2)\ cm^2 = 88\ cm^2$$

$$n = hA \frac{(mc)_1 + (mc)_2}{(mc)_1 (mc)_2} = \frac{350\ W}{m^2K} 88 \times 10^{-4} m^2 \frac{0.5 + 0.84}{0.5 \times 0.84} \frac{K}{kJ}$$

$$= 0.0098\ s^{-1}$$

$$1 - \exp(-nt) = 1 - \exp\left(- \frac{0.0098}{s} 60s\right) = 0.445$$

$$T_1 = 60°C - \frac{(60 - 10)°C}{1 + \frac{0.5}{0.84}} 0.445 \cong 46.1°C$$

$$T_2 = 10°C + \frac{(60 - 10)°C}{1 + \frac{0.84}{0.5}} 0.445 \cong 18.3°C$$

b) If the amount of cold water is assumed to remain at 10°C, the problem reduces to the single lumped capacitance of section 4.2. Alternatively, the corresponding T_1 expression can be obtained by setting $(mc)_2 \rightarrow \infty$ in the $T_1(t)$ formula given in the preceding problem statement, hence

$$T_1 = T_{1,0} - (T_{1,0} - T_{2,0})\ (1 - e^{-nt})$$

where this time

$$n = \frac{hA}{(mc)_1} = \frac{350\ W}{m^2K} 88 \times 10^{-4}\ m^2 \frac{K}{0.5 \times 10^3 J}$$

$$= 0.0062\ s^{-1}$$

Numerically, the new T_1 estimate at t = 60s is

$$T_1 = 60°C - (60 - 10)°C\ (1 - e^{-0.0062 \times 60})$$

$$\cong 44.5°C$$

This estimate is approximately 1.5°C lower than the correct T_1 value calculated in part (a).

Problem 4.4. The temperature history of the lumped thermal capacitance is expressed by eq. (4.17), in which $T_\infty = T_0 + \Delta T$, and the initial temperature is T_0 at $t = 0$,

$$\frac{T-(T_0+\Delta T)}{T_0-(T_0+\Delta T)} = \exp\left(-\frac{hA}{\rho c V}t\right) \tag{1}$$

It is convenient to express the temperature history T(t) in terms of the following dimensionless variables, i.e. as $\theta(\tau)$,

$$\theta = \frac{T-T_0}{\Delta T} \tag{2}$$

$$\tau = \frac{hA}{\rho c V}t \tag{3}$$

Equation (1) becomes

$$-\theta + 1 = \exp(-\tau) \tag{4}$$

and, at $t = t_1$, when $T = T_1$,

$$\theta_1 = 1 - \exp(-\tau_1) \tag{5}$$

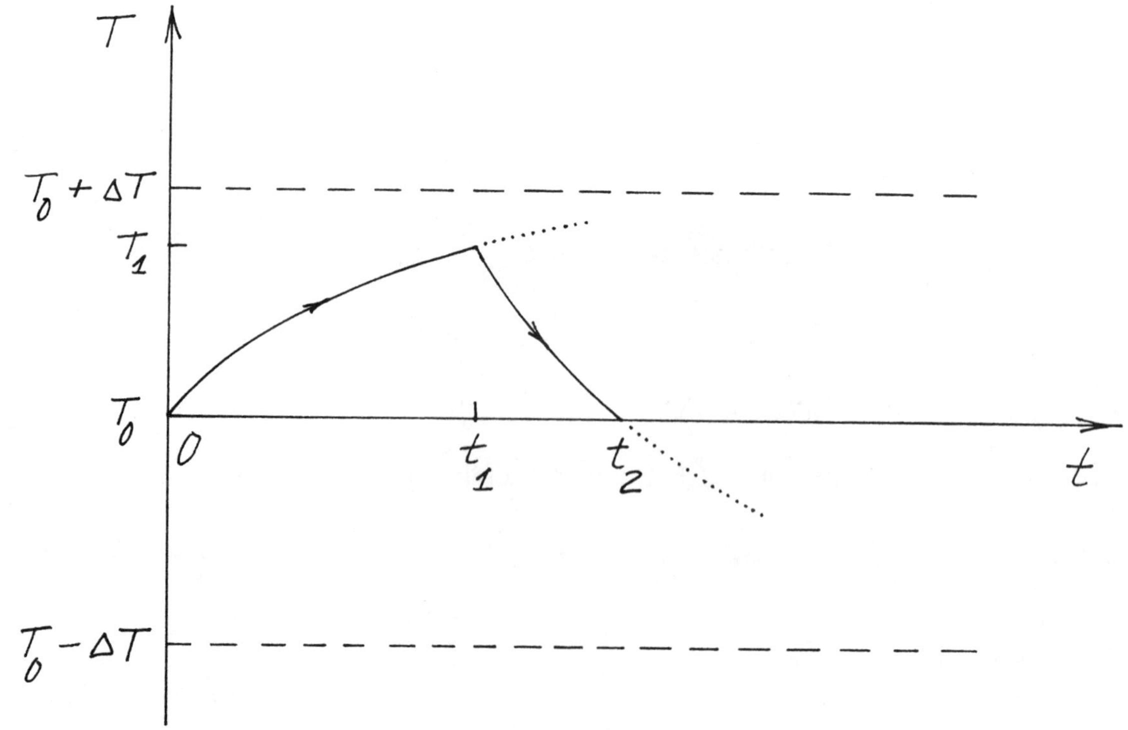

In the second phase of the process, the lumped capacitance is plunged into the cold bath $T_\infty = T_0 - \Delta T$, having $T = T_1$ at $t = t_1$, as initial condition. Equation (4.17) states that

$$\frac{T-(T_0-\Delta T)}{T_1-(T_0-\Delta T)} = \exp\left[-\frac{hA}{\rho c V}(t-t_1)\right] \tag{6}$$

or in terms of θ and τ,

$$\frac{\theta+1}{\theta_1+1} = \exp(-\tau+\tau_1) \tag{7}$$

Finally, we eliminate θ_1 between eqs. (5) and (7), and find that the t_2/t_1 ratio depends on the first dimensionless time,

$$\frac{t_2}{t_1} = \frac{\tau_2}{\tau_1} = 1 + \frac{\ln[2-\exp(-\tau_1)]}{\tau_1} \tag{8}$$

The asymptotic value of this ratio approaches 2 when $\tau_1 << 1$, and 1 when $\tau_1 >> 1$. This means that the ratio $(t_2 - t_1)/t_1$ has values between 1 and 0, i.e. that the time spent in the cold bath is always shorter than the time spent in the hot fluid.

Problem 4.5. In accordance with eqs. (4.12)-(4.14), the first law of thermodynamics for the volume V is

$$\frac{\rho c V}{hA}\frac{dT}{dt} = T_\infty - T \tag{1}$$

where

$$T_\infty(t) = T_0 + a \sin bt$$

Equation (1) can be arranged as follows,

$$\frac{d\theta}{d\tau} + \theta - a \sin(B\tau) = 0 \tag{2}$$

by defining

$$\theta = T - T_0$$

$$\tau = t\frac{hA}{\rho c V}$$

$$B = b\frac{\rho c V}{hA}$$

Equation (2) is a linear ordinary differential equation of the type

$$\frac{dy}{dx} + P(x)\cdot y + Q(x) = 0$$

with the general solution

$$y = e^{-\int P dx}\left(C - \int Q\, e^{\int P dx}\, dx\right)$$

In the present problem, $y = \theta$, $x = \tau$, $P = 1$ and $Q = -a \sin B\tau$, therefore the solution to eq. (2) is

$$\theta = e^{-\tau}\left[C + a \int e^{\tau} \sin(B\tau)\, d\tau\right]$$

$$= e^{-\tau}\left\{C + a \frac{e^{\tau}}{1 + B^2}[\sin(B\tau) - B\cos(B\tau)]\right\}$$

$$= C\, e^{-\tau} + \frac{a}{1 + B^2}[\sin(B\tau) - B\cos(B\tau)] \tag{3}$$

The term $Ce^{-\tau}$ vanishes after a sufficiently long time, or number of daily cycles. That happens when $\tau >> 1$, i.e. when

$$t >> \frac{\rho c V}{hA}$$

The solution (3) reduces to

$$\theta = \frac{a}{1 + B^2}[\sin(B\tau) - B\cos(B\tau)] \tag{4}$$

The quantity in the square brackets is a periodic function that can be seen better by remembering the trigonometry formulas

$$\sin(\alpha + \beta) = \sin\alpha \cos\beta + \cos\alpha \sin\beta$$

$$\frac{\sin(\alpha + \beta)}{\cos\beta} = \sin\alpha + \cos\alpha \tan\beta \tag{5}$$

By comparing the right-hand side of eq. (5) with the quantity in square brackets in eq. (4), we conclude that

$$\alpha = B\tau$$

$$\tan\beta = -B$$

$$-\beta = \tan^{-1} B = \varphi > 0$$

and that eq. (4) can be rewritten as

$$\theta = \frac{a}{\left(1 + B^2\right)\cos\varphi}\sin(B\tau - \varphi)$$

$$= \frac{a}{\left(1 + B^2\right)\cos\varphi}\sin\left[b\left(t - \frac{\varphi}{b}\right)\right]$$

This shows that the body temperature oscillation lags behind the ambient temperature oscillation by the time interval φ/b. Relative to the ambient amplitude a, the amplitude of the body temperature oscillation is attenuated by the dimensionless factor $(1 + B^2)\cos\varphi$, which depends only on B.

<u>Problem 4.6.</u> The order of magnitude equivalent of the boundary condition (4.47) is

$$h(T_\infty - T_0) \sim -k\frac{T_i - T_0}{\delta} \tag{1}$$

in which $\delta \sim (\alpha t)^{1/2}$. This equality of "scales" must be accurate within a factor the value of which is comparable with 1. Let C be that factor. Multiplying one side of eq. (1) by C, we obtain an <u>equation</u> instead of just an equivalence of scales,

$$Ch\,(T_\infty - T_0) = -k\frac{T_i - T_0}{(\alpha t)^{1/2}} \tag{2}$$

This equation can be rearranged so that the left side shows the dimensionless temperature difference ratio used as ordinate in Fig. 4.4 (right):

$$\frac{T_0 - T_\infty}{T_i - T_\infty} = \frac{1}{1 + C\,Bi_\delta}, \qquad \text{where } Bi_\delta = \frac{h}{k}(\alpha t)^{1/2} \tag{3}$$

This expression can be "fitted" to the exact curve by choosing a C value such that both eq. (3) and the exact curve coincide. The equation for the exact curve is obtained by setting x = 0 in eq. (4.48),

$$\frac{T_0 - T_\infty}{T_i - T_\infty} = \exp\left(Bi_\delta^2\right)\operatorname{erfc}\left(Bi_\delta\right) \tag{4}$$

If eqs. (3) and (4) are made to coincide at $Bi_\delta = 1$ (i.e. roughly in the middle of the curve), then C = 1.339. The following table shows how well eq. (3) approximates the exact curve.

$\frac{h}{k}(\alpha t)^{1/2}$	$\left(\frac{T_0 - T_\infty}{T_i - T_\infty}\right)_{\text{exact, eq. (4)}}$	$\left(\frac{T_0 - T_\infty}{T_i - T_\infty}\right)_{\text{approximate, eq. (3)}}$ with C = 1.339
0	1	1
0.1	0.897	0.882
0.3	0.735	0.714
1	0.428	0.428
3	0.179	0.199
∞	0	0

Problem 4.7. a) Assuming that the surface temperature is constant and equal to the liquid temperature T_∞ = 70°C, we use eq. (4.42) in which T_i = 25°C, x = 2 mm, t = 3s, and α = 0.004 cm²/s:

$$\frac{T - T_\infty}{T_i - T_\infty} = \text{erf}\left[\frac{x}{2(\alpha t)^{1/2}}\right]$$

$$\frac{T - 70°C}{(25 - 70)°C} = \text{erf}\left[\frac{0.2\text{ cm}}{2\left(0.004\,\frac{\text{cm}^2}{\text{s}}\,3\text{s}\right)^{1/2}}\right]$$

$$= \text{erf}\,(0.909)$$

$$\cong 0.801$$

$$T = 70°C - 0.801 \times 45°C$$

$$\cong 34°C$$

b) If the liquid side heat transfer coefficient is h = 1200 W/m²K, the Biot number based on the thermal penetration thickness $(\alpha t)^{1/2}$ is

$$\text{Bi} = \frac{h}{k}(\alpha t)^{1/2} = \frac{1200\text{ W}}{\text{m}^2\text{K}}\,\frac{\text{m}\cdot\text{K}}{1.03\text{ W}}\left(0.004\,\frac{\text{cm}^2}{\text{s}}\,3\text{s}\right)^{1/2}$$

$$= 1.28$$

in which k = 1.03 W/m·K is the thermal conductivity of the solid (porcelain). This Biot number appears as a parameter on the curves drawn in Fig. 4.4, left. The value of the abscissa parameter in the same figure is 0.909 [see the argument of erf in part (a) of this problem], therefore from Fig. 4.4, left, we read approximately

$$\frac{T - T_\infty}{T_i - T_\infty} \cong 0.9$$

$$T = 70°C + 0.9\ (25 - 70)°C$$

$$= 24.5°C$$

The reading of Fig. 4.4, left, was "approximate" because the point of interest is located somewhere between the two curves drawn for Bi = 1 and Bi = 3. We get a better idea of how close the Bi = 1.28 curve would have been relative to the Bi = 1 curve, by using the right side of Fig. 4.4 and calculating the relative surface temperature at the moment of interest (Bi = 1.28):

$$\frac{T_0 - T_\infty}{T_i - T_\infty} \cong 0.36$$

We now place this value on the left side of Fig. 4.4, left (i.e. at x = 0), and see that the Bi = 1.28 curve is located much closer to the Bi = 1 curve than to the Bi = 3 curve. It is safe to assume that the relative positioning of these Bi = constant curves is preserved all the way to the abscissa parameter of interest, $x/2(\alpha t)^{1/2} = 0.909$: this is how I read the 0.9 value for $(T - T_\infty)/(T_i - T_\infty)$ at the start of part (b).

c) A more precise calculation is made possible by eq. (4.48), for which we have all the ingredients:

$$\frac{T - T_\infty}{T_i - T_\infty} = \text{erf}\left[\frac{x}{2\,(\alpha t)^{1/2}}\right] + \exp\left(\frac{hx}{k} + \frac{h^2 \alpha t}{k^2}\right) \cdot \text{erfc}\left[\frac{x}{2\,(\alpha t)^{1/2}} + \frac{h}{k}(\alpha t)^{1/2}\right]$$

$$= \text{erf}\ (0.909) + \exp\ (0.909 \times Bi \times 2 + Bi^2) \cdot \text{erfc}\ (0.909 + Bi)$$

$$= 0.801 + 52.98 \times 0.001975 = 0.906$$

$$T = 70°C + 0.906\ (25 - 70)°C$$

$$= 29.2°C$$

In conclusion, the temperature estimate based on the analytical solution (4.48) is nearly the same as that provided by the reading of Fig. 4.4, left [see part (b)]. Both estimates are lower than that of part (a),which was based on the erroneous assumption that the surface temperature is fixed (Bi = ∞).

Problem 4.8. a) According to eq. (4.46), the surface temperature at t = 3600s is

$$T_0 = T_i + 2\frac{q''}{k}\left(\frac{\alpha t}{\pi}\right)^{1/2}$$

$$= 10°C + 2\,\frac{100\ W}{m^2}\,\frac{m \cdot K}{0.5\ W}\left(0.005\,\frac{cm^2}{s}\,\frac{3600s}{\pi}\right)^{1/2}$$

$$= 10°C + 9.6°C = 19.6°C$$

b) The temperature at the distance x = 10 cm inside the brick wall can be calculated using eq. (4.44), in which

$$\frac{x}{2\,(\alpha t)^{1/2}} = \frac{10\ cm}{2\left(0.005\,\frac{cm^2}{s}\,3600s\right)^{1/2}} \cong 1.18$$

$$\frac{x^2}{4\alpha t} = (1.18)^2 \cong 1.39$$

$$2\frac{q''}{k}\left(\frac{\alpha t}{\pi}\right)^{1/2} \cong 9.6°C, \qquad \text{[see part (a)]}$$

$$\frac{q''}{k}x = 2\frac{q''}{k}(\alpha t)^{1/2}\,\frac{x}{2\,(\alpha t)^{1/2}}$$

$$= 9.6°C \times \pi^{1/2} \times 1.18 \cong 20°C$$

Substituting all these quantities in eq. (4.44) we obtain

$$T - T_i = 9.6°C \exp(-1.39) - 20°C \underbrace{\text{erfc}\,(1.18)}_{0.0962}$$

$$\cong 0.5°C$$

which means that $T \cong 10.5°C$, and that the heating effect of q" has not been really felt yet at x = 10 cm.

Problem 4.9. If the outer surface of the porcelain wall can be modelled as an adiabatic surface, then it behaves the same as the midplane of a twice thicker porcelain slab exposed to hot coffee on both sides. Therefore we can rely on Fig. 4.7 in which L = 0.5 cm, and

$$Bi = \frac{hL}{k} = \frac{700\ W}{m^2K}\ \frac{0.005\ m}{1.03\ W/m \cdot K} = 3.4$$

$$\frac{1}{Bi} = \frac{1}{3.4} \cong 0.3$$

$$\frac{T_c - T_\infty}{T_i - T_\infty} = \frac{(50 - 90)°C}{(10 - 90)°C} = 0.5$$

These numerical values pinpoint the abscissa parameter

$$\frac{\alpha t}{L^2} \cong 0.61$$

which leads to the time interval

$$t = 0.61\ \frac{L^2}{\alpha} = 0.61\ \frac{0.25\ cm^2}{0.004\ cm^2/s} \cong 38\ s$$

Problem 4.10. We treat the icicle as a long cylinder of radius r_o = 1 cm and initial temperature T_i = 0°C. In order to use Fig. 4.10 to calculate the centerline temperature, we must first determine:

$$Fo = \frac{\alpha t}{r_o^2} = 0.012\ \frac{cm^2}{s}\ \frac{10 \times 60\ s}{1\ cm^2} = 7.2$$

$$Bi = \frac{hr_o}{k} = \frac{10\ W}{m^2K}\ \frac{0.01\ m}{2.25\ W/m \cdot K} = 0.044$$

$$\frac{1}{Bi} = 22.5$$

In line with these Fo and 1/Bi values, Fig. 4.10 recommends

$$\frac{T_c - T_\infty}{T_i - T_\infty} \cong 0.54$$

which corresponds to the centerline temperature

$$T_c = T_\infty + 0.54\,(T_i - T_\infty)$$

$$= -5°C + 0.54\,(0 + 5)°C = -2.3°C$$

An alternative way to calculate T_c is by relying on eq. (4.69), in which the first term is a good substitute for the sum of the entire series

$$\frac{T(r,t) - T_\infty}{T_i - T_\infty} \cong K_1\, J_0\left(b_1 \frac{r}{r_o}\right) \exp\left(-\, b_1^2\, Fo\right)$$

From Table 4.1 we learn that at Bi = 0.044 the b_1 and K_1 constants are, approximately,

$$b_1 \cong 0.28 \qquad \text{and} \qquad K_1 \cong 1.011$$

Furthermore, on the centerline (r = 0) the value of the Bessel function is $J_0(0) = 1$, therefore

$$\frac{T_c - T_\infty}{T_i - T_\infty} \cong 1.011\, \exp\left(-\, 0.28^2 \times 7.2\right) = 0.575$$

$$T_c = T_\infty + 0.575\,(T_i - T_\infty)$$

$$= -5°C + 0.575\,(0 + 5)°C = -2.1°C$$

To verify that the sum of the series on the right side of eq. (4.69) is approximated well by the first term, we calculate the second term, for which Table 4.1 suggests

$$b_2 \cong 3.84 \qquad \text{and} \qquad K_2 = -\,0.0148$$

On the centerline, r = 0, the second term of the series reduces to

$$K_2 \exp\left(-\, b_2^2\, Fo\right) = -\,0.0148\, \exp\left(-\, 3.84^2 \times 7.2\right)$$

$$\cong -\,10^{-48}$$

This term is obviously negligible relative to the first term, whose value (0.575) was calculated earlier.

Problem 4.11. We approximate the properties of window glass as being equal to those listed for 20°C in Appendix B:

$$k = 0.81 \frac{W}{m \cdot K} \qquad \alpha = 0.0034 \frac{cm^2}{s}$$

In order to use Fig. 4.13 to determine the cooling time (Fo) of the glass sphere, we must first calculate

$$Bi = \frac{hr_o}{k} = \frac{324\ W}{m^2K} \frac{(0.25)\ 10^{-3} m}{0.81\ W/m \cdot K} = 0.1$$

$$\frac{1}{Bi} = 10$$

$$\frac{T_c - T_\infty}{T_i - T_\infty} = \frac{(40 - 20)°C}{(500 - 20)°C} = 0.042$$

Entering these last two values in the reading of Fig. 4.13, on the abscissa we find Fo ≅ 12, which means that

$$t \cong 12 \frac{r_o^2}{\alpha} = 12 \frac{(0.025)^2\ cm^2}{0.0034\ cm^2/s} = 2.2\ s$$

The height from which the glass spheres fall must be at least as great as

$$H = Ut = 3.8 \frac{m}{s}\ 2.2\ s \cong 8.4\ m$$

For the surface temperature at t = 2.2 s, we identify the curve labeled $r/r_o = 1$ in Fig. 4.14. In combination with 1/Bi = 10, this curve shows that

$$\frac{T - T_\infty}{T_c - T_\infty} \cong 0.95$$

$$T = 20°C + 0.95\ (40 - 20)°C = 39°C$$

The surface temperature is nearly the same as the temperature in the center of the sphere. In conclusion, in these late stages the cooling of the glass bead would be described adequately by the lumped-capacitance model. This conclusion is consistent with the lower-right quadrant of Fig. 4.5, because the present calculations (Bi < 1, Fo > 1) fell marginally in that domain.

Problem 4.12. From the outset, we expect that the various potato shapes will have transversal (e.g. radial) dimensions of the order of 1 cm, therefore

$$Bi = \frac{hr_o}{k} \sim \frac{2 \times 10^4 W}{m^2 K} \frac{0.01\ m}{0.6\ W/m \cdot K}$$

$$= 333$$

Since Bi >> 1, we can rely on Fig. 4.16, where the ordinate is set regardless of shape at

$$\frac{\overline{T} - T_\infty}{T_i - T_\infty} = \frac{(65 - 100)°C}{(30 - 100)°C} = 0.5$$

The three curves indicate the following readings on the abscissa:

$$Fo_{sphere} \cong 0.031 = \frac{\alpha t}{r_o^2} \quad (1)$$

$$Fo_{cylinder} \cong 0.064 = \frac{\alpha t}{r_o^2} \quad (2)$$

$$Fo_{plate} \cong 0.196 = \frac{\alpha t}{L^2} \quad (3)$$

1) In the case of the spherical shape, the radius is fixed by the mass m = 5 grams,

$$m = \rho \frac{4\pi}{3} r_o^3$$

$$r_o^3 = \frac{5g}{0.9 \frac{g}{cm^3}} \frac{3}{4\pi} = 1.33\ cm^3$$

$$r_o = 1.1\ cm$$

and eq. (1) pinpoints the warming time,

$$t = 0.031 \frac{(1.1)^2\ cm^2}{0.0017\ cm^2/s} \cong 22\ s$$

2) For a cylinder of length Z = 6 cm, the radius is

$$m = \rho Z \pi r_0^2$$

$$r_0^2 = \frac{5g}{0.9 \frac{g}{cm^3}} \frac{1}{\pi\, 6\ cm} = 0.29\ cm^2$$

$$r_0 = 0.54\ cm$$

and eq. (2) recommends

$$t = 0.064 \frac{(0.54)^2\ cm^2}{0.0017\ cm^2/s} \cong 11\ s$$

3) Finally, in the case of a disc of diameter D = 4 cm, the disc half-thickness L is

$$m = \rho \frac{\pi D^2}{4} 2L$$

$$L = \frac{5g}{0.9 \frac{g}{cm^3}} \frac{2}{\pi\, 16\ cm^2} = 0.22\ cm$$

with the corresponding warming time, cf. eq. (3),

$$t = 0.196 \frac{(0.22)^2\ cm^2}{0.0017\ cm^2/s} \cong 6\ s$$

Comparing these three time intervals (22 s, 11 s, 6 s) we see that the disc shape (the potato "slice") promises to cook much faster than the other two shapes.

Problem 4.13. a) With regard to the phenomenon of time-dependent conduction in a slab, eq. (4.62), we know the following:

$$L = 0.02\ m \qquad h = 60 \frac{W}{m^2\,K}$$

$$T_\infty = 150°C \qquad k = 0.4 \frac{W}{m \cdot K}$$

$$T_i = 25°C$$

$$T(0,t) = 80°C \qquad \alpha = 1.25 \times 10^{-7} \frac{m^2}{s}$$

We calculate in order

$$\frac{hL}{k} = 60 \frac{W}{m^2\,K} 0.02\,m \frac{m{\cdot}K}{0.4\,W} = 3$$

$a_1 = 1.1925$ (from Table 3.1)

$$\frac{T(0,t) - T_\infty}{T_i - T_\infty} = \frac{80 - 150}{25 - 150} = 0.56$$

so that we can finally use eq. (4.62), with only the first term on the right side,

$$0.56 \cong 2 \frac{\sin(a_1)}{a_1 + \sin(a_1)\cos(a_1)} \cdot 1 \cdot \exp\left(- a_1^2\,Fo\right)$$

$$0.4627 \cong \exp\left(- a_1^2\,Fo\right)$$

$$Fo \cong \frac{-\,0.7706}{-\,(1.1925)^2} = 0.5419 \cong \frac{\alpha t}{L^2}$$

$$t \cong 0.5419 \frac{(0.02)^2\,m^2}{1.25 \times 10^{-7}\,m^2/s} = 1734s$$

$$= 28.9 \text{ minutes}$$

The order of magnitude of each term in the series on the right side of eq. (4.62) is dictated by the exponential. The first term is "of order 1" (note that the exponential calculated above was 0.4627). Let us assume that the "exact" Fo is indeed equal to 0.5419, as calculated above. From Table 3.1 we select $a_2 = 3.8088$, and learn that the exponential factor of the second term is

$$\exp\left(- a_2^2\,Fo\right) = 0.00039$$

i.e. of order 10^{-4}. This means that the first term of the series is an adequate approximation for the right side of eq. (4.62), and that Fo = 0.5419 is a good estimate of the true Fourier number.

b) In order to obtain an upper bound estimate for the effect of meat shrinkage on the cooking time, we repeat the above calculations for the "shrunk" meat slab:

$$L = 0.02\,m\ 0.75 = 0.015\,m$$

$$\frac{hL}{k} = 2.25$$

$a_1 \cong 1.068$, by interpolating linearly between hL/k = 1 and hL/k = 3 in Table 3.1

$$0.4762 = \exp\left(-a_1^2 \, Fo\right), \qquad \text{eq. (4.62)}$$

$$Fo = 0.6504$$

$$t = 0.6504 \frac{L^2}{\alpha} = 1171\,s = 19.5 \text{ minutes}$$

c) In conclusion, the shrinking of meat shortens the cooking time. The actual cooking time will be somewhere between these two estimates, 28.9 minutes and 19.5 minutes, because during cooking the half thickness L decreases continuously from 2 cm to 1.5 cm.

Problem 4.14. The conservation of energy at the interface of the two rubbing surfaces requires

$$q'' = q_1'' + q_2'' \tag{1}$$

where both q_1'' and q_2'' are defined as positive when they enter the respective solid blocks. According to eq. (4.43), the q_1'' part is

$$q_1'' = k_1 \frac{T_0 - T_\infty}{(\pi\alpha_1 t)^{1/2}} \tag{2}$$

Similarly, after noting the sign change required by the definition of q_2'', eq. (4.43) yields also

$$q_2'' = k_2 \frac{T_0 - T_\infty}{(\pi\alpha_2 t)^{1/2}} \tag{3}$$

Equations (1) - (3) yield, in order,

$$\frac{q_2''}{q_1''} = \left[\frac{(k\rho c)_2}{(k\rho c)_1}\right]^{1/2} = R$$

$$q_1'' = \frac{1}{1+R} q'', \qquad q_2'' = \frac{R}{1+R} q''$$

$$T_0 - T_\infty = \frac{(\pi t)^{1/2}\, q''}{(k\rho c)_1^{1/2} + (k\rho c)_2^{1/2}}$$

In conclusion, the block with the higher (kρc) value absorbs the larger share of the heat flux q" generated by friction at the interface.

Problem 4.15. With reference to eq. (4.62), we recognize the following quantities

$T_i = -10°C$ $\qquad$ $x = 0$

$T_\infty = 0°C$ $\qquad$ $\alpha = 1.15 \times 10^{-6} m^2/s$

$T(0,t) = -0.1°C$ $\qquad$ $a_1 = \frac{\pi}{2}$, (Table 3.1)

Taking only the first term in the series on the right side of eq. (4.62), we write

$$\frac{-0.1 - 0}{-10 - 0} = 2 \frac{1}{\pi/2 + 0} \exp\left(- \frac{\pi^2}{4} Fo\right)$$

leading to

$$Fo = 1.964 = \frac{\alpha t}{L^2}$$

where L = 0.5 cm. The time t required by the warm up of the slab centerplane is therefore

$$t = 1.964 \frac{L^2}{\alpha} = 1.964 \frac{0.25\ cm^2}{(1.15)\ 10^{-2} cm^2/s}$$

$$= 42.7\ s$$

Problem 4.16. a) We estimate, in order, the inverse of the Biot number,

$$\frac{1}{Bi} = \frac{k}{h\, r_o}$$

$$= 0.4 \frac{W}{m \cdot K} \frac{m^2 K}{200 W} \frac{1}{0.01 m} = 0.2$$

the ordinate of Fig. 4.10,

$$\frac{T_c - T_\infty}{T_i - T_\infty} = \frac{65 - 95}{20 - 95} = 0.4$$

the abscissa of Fig. 4.10,

$$\frac{\alpha t}{r_o^2} \cong 0.33$$

and the elapsed time,

$$t = 0.33 \frac{r_o^2}{\alpha}$$

$$= 0.33\ (1\ \text{cm})^2 \frac{\text{s}}{0.0014\ \text{cm}^2} \cong 4 \text{ minutes}$$

b) For the hot dog surface, in Fig. 4.11 we set $r/r_o = 1$, and $1/Bi = 0.2$ on the abscissa. On the ordinate we read

$$\frac{T - T_\infty}{T_c - T_\infty} \cong 0.25$$

$$T \cong T_\infty + 0.25\left(T_c - T_\infty\right)$$

$$= 95°\text{C} + 0.25\ (65 - 95)°\text{C}$$

$$= 87.5°\text{C}$$

Problem 4.17. The properties of 304 stainless steel near 600°C are

$$k = 25 \frac{\text{W}}{\text{m·K}}$$

$$\rho = 7817 \frac{\text{kg}}{\text{m}^3}$$

$$c_P = 0.46 \frac{\text{kJ}}{\text{kg K}}$$

$$\alpha = \frac{k}{\rho\, c_P} \cong 7 \times 10^{-6} \frac{\text{m}^2}{\text{s}}$$

The requirement that the wire store 90 percent of the maximum ($t \rightarrow \infty$) heat transfer means that the parameter on the ordinate of Fig. 4.12 is fixed:

$$\frac{Q(t)}{Q_i} = 0.9 \qquad (1)$$

It is fortuitous that the heat transfer coefficient varies inversely with the wire diameter, because it means that the Biot number is fixed regardless of the D value:

$$Bi = \frac{h\, r_o}{k} = \frac{hD}{2k}$$

$$= \frac{1 \dfrac{W}{m \cdot K}}{2 \times 25 \dfrac{W}{m \cdot K}} = 0.02 \qquad (2)$$

Intersecting lines (1) and (2) on Fig. 4.12 we find, on the abscissa,

$$Bi^2\, Fo \cong 0.025$$

in other words,

$$Fo = \frac{\alpha t}{r_o^2} \cong \frac{0.025}{Bi^2}$$

$$= \frac{0.025}{(0.02)^2} = 62.5$$

$$r_o \cong \left(\frac{\alpha t}{62.5}\right)^2 = \left(\frac{7 \times 10^{-6} \frac{m^2}{s}\, 0.01\ s}{62.5}\right)^{1/2}$$

$$\cong 33.5\ \mu m$$

In conclusion, the needed wire diameter is about 67 μm.

Problem 4.18. The function f(x,t) is a solution of the conduction equation (4.87), therefore

$$\frac{\partial^2 f}{\partial x^2} = \frac{1}{\alpha} \frac{\partial f}{\partial t} \qquad (1)$$

To see whether the new function

$$g(x,t) = \frac{\partial f}{\partial x} \qquad (2)$$

satisfies the same conduction equation, all we have to do is substitute the expression (2) in eq. (4.87). We obtain, in order

$$\frac{\partial^2 g}{\partial x^2} - \frac{1}{\alpha} \frac{\partial g}{\partial t} = 0$$

$$\frac{\partial^2}{\partial x^2}\left(\frac{\partial f}{\partial x}\right) - \frac{1}{\alpha} \frac{\partial g}{\partial t}\left(\frac{\partial f}{\partial x}\right) = 0$$

$$\frac{\partial}{\partial x}\left(\frac{\partial^2 f}{\partial x^2} - \frac{1}{\alpha}\frac{\partial f}{\partial t}\right) = 0 \qquad (3)$$

The expression in parentheses in eq. (3) vanishes cf. eq. (1), therefore, eq. (3) is indeed satisfied. In conclusion, the function g(x,t) satisfies the conduction equation (4.87).

One example of a function f(x,t) that satisfies eq. (4.87) [actually, eq. (4.22)] is the error-function solution (4.41):

$$f(x,t) = C \operatorname{erf}\left[\frac{x}{2(\alpha t)^{1/2}}\right] \qquad (4)$$

The g(x,t) function that corresponds to this example is

$$g = \frac{\partial f}{\partial x} = \frac{C}{2(\alpha t)^{1/2}} \exp\left(-\frac{x^2}{4\alpha t}\right) \qquad (5)$$

This is the expression chosen in eq. (4.88), as a solution to eq. (4.87).

<u>Problem 4.19</u>. The time of maximum temperature at a distance x away from the instantaneous plane source is determined by solving $\partial\theta/\partial t = 0$, for which $\theta(x,t)$ is reported in eq. (4.93). We obtain in this way

$$-\frac{1}{2} t^{-3/2} \exp(\,) + \frac{1}{t^{1/2}} \exp(\,) \frac{x^2}{4\alpha t^2} = 0$$

in other words,

$$t = \frac{x^2}{2\alpha}$$

Substituting this special time value in the $\theta(x,t)$ expression (4.93), we obtain the maximum temperature that will be registered in the constant-x plane

$$\theta_{max}(x) = \frac{1}{(2\pi e)^{1/2}} \frac{Q''}{\rho c |x|}$$

Problem 4.20. The time of the temperature maximum is obtained by solving $\partial\theta/\partial t = 0$ in conjunction with eq. (4.94),

$$-\frac{1}{t^2}\exp(\,) + \frac{1}{t}\exp(\,)\frac{r^2}{4\alpha t^2} = 0$$

This equation yields

$$t = \frac{r^2}{4\alpha}$$

The corresponding maximum temperature in the cylindrical surface of radius r (fixed) is

$$\theta_{max}(r) = \frac{1}{\pi e}\frac{Q'}{\rho c r^2}$$

Problem 4.21. The time of maximum temperature at the distance r away from the instantaneous point source is determined by solving $\partial\theta/\partial t = 0$, where $\theta(r,t)$ is the expression listed in eq. (4.95):

$$-\frac{3}{2}t^{-5/2}\exp(\,) + t^{-3/2}\exp(\,)\frac{r^2}{4\alpha t^2} = 0$$

The time that satisfies this condition is

$$t = \frac{r^2}{6\alpha}$$

The corresponding maximum temperature on the constant-r spherical surface is

$$\theta_{max}(r) = \left(\frac{3}{2\pi e}\right)^{3/2}\frac{Q}{\rho c r^3}$$

Problem 4.22. With reference to the steady (long times) temperature distribution around a continuous point source, eq. (4.107), the temperature T_1 on the spherical surface of radius r_1 is

$$\theta(r_1, \infty) = T_1 - T_\infty = \frac{q}{4\pi k\, r_1}$$

The formula for the steady heat transfer rate q between an isothermal sphere (r_1, T_1) and an infinite conducting medium is therefore

$$q = (4\pi\, r_1)\, k\, (T_1 - T_\infty)$$

Recalling the definition of conduction shape factor, $q = Sk(T_1 - T_\infty)$, we conclude that

$$S = 4\pi r_1$$

This result can be deduced also from the first two "spherical surfaces" entries in Table 3.3, by letting $D/4H \to 0$.

Problem 4.23. Our first task is to transform the T(x,y) problem (4.108)-(4.110) into a problem for $\theta(\eta)$, where

$$T - T_\infty = \frac{q'/\rho c}{(U\alpha x)^{1/2}}\,\theta(\eta)$$

$$\eta = y\left(\frac{U}{\alpha x}\right)^{1/2}$$

The left and right sides of the energy eq. (4.108) yield, respectively,

$$U\frac{\partial T}{\partial x} = U\frac{q'/\rho c}{(U\alpha)^{1/2}}\left(-\frac{1}{2}\right)x^{-3/2}\theta + U\frac{q'/\rho c}{(U\alpha x)^{1/2}}\frac{d\theta}{d\eta}\,y\left(\frac{U}{\alpha}\right)^{1/2}\left(-\frac{1}{2}\right)x^{-3/2}$$

$$\alpha\frac{\partial^2 T}{\partial y^2} = \alpha\frac{q'/\rho c}{(U\alpha x)^{1/2}}\frac{d^2\theta}{d\eta^2}\frac{U}{\alpha x}$$

Substituted back into eq. (4.108), these lead to the new form of the energy equation

$$-\frac{1}{2}\left(\theta + \eta\,\theta'\right) = \theta'' \qquad \text{(a)}$$

It is easy to show that the $\theta(\eta)$ form of the boundary conditions (4.109) and the energy integral (4.110) is

$$\theta = 0 \qquad \text{at} \qquad \eta = \pm\infty \qquad \text{(b)}$$

$$\int_{-\infty}^{\infty}\theta\, d\eta = 1 \qquad \text{(c)}$$

The new problem is represented by eqs. (a)-(c). The energy eq. (a) can be integrated in the following sequence:

$$-\frac{1}{2}(\eta\,\theta)' = \theta''$$

$$-\frac{1}{2}\eta\,\theta = \theta' + \underbrace{\text{constant}}$$

zero, because of the symmetry about the $\eta = 0$ plane

$$-\frac{1}{2}\eta\, d\eta = \frac{d\theta}{\theta}$$

$$-\frac{\eta^2}{4} = \ln\theta + \text{constant}$$

This last expression is equivalent to

$$\theta = C\exp\left(-\frac{\eta^2}{4}\right)$$

which substituted into eq. (c) pinpoints the value of the constant coefficient C:

$$C = \frac{1}{2\pi^{1/2}}$$

In conclusion, the $\theta(\eta)$ solution is

$$\theta(\eta) = \frac{1}{2\pi^{1/2}}\exp\left(-\frac{\eta^2}{4}\right)$$

which means that the temperature distribution in the two-dimensional wake is

$$T(x,y) - T_\infty = \frac{q'/\rho c}{(4\pi\alpha x)^{1/2}}\exp\left(-\frac{U\,y^2}{4\alpha x}\right)$$

Problem 4.24. a) The thermal diffusion effect penetrates a rock layer of thickness

$$\delta \sim (\alpha t)^{1/2}$$

where $\alpha = 0.012\ \text{cm}^2/\text{s}$ and $t = 10$ days:

$$\delta \sim \left(0.012\,\frac{\text{cm}^2}{\text{s}}\,10 \times 24 \times 3600\text{s}\right)^{1/2} = 102\ \text{cm} \sim 1\text{m}$$

Since δ (~ 1m) is much smaller than the crack diameter D(= 10m), the cooled zone is a "pancake" of thickness ~ 2m in which the heat transfer is mainly unidirectional (perpendicular to the plane of the crack).

The instantaneous heat flux at t = 10 days is, cf. eq. (4.43'),

$$q'' = \left(T_\infty - T_c\right)\frac{k}{(\pi\alpha t)^{1/2}}$$

$$= (200 - 25)\text{ K}\,\frac{2.9\,\frac{\text{W}}{\text{m·K}}}{\pi^{1/2}\ 1.02\text{m}} = 280.1\,\frac{\text{W}}{\text{m}^2}$$

The total heat transfer area is

$$A = 2 \times \frac{\pi D^2}{4} = 157.1\text{ m}^2$$

where we made use of the observation that the water-cooled space has two sides. The corresponding total heat transfer rate is

$$q = q''A = 280.1\,\frac{\text{W}}{\text{m}^2}\,157.1\text{ m}^2 \cong 44\text{ kW}$$

b) By setting δ ~ 10m in the expression $\delta \sim (\alpha t)^{1/2}$ we find the thermal diffusion penetration time

$$t \sim \frac{\delta^2}{\alpha} = \frac{10^2\text{ m}^2}{0.012\text{ cm}^2/\text{s}} = 965\text{ days} = 2.64\text{ years}$$

We assume that after a time of order 2 years (or longer) the rock zone cooled by the water stream can be modeled as a sphere of diameter 10m and temperature 25°C. The steady-state heat current that is sucked by this sphere from the hot-rock (infinite medium) that surrounds it can be calculated with eq. (4.107):

$$q_{sink} = 4\pi\,kr\left(T_\infty - T_c\right)$$

$$= 4\pi\ 2.9\,\frac{\text{W}}{\text{m·K}}\ 5\text{m}\ (200 - 25)\text{ k}$$

$$= 31\ 887\text{ W}$$

In conclusion, the long-time power output from this well would be of the order of 30 kW. This power level is low primarily because of the smallness of the cooled zone. The design of pilot plants for the extraction of energy from hot-dry-rock beds calls for much larger cracks (D ~ 1 km) and shorter energy extraction periods.

Problem 4.25. The quasi-steady temperature distribution in the liquid layer of thickness δ is linear, cf. eq. (4.114),

$$\frac{T - T_m}{T_0 - T_m} = 1 - \frac{x}{\delta} \tag{1}$$

Likewise, the temperature distribution in the solid layer of thickness $L - \delta$ can be written as

$$\frac{T - T_m}{T_L - T_m} = \frac{x - \delta}{L - \delta} \tag{2}$$

The conservation of energy at the liquid-solid interface requires

$$q''_{x,f} - q''_{x,s} = \rho h_{sf} \frac{d\delta}{dt} \tag{3}$$

where the two heat fluxes are defined as positive in the x direction,

$$q''_{x,f} = - k_f \left(\frac{\partial T}{\partial x}\right)_{x = \delta} = \frac{k_f}{\delta} (T_0 - T_m) \tag{4}$$

$$q''_{x,s} = - k_s \left(\frac{\partial T}{\partial x}\right)_{x = \delta} = \frac{k_s}{L - \delta} (T_m - T_L) \tag{5}$$

and where "f" and "s" indicate the liquid and solid sides of the interface. Combining eqs. (3) - (5), and using the dimensionless notation defined in the problem statement, we obtain

$$\frac{1}{X} - \frac{B}{1 - X} = \frac{dX}{d\tau} \tag{6}$$

The true steady state corresponds to $dX/d\tau = 0$, therefore the steady-state value of the dimensionless thickness X is

$$X = \frac{1}{1 + B}, \qquad \text{as } \tau \rightarrow \infty \tag{7}$$

Note that in the steady state the system is a sandwich of liquid and solid, because the value $X = (1 + B)^{-1}$ is less than 1.

For the more general case of a still moving melting front, we integrate eq. (6) from the initial condition $X(\tau = 0) = 0$, and obtain

$$\tau = \frac{X^2}{2(1+B)} - \frac{BX}{(1+B)^2} - \frac{B}{(1+B)^3} \ln[1 - (1+B)X] \quad (8)$$

Problem 4.26. The continuity of heat flux through the $x = 0$ surface requires that

$$h(T_\infty - T_0) = k\frac{T_0 - T_m}{\delta} \quad (1)$$

In accordance with eq. (4.113), the conservation of energy at the melting front is written as

$$k\frac{T_0 - T_m}{\delta} = \rho\, h_{sf}\frac{d\delta}{dt} \quad (2)$$

Equation (1) can be rearranged to obtain an expression for the temperature of the $x = 0$ surface,

$$T_0 - T_m = (T_\infty - T_m)\frac{\frac{h\delta}{k}}{1 + \frac{h\delta}{k}} \quad (3)$$

We substitute this expression on the left side of equation (2) and obtain

$$\frac{h}{\rho\, h_{sf}}(T_\infty - T_m) = \left(1 + \frac{h\delta}{k}\right)\frac{d\delta}{dt} \quad (4)$$

Integrated from $t = 0$, where $\delta = 0$, equation (4) yields

$$\frac{h}{\rho\, h_{sf}}(T_\infty - T_m)\,t = \delta + \frac{h}{k}\cdot\frac{\delta^2}{2} \quad (5)$$

Equation (5) shows that in the very beginning the liquid layer thickness δ increases proportionally with the time t. In other words, when δ is sufficiently small the second term on the right-hand side is negligible. Later, the second term takes over, and δ increases more like $t^{1/2}$. This second regime corresponds to eq. (4.116) in the text. The transition to the $\delta \sim t^{1/2}$ regime occurs when the second term dominates the right-hand side of eq. (5),

$$\frac{h}{k}\cdot\frac{\delta^2}{2} > \delta \quad (6)$$

i.e. when the liquid film has become thick enough so that

$$\delta > 2\frac{k}{h} \tag{7}$$

Equation (5) can be rewritten in nondimensional terms by using the definitions

$$\mathrm{Ste} = \frac{c(T_\infty - T_m)}{h_{sf}}, \qquad \lambda = \frac{\delta}{2(\alpha t)^{1/2}} \tag{8}$$

The resulting expression is

$$\frac{1}{2}\,\mathrm{Ste}\,\mathrm{Bi} = \lambda + \lambda^2\,\mathrm{Bi} \tag{9}$$

where Bi is a Biot number based on the conduction thickness $(\alpha t)^{1/2}$ as length scale,

$$\mathrm{Bi} = \frac{h}{k}(\alpha t)^{1/2} \tag{10}$$

Problem 4.27. We approximate the popsicle properties by using those of ice at 0°C,

$$h_{sf} = 333.4\,\frac{\mathrm{kJ}}{\mathrm{kg}} \qquad k = 2.25\,\frac{\mathrm{W}}{\mathrm{m\cdot K}}$$

$$c_P = 2.04\,\frac{\mathrm{kJ}}{\mathrm{kg\,K}} \qquad \rho = 917\,\frac{\mathrm{kg}}{\mathrm{m}^3}$$

The first question is whether eq. (4.116) is accurate, or we must use Fig. 4.26. For this we calculate the Stefan number for solidification [compare with eq. (4.119)]:

$$\mathrm{Ste} = \frac{c_P(T_m - T_0)}{h_{sf}}$$

$$= 2.04\,\frac{\mathrm{kJ}}{\mathrm{kg\,K}}\,5\mathrm{K}\,\frac{\mathrm{kg}}{333.4\,\mathrm{kJ}} = 0.03$$

The Stefan number is very small: Fig. 4.26 shows that the equation to use is the "solidification" version of eq. (4.116):

$$\delta = \left[2\,\frac{kt}{\rho\,h_{sf}}(T_m - T_0)\right]^{1/2}$$

$$t = \frac{\delta^2 \rho h_{sf}}{2k(T_m - T_0)}$$

$$= \frac{(0.007\text{m})^2\ 917\ \frac{\text{kg}}{\text{m}^3}\ 333.4\ \frac{10^3\text{K}}{\text{kg}}}{2 \times 2.25\ \frac{\text{J/s}}{\text{m·K}}\ 5\text{K}}$$

$$= 666\text{s} \cong 11 \text{ minutes}$$

Problem 4.28. The relevant properties of frozen fish are

$$h_{sf} = 235\ \frac{\text{kJ}}{\text{kg}} \qquad k \cong 1\ \frac{\text{W}}{\text{m·K}}$$

$$c = 1.7\ \frac{\text{kJ}}{\text{kg K}} \qquad \rho \cong 1000\ \frac{\text{kg}}{\text{m}^3}$$

We calculate, in order,

$$\text{Ste} = \frac{c(T_m - T_0)}{h_{sf}}$$

$$= 1.7\ \frac{\text{kJ}}{\text{kg K}}\ \frac{(-2.2 + 20)\ \text{K}}{235\ \text{kJ/kg}} = 0.13$$

$$\frac{\delta}{(\alpha t)^{1/2}} \cong 0.51 \qquad \text{(Fig. 4.26, at Ste = 0.13)}$$

$$\alpha = \frac{k}{\rho c} \cong 1\ \frac{\text{W}}{\text{m·K}}\ \frac{\text{m}^3}{10^3\ \text{kg}}\ \frac{\text{kg K}}{1.7 \times 10^3\ \text{J}}$$

$$\cong 0.0059\ \frac{\text{cm}^2}{\text{s}}$$

$$t = \frac{\delta^2}{\alpha}\ \frac{1}{(0.51)^2} = \frac{1\ \text{cm}^2}{0.0059\ \frac{\text{cm}^2}{\text{s}}}\ \frac{1}{0.26}$$

$$= 652\text{s} \cong 11 \text{ minutes}$$

For the complete freezing of each fish, the conveyor belt speed must not exceed

$$U = \frac{L}{t} = \frac{30\text{m}}{652\text{s}} = 4.6\ \frac{\text{cm}}{\text{s}}$$

Problem 4.29. With reference to the $\Delta x/2$-wide control volume shown in the last drawing of Table 4.4, we write the first law of thermodynamics for a closed system that operates in a time-dependent manner,

$$\rho \frac{\Delta x}{2} c \frac{\partial T}{\partial t} = \underbrace{k \frac{T_{i-1}^m - T_i^m}{\Delta x}}_{\substack{\text{entering}\\ \text{from the}\\ \text{left}}} + \underbrace{q''}_{\substack{\text{entering}\\ \text{from the}\\ \text{right}}}$$

in other words

$$\frac{\partial T}{\partial t} = 2 \frac{\alpha}{(\Delta x)^2}\left(T_{i-1}^m - T_i^m\right) + \frac{2q''}{\rho c\, \Delta x} \qquad (1)$$

In order to obtain the explicit finite-difference equation, we make use of the forward difference

$$\frac{\partial T}{\partial t} \cong \frac{T_i^{m+1} - T_i^m}{\Delta t}$$

and eq. (1) becomes

$$T_i^{m+1} = (1 - 2\,\text{Fo})\, T_i^m + 2\,\text{Fo}\left(T_{i-1}^m + \frac{q''\,\Delta x}{k}\right)$$

The condition for numerical stability is $(1 - 2\,\text{Fo}) \geq 0$, or

$$\text{Fo} \leq \frac{1}{2}$$

For the implicit finite-difference equation, we rely on the backward difference

$$\frac{\partial T}{\partial t} \cong \frac{T_i^m - T_i^{m-1}}{\Delta t}$$

and, after combining it with eq. (1), we obtain

$$(1 + 2\,\text{Fo})\, T_i^m - 2\,\text{Fo}\, T_{i-1}^m = T_i^{m-1} + 2\,\text{Fo}\frac{q''\,\Delta x}{k}$$

Or, if we use (m + 1) as superscript for the unknown temperatures, and m for the known temperature, the implicit equation can be written also as

$$(1 + 2\,Fo)\,T_i^{m+1} - 2\,Fo\,T_{i-1}^{m+1} = T_i^m + 2\,Fo\,\frac{q''\,\Delta x}{k}$$

Problem 4.30. The first-law statement for the boundary control volume of height Δx and thickness $\Delta x/2$ is

$$\rho\,\frac{\Delta x}{2}\,\Delta x\,c\,\frac{\partial T}{\partial t} = k\,\Delta x\,\frac{T_{i-1,j}^m - T_{i,j}^m}{\Delta x} + k\,\frac{\Delta x}{2}\,\frac{T_{i,j+1}^m - T_{i,j}^m}{\Delta x}$$

$$+ k\,\frac{\Delta x}{2}\,\frac{T_{i,j-1}^m - T_{i,j}^m}{\Delta x} + q_j''\,\Delta x$$

or, after dividing by $\rho c(\Delta x)^2/2$,

$$\frac{\partial T}{\partial t} = \frac{\alpha}{(\Delta x)^2}\left(2\,T_{i-1,j}^m + T_{i,j+1}^m + T_{i,j-1}^m - 4\,T_{i,j}^m\right) + q_j''\,\frac{2}{\rho c\,\Delta x} \qquad (1)$$

For the explicit equation, we use the forward difference

$$\frac{\partial T}{\partial t} \cong \frac{T_{i,j}^{m+1} - T_{i,j}^m}{\Delta t}$$

and eq. (1) becomes

$$T_{i,j}^{m+1} = (1 - 4\,Fo)\,T_{i,j}^m + Fo\left(2\,T_{i-1,j}^m + T_{i,j+1}^m + T_{i,j-1}^m\right) + 2\,Fo\,\frac{q_j''\,\Delta x}{k}$$

The stability requirement is $(1 - 4\,Fo) \geq 0$, or

$$Fo \leq \frac{1}{4}$$

For the implicit form of the finite-difference equation, we use the backward difference

$$\frac{\partial T}{\partial t} \cong \frac{T_{i,j}^m - T_{i,j}^{m-1}}{\Delta t}$$

and, after some manipulation, eq. (1) reduces to

$$(1 + 4\,Fo)\,T_{i,j}^{m} + Fo\left(2\,T_{i-1,j}^{m} + T_{i,j+1}^{m} + T_{i,j-1}^{m}\right)$$

$$= T_{i,j}^{m-1} + 2\,Fo\,\frac{q_j''\,\Delta x}{k}$$

This equation is analogous to the one listed in the problem statement, in which (m+1) denotes the new (unknown) temperatures.

Problem 4.31. i) It is advisable to solve the problem in dimensionless form, so that the results (dimensionless) may be relevant to more than one numerical application. Because of symmetry, we consider only the semiinfinite domain $x > 0$ in which the problem can be stated as follows.

Equation: $$\frac{1}{\alpha}\frac{\partial T}{\partial t} = \frac{\partial^2 T}{\partial x^2} \qquad (1)$$

Initial condition: $$t = 0, \quad T = T_i \quad \text{for} \quad 0 < x \le L \qquad (2)$$

$$T = T_\infty \quad \text{for} \quad x > L$$

Boundary conditions: $$x = 0, \quad \frac{\partial T}{\partial x} = 0 \qquad (3)$$

$$x \to \infty, \quad T \to T_\infty \qquad (4)$$

If we define the dimensionless variables

$$\theta = \frac{T - T_\infty}{T_i - T_\infty} \qquad (5)$$

$$\xi = \frac{x}{L} \qquad (6)$$

$$\tau = \frac{\alpha t}{L^2} \qquad (7)$$

the problem statement (1)-(4) assumes the form

$$\frac{\partial \theta}{\theta \tau} = \frac{\partial^2 \theta}{\partial \xi^2} \tag{8}$$

$$\tau = 0, \qquad \theta = 1 \qquad \text{for} \qquad 1 < \xi \le 1$$
$$\qquad\qquad \theta = 0 \qquad \text{for} \qquad \xi > 1 \tag{9}$$

$$\xi = 0, \qquad \frac{\partial \theta}{\theta \xi} = 0 \tag{10}$$

$$\xi \to \infty, \qquad \theta \to 0 \tag{11}$$

We cover the conducting medium with slices (control volumes) of equal thickness, $\Delta x = L/N$. The first control volume has the plane $x = 0$ as plane of symmetry. The $(N + 1)^{th}$ control volume is centered about the $x = L$ plane. This also means that $\Delta\xi = 1/N$, $i = 1$ for the plane of symmetry ($x = 0$), $i = N + 1$ for the edge of the original hot zone ($x = L$), and $i > N + 1$ for $x > L$.

To the time step Δt corresponds the dimensionless time step

$$\Delta\tau = \frac{\alpha}{L^2} \Delta t \tag{12}$$

The grid Fourier number (4.125) becomes

$$Fo = \frac{\Delta\tau}{(\Delta\xi)^2} = N^2 \, \Delta\tau \tag{13}$$

The implicit method begins with selecting out of Table 4.4 the proper equations for the symmetry (adiabatic) plane ($i = 1$)

$$(1 + 2\,Fo)\, T_1^{m+1} - 2\,Fo\, T_2^{m+1} = T_i^m \tag{14}$$

and for any internal plane ($i > 1$)

$$(1 + 2\,Fo)\, T_1^{m+1} - Fo\left(T_{i-1}^{m+1} + T_{i+1}^{m+1}\right) = T_i^m \tag{15}$$

Next, the dimensionless counterparts of eqs. (14,15) are arranged in the form required by the Gauss-Seidel iteration method

$$\theta_1^{m+1} = \frac{\theta_1^m}{1 + 2\,Fo} + \frac{2\,Fo}{1 + 2\,Fo}\,\theta_2^{m+1} \tag{16}$$

$$\theta_i^{m+1} = \frac{\theta_i^m}{1 + 2\,Fo} + \frac{Fo}{1 + 2\,Fo}\left(\theta_{i-1}^{m+1} + \theta_{i+1}^{m+1}\right) \tag{17}$$

The overall thickness of the conduction (computational) domain must be chosen in such a way (i.e. the thickness must be large enough) that at the edge of the domain $T = T_\infty$ for all the time steps. This thickness corresponds to the highest i, and increases from one time step to the next as conduction penetrates into the unheated portion of the domain. It becomes evident as we set up the following algorithm.

We begin with the initial time t = 0 (or m = 0), and initialize all the node temperatures (θ_i^0) in accordance with eqs. (9). This means that

$$\theta_i^0 = \begin{cases} 1, & 1 \le i \le N \\ \frac{1}{2}, & i = N + 1 \\ 0, & N + 1 < i \le M \end{cases} \tag{18}$$

The initial condition $\theta_{N+1}^0 = 1/2$ accounts for the fact that only half of the (N + 1)th control volume is initially hot. To initiate a wide enough domain we set M = 10 (N + 1). We then increase the time by $\Delta\tau$, and sweep the domain from left to right, by using eq. (16) followed by eqs. (17). We do not have to go all the way to i = M, because the boundary condition (11) is met by a θ_i^1 much closer to the heated zone. That location (i = K) is determined by a criterion of the type

$$\theta_K^m < \varepsilon \tag{19}$$

in which ε is a number close to zero. I used $\varepsilon = 10^{-3}$. The time is increased again by $\Delta\tau$, and the sweep is repeated to a new (larger) K, i.e. to a more distant slice. The elapsed time is

$$\tau = m\,\Delta\tau \tag{20}$$

ii) The accuracy test refers to the values that are assigned to $\Delta\xi$ and $\Delta\tau$, or N and $\Delta\tau$. The effect of (N, $\Delta\tau$) on the numerical results becomes clear as we consider the time evolution of the midplane temperature $\theta_1(\tau)$. This decreases from 1 at $\tau = 0$, to 0 as $\tau \to \infty$. The time τ^* when θ_1 drops to an intermediate value, say $\theta_1{}^* = 2/3$, depends on N and $\Delta\tau$,

$$\tau^* = \text{function}\ (N, \Delta\tau) \tag{21}$$

This function (the value of τ^*) is presented in the following table, which shows that the grid N = 10 and $\Delta\tau = 0.0001$ is sufficiently fine to make τ^* relatively insensitive to changes in N and $\Delta\tau$.

$\Delta\tau$	N = 5	N = 10	N = 15
0.01	0.67	1.07	1.74
0.001	0.55	0.59	0.654
0.0001	0.54	0.54	0.546

iii) The following results for the centerplane temperature $\theta_1(\tau)$ were obtained using N = 5 and $\Delta\tau = 0.001$, which is rough but adequate for the following comparison. The plane instantaneous source eq. (4.93) recommends the following expression for the center plane temperature (x = 0)

$$T(x=0) - T_\infty = \frac{Q''}{2\,\rho c\,(\pi \alpha t)^{1/2}} \tag{22}$$

The source strength [J/m^2] is

$$Q'' = 2L\,\rho c\,(T_i - T_\infty) \tag{23}$$

therefore, in view of definitions (5) and (7), eq. (22) becomes

$$\theta(0,\tau) = (\pi\tau)^{-1/2} \tag{24}$$

The following table shows that eq. (24) anticipates well the numerical results (θ_1) when $\tau > 1$. This conclusion makes sense, because when τ exceeded 1 conduction has had enough time to travel (penetrate) the half-thickness L, and to smooth the step-shape of the original temperature distribution (i.e. to make it look bell-shaped like in Fig. 4.21)

τ	θ_1	$(\pi\tau)^{-1/2}$
0	1	∞
0.5	0.69	0.80
1	0.53	0.56
2	0.39	0.40
3	0.32	0.33
4	0.28	0.28

Project 4.1. The following solution was performed using the explicit method. With reference to the drawing that accompanies Example 4.4 in the text, the explicit equations for the internal planes and the adiabatic midplane are the same as eqs. (iii) and (iv). The equation for the left surface is collected from Table 4.4,

$$T_1^{m+1} = 2\,\text{Fo}\left(T_2^m + \text{Bi}\,T_\infty\right) + (1 - 2\,\text{Fo} - 2\,\text{Bi}\,\text{Fo})\,T_1^m \qquad \text{(a)}$$

in which

$$\text{Bi} = \frac{h\,\Delta x}{k} = \frac{hL}{k}\frac{1}{n} = \frac{10^3 \text{W}}{\text{m}^2\text{K}}\,\frac{0.02\text{ m}}{1 \text{W/m·K}}\,\frac{1}{n}$$

$$= \frac{20}{n} \qquad \text{(b)}$$

The actual expression for the fluid temperature $T_\infty(t)$ depends on the particular time "regime". There are three such regimes. In the earliest, when $0_s < t \le 30$ s, the fluid temperature rises linearly:

$$T_\infty^{I} = \frac{100°\text{C}}{30\text{ s}}\,t = \frac{100°\text{C}}{30\text{ s}}\,m\,\Delta t = \ldots.$$

$$= 266.67°\text{C}\,\frac{m\,\text{Fo}}{n^2} \qquad \text{(c)}$$

This regime expires when $T_1 = 30$ s, which corresponds to the time frame numbered

$$m_1 = \frac{t_1}{\Delta t} = t_1\,\frac{n^2}{\text{Fo}\,L^2\,\alpha} = 0.375\,\frac{n^2}{\text{Fo}} \qquad \text{(d)}$$

In the intermediate regime, $30\text{ s} < t \le 60$ s, the fluid temperature decreases linearly,

$$T_\infty^{II} = -\frac{100°\text{C}}{30\text{ s}}\,t + 200°\text{C}$$

$$= -266.67°\text{C}\,\frac{m\,\text{Fo}}{n^2} + 200\ °\text{C} \qquad \text{(e)}$$

This second regime expires at $t_2 = 60$ s, which in terms of m means

$$m_2 = \frac{t_2}{\Delta t} = 0.750\,\frac{n^2}{\text{Fo}} \qquad \text{(f)}$$

Equations (c) - (f) were derived by using the definitions $n = L/\Delta x$ and $\text{Fo} = \alpha\Delta t/(\Delta x)^2$. Finally, during the third regime $t > 60$ s (or $m > m_2$), the fluid temperature remains constant,

$$T_{\infty}^{III} = 0°C \tag{g}$$

For numerical stability, the most stringent requirement is posed by eq. (a),

$$Fo \le \frac{1}{2\,(1 + Bi)} = \frac{1}{2\left(1 + \frac{20}{n}\right)} \tag{h}$$

The maximum allowable Fo is therefore a function of the spatial division,

n	3	4	5	6
Fo_{max}	0.0652	0.0833	0.1	0.1154

We know from Example 4.4 that in the present geometry n = 5 provides a reasonably fine grid. Therefore we set n = 5, and choose a safe Fourier number, say

$$Fo = 0.09375$$

so that the dimensionless times m_1 and m_2 are integers, $m_1 = 100$ and $m_2 = 200$. A sample of the temperature histories of the outer surface (T_1) and the midplane of the slab (T_6) is presented below. In particular, the midplane temperature reaches the maximum 45.4°C at the time $t \cong 54.6$ s.

m	t (s)	T_∞ (°C)	T_1 (°C)	T_6 (°C)
0	0	0	0	0
50	15	50	44.012	2.714
100	30	100	91.481	18.787
150	45	50	51.810	40.961
200	60	0	5.787	44.079
250	75	0	2.382	31.898
300	90	0	1.550	21.066

With the grid fixed at n = 5, a more refined solution can be obtained by using a smaller time step, i.e. a smaller Fourier number,

$$Fo = 0.0625$$

In this case, the three time regimes are separated by $m_1 = 150$ and $m_2 = 300$. The following table shows that the solution is nearly identical to the first. The maximum midplane temperature is 45.32°C, and it occurs at $t \cong 54.4$ s.

m	t (s)	T_∞ (°C)	T_1 (°C)	T_6 (°C)
0	0	0	0	0
75	15	50	44.017	2.766
150	30	100	91.487	18.864
225	45	50	51.805	40.933
300	60	0	5.780	43.993
375	75	0	2.380	31.853
450	90	0	1.549	21.053

Project 4.2. We adopt the dimensionless notation

$$\theta = \frac{T - T_\infty}{T_i - T_\infty}, \qquad \xi = \frac{x}{L}, \qquad \tau = \frac{\alpha t}{L^2}$$

so that the half-thickness extends from $\xi = 0$ (the midplane) to $\xi = 1$ (the surface with convection). Next, we divide this ξ interval into N equal slices,

$$\Delta\xi = \frac{1}{N}$$

where $\Delta\xi$ is the thickness of one slice. We assign $i = 1$ to the midplane (adiabatic), and $i = N + 1$ to the surface with convection. The following solution is based on the explicit method.

The internal planes ($i = 2,\ldots,N$) are accounted for by the first entry in Table 4.4, which after using the θ notation becomes

$$\theta_i^{m+1} = \text{Fo}\left(\theta_{i-1}^m + \theta_{i+1}^m\right) + (1 - 2\,\text{Fo})\,\theta_i^m \qquad (1)$$

The grid Fourier number Fo is in this case

$$Fo = \frac{\alpha \Delta t}{(\Delta x)^2} = \frac{\Delta \tau}{(\Delta \xi)^2} =$$

$$= N^2 \Delta \tau \tag{2}$$

and the associated stability condition is

$$N^2 \, \Delta \tau \leq \frac{1}{2} \tag{3}$$

The surface with convection (i = N + 1) is described by the second entry in Table 4.4. In terms of θ, this reduces to

$$\theta_{N+1}^{m+1} = 2 \, Fo \, \theta_N^m + (1 - 2 \, Fo - 2 \, Bi \, Fo) \, \theta_{N+1}^m \tag{4}$$

where the grid Biot number is

$$Bi = \frac{h \Delta x}{k} = \frac{hL}{k} \Delta \xi = \frac{\overline{Bi}}{N} \tag{5}$$

And, since the "macroscopic" Biot number has been specified,

$$\overline{Bi} = 1 \tag{6}$$

the grid Biot number is given by

$$Bi = \frac{1}{N} \tag{7}$$

The stability condition is Fo (1 + Bi) ≤ 0.5, which means

$$N^2 \, \Delta \tau \left(1 + \frac{1}{N}\right) \leq \frac{1}{2} \tag{8}$$

or

$$\Delta \tau \leq \frac{0.5}{N \, (N + 1)} \tag{9}$$

The adiabatic midplane (i = 1) is represented by eq. (4), after setting Bi = 0, and after replacing i = N + 1 with i = 1:

$$\theta_1^{m+1} = 2 \, Fo \, \theta_0^m + (1 - 2 \, Fo) \, \theta_1^m \tag{10}$$

In this, we note that because of symmetry the temperature θ_0^m (of the plane situated immediately to the left of the midplane) is the same as the temperature in the plane situated to the right of the midplane, θ_2^m. The finite difference equation for the midplane is therefore

$$\theta_1^{m+1} = 2\,Fo\,\theta_2^m + (1 - 2\,Fo)\,\theta_1^m \tag{11}$$

with the same stability condition as in eq. (3).

In summary, the most stringent stability condition is eq. (9): this was used in the numerical solution that is described next.

<u>Numerical results</u>. The algorithm consisted of first choosing N and $\Delta\tau$ (based on eq. 9), and then calculating Bi and Fo. All the slab temperatures were initialized by setting

$$\theta_i^0 = 1 \qquad \text{for } i = 1,2,\ldots,N+1$$

The calculations for one particular time step consisted of sweeping through the finite difference equations once, in the following order:

eq. (11)	$\rightarrow$	eqs. (1)	$\rightarrow$	eq. (4)
$i = 1$		$i = 2,\ldots,N$		$i = N+1$

The index m keeps track of the number of sweeps of this kind. The "current elapsed time" that corresponds to a particular time step m is

$$\tau = m \cdot \Delta\tau \tag{12}$$

The attached table shows the results obtained for the midplane and surface temperatures, as a function of time (τ) and grid fineness (a,b,c):

Grid	N	$\Delta\tau$
a	5	0.01
b	10	0.0025
c	15	0.002

The numerical values show that the grid (b) is fine enough, because the results obtained with it are nearly the same as those obtained with the finer grid (c).

Time τ	Midplane ($\xi = 0$) θ_1			Surface ($\xi = 1$) θ_{N+1}		
	a	b	c	a	b	c
0.1	0.997	0.994	0.994	0.733	0.726	0.725
0.2	0.957	0.952	0.952	0.648	0.645	0.644
0.3	0.899	0.894	0.893	0.592	0.590	0.590
0.4	0.837	0.833	0.832	0.547	0.545	0.545

Project 4.3. The following information is known:

ash properties

$$\rho = 1900 \frac{kg}{m^3}$$

$$c = 1 \frac{kJ}{kg \cdot K}$$

$$k = 1 \frac{W}{m \cdot K}$$

$$\alpha = \frac{k}{\rho c} = 5.26 \times 10^{-7} \frac{m^2}{s}$$

ash hopper features

$$\dot{m} = 500 \frac{kg}{h}$$

$$A = 6 \ m^2$$

$$H = 1 m$$

$$T_i = 900°C$$

$$T_\infty = 30°C$$

$$h = 30 \frac{W}{m^2 K}$$

The time needed to accumulate the ash for filling the hopper once is

$$t_1 = \frac{\rho AH}{\dot{m}} = 1900 \frac{kg}{m^3} \frac{6 \ m^2 \ 1m}{500 \ kg/h}$$

$$= 22.8h \cong 82080 \ s$$

a) After the hopper is filled by one large discharge (at t = 0), the 1m-deep ash layer sits in the hopper for 22.8h while its upper surface is cooled by convection.

During this time interval, conduction penetrates beneath the top surface to a depth of order

$$\delta \sim (\alpha\, t_1)^{1/2} = \left(5.26 \times 10^{-7}\ \frac{m^2}{s}\ 82080\ s\right)^{1/2}$$

$$\sim 0.21\ m$$

This penetration length is much smaller than the depth of the ash pile (1m). In conclusion, at the end of the residence time of 22.8h, when the ash hopper must be emptied, the temperature at the bottom of the ash pile is the same as the initial temperature

$$T = T_i = 900°C$$

b) In the opposite extreme, when the filling of the hopper is based on a large number of discharges of infinitesimal size, we can model the filling process as "steady" with the mass flowrate 500 kg/h. For the flat control volume drawn around the rising top surface of the ash pile, the first law of thermodynamics requires (see the figure)

$$\dot{m}\, i_{in} - \dot{m}\, i_{out} = hA\,(T - T_\infty)$$

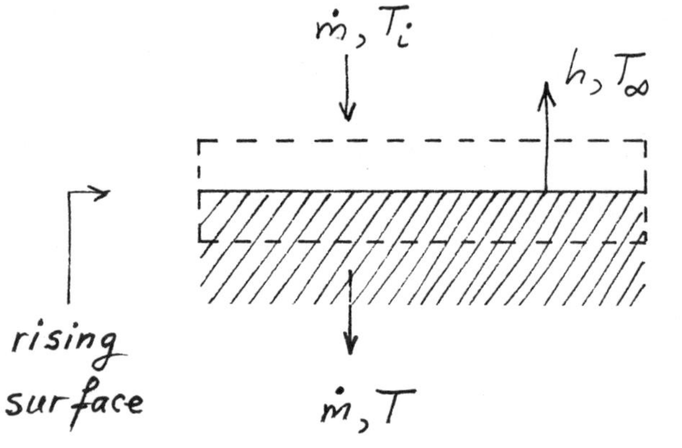

For ash as an incompressible substance at constant pressure, the specific enthalpy drop ($i_{in} - i_{out}$) is the same as the drop in its specific internal energy,

$$i_{in} - i_{out} = c\,(T_i - T)$$

therefore the first law becomes

$$\dot{m}c\,(T_i - T) = hA\,(T - T_\infty)$$

or

$$\frac{T - T_\infty}{T_i - T} = \frac{\dot{m}c}{hA} = \ldots. = 0.772$$

and, finally

$$T = 409°C$$

c) The safe temperature for the bottom of the ash pile (550°C) falls between the temperatures estimated in part (a) and part (b). This means that a bottom temperature of 550°C (or lower) can be achieved while filling the hopper intermittently, using only a few equal discharges that are equally spaced in time.

Let n be the number of discharges. Each discharge adds a fresh ash layer of thickness

$$L = \frac{H}{n}$$

to the pile. The residence time of the fresh layer on top of the pile is clearly

$$t_n = \frac{t_1}{n}$$

A simple way of estimating the temperature at the bottom of the hopper consists of viewing the bottom layer (the first discharge) as half of a slab of thickness 2L with convection on both sides. The "bottom temperature" registered after the time interval t_n is the same as the temperature in the midplane of the slab, and can be calculated (approximately of course) with the Heisler chart of Fig. 4.7.

For example, if n = 10 the calculation involves the following steps:

$$L = 0.1 \text{m}$$

$$t_{10} = 8208 \text{ s}$$

$$Fo = \frac{\alpha\, t_{10}}{L^2} = \ldots = 0.43$$

$$\frac{k}{hL} = 1 \frac{W}{m \cdot K} \frac{m^2 K}{30\, W} \frac{1}{0.1\, m} = \frac{1}{3}$$

$$\frac{T - T_\infty}{T_i - T_\infty} \cong 0.66$$

$$T \cong 30°C + 0.66\,(900°C - 30°C) = 604°C$$

The bottom temperature of 604°C occurs at $t = t_{10}$ into the filling process. The second layer will fall on top of the first, and will experience a similar degree of convection cooling as the first. At the end of this second time interval, $t = 2t_{10}$, the first layer will be "insulated" from above by a layer of similar temperature. In this way, discharge after discharge the hopper fills with ash whose temperature resembles the 604°C calculated above. This estimate is in fact conservative (i.e. the average ash temperature will be somewhat lower than 604°C), because at the end of its t_{10}-life on top of the pile, the average temperature of the top layer is lower than its bottom temperature (lower because its top surface is being cooled by convection).

A bottom temperature lower than 604°C can be achieved by using more than 10 discharges. Here are two additional calculations

$n = 14$	$n = 20$
$L = 0.071$ m	$L = 0.05$ m
$t_{14} = 5863$ s	$t_{20} = 4104$ s
$Fo = 0.612$	$Fo = 0.864$
$\frac{k}{hL} = 0.47$	$\frac{k}{hL} = 0.67$
$\frac{T - T_\infty}{T_i - T_\infty} \cong 0.57$	$\frac{T - T_\infty}{T_i - T_\infty} \cong 0.5$
$T \cong 526°C$	$T \cong 465°C$

The last line shows that the bottom temperature of the ash pile can be maintained below 550°C by using 14 or more equidistant discharges.

d) The following numerical results were obtained for 15 discharges (n = 15), using the implicit method and a grid consisting of:

- 50 equal time steps for the life of each new layer at the top of the pile, Δt = 82080s/15/50 = 109.44s

- 60 spatial steps per layer, i.e. grid spaces in each new layer, Δx = H/n/60 = 1.11 mm

The figure shows the history of the temperature at the very bottom of the pile. Each circle indicates the addition of a fresh 900°C-layer at the top of the pile. It is now clear that the bottom temperature approaches its final temperature within no more than 5 discharges. Almost all the cooling that is experienced by the bottom layer occurs while that layer is alone in the hopper.

Here are the two tests that demonstrate that the (Δx,Δt) grid is fine enough. First, by fixing Δt = 109.44s (i.e. 50 time steps per layer), we determined the effect of Δx on the final bottom temperature:

number of steps Δx per layer	5	10	20	30	40	50	60	70
bottom temperature at t = 82080s, T(°C)	448.7	472.0	483.4	487.2	489.0	490.1	490.9	491.4

Next by fixing Δx = 1.11 mm (i.e. 60 steps Δx per layer), we found the effect of the time step on the final bottom temperature:

number of steps Δx per layer	5	10	20	30	40	50
bottom temperature at t = 82080s, T(°C)	499.4	494.7	492.3	491.5	491.1	490.9

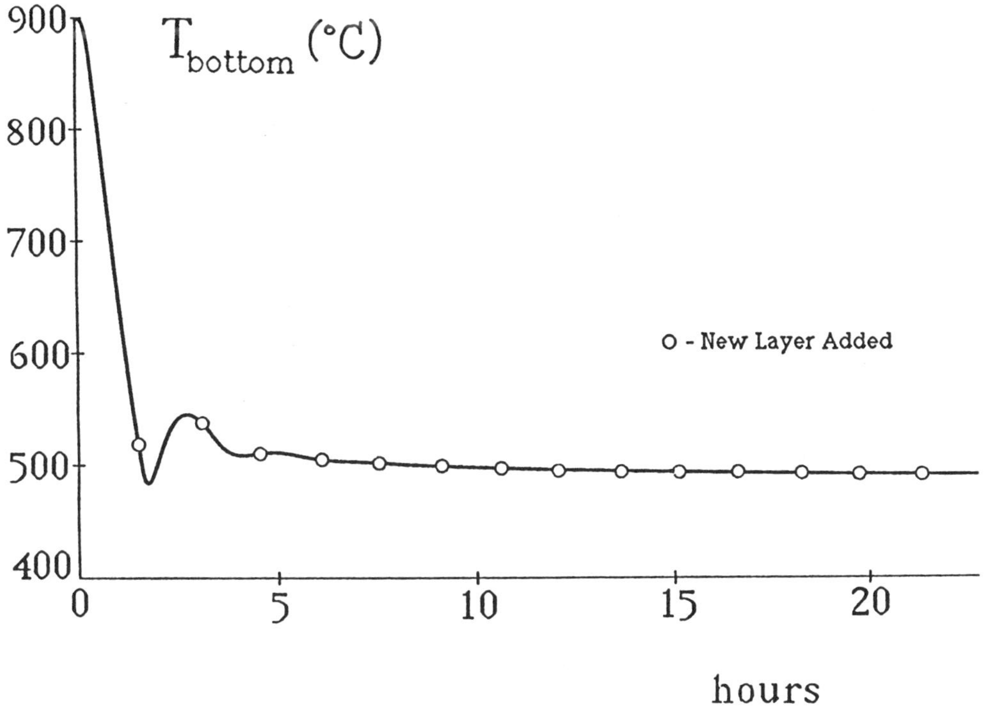

Project 4.4. Symmetry permits us to work only with the domain $x > 0$, in which the problem is stated as follows:

Equation:
$$\frac{1}{\alpha}\frac{\partial T}{\partial t} = \frac{\partial^2 T}{\partial x^2} + \frac{\rho_e J^2}{k} \tag{1}$$

Initial condition:
$$t = 0, \qquad T = T_i \quad \text{for} \quad 0 < x \le L$$
$$T = T_\infty \quad \text{for} \quad x > L \tag{2}$$

Boundary conditions:
$$x = 0, \qquad \frac{\partial T}{\partial x} = 0 \tag{3}$$

$$x \to \infty, \qquad T \to T_\infty \tag{4}$$

The source term $\rho_e J^2/k$ is present in eq. (1) only if the local temperature is greater than the critical temperature T_c.

We adopt the dimensionless notation

$$\theta = \frac{T - T_\infty}{T_c - T_\infty} \tag{5}$$

$$\xi = \frac{x}{L} \tag{6}$$

$$\tau = \frac{\alpha t}{L^2} \tag{7}$$

and the problem statement (1)-(4) becomes

$$\frac{\partial \theta}{\partial \tau} = \frac{\partial^2 \theta}{\partial \xi^2} + Q \tag{8}$$

$$\tau = 0, \qquad \theta = B > 1 \quad \text{for} \quad 1 < \xi \le 1$$
$$\theta = 0 \quad \text{for} \quad \xi > 1 \tag{9}$$

$$\xi = 0, \qquad \frac{\partial \theta}{\partial \xi} = 0 \tag{10}$$

$$\xi \to \infty, \qquad \theta \to 0 \tag{11}$$

where B and Q are defined by

$$B = \frac{T_i - T_\infty}{T_c - T_\infty} \tag{12}$$

$$Q = \frac{\rho_e J^2 L^2}{k\left(T_c - T_\infty\right)} \tag{13}$$

Again, note that Q = 0 if $\theta < 1$, i.e. if the local temperature is below critical.

We divide the conductor into control volumes (slices) of equal thickness, $\Delta x = L/N$. The first control volume (node i = 1) has the x = 0 plane as plane of symmetry. The $(N + 1)^{th}$ control volume is centered about the x = L plane. The dimensionless time step

$$\Delta\tau = \frac{\alpha}{L^2}\Delta t \tag{14}$$

and the grid Fourier number (4.125) are related in this way

$$Fo = \frac{\Delta\tau}{(\Delta\xi)^2} = N^2\,\Delta\tau \tag{15}$$

We use Table 4.4 as a guide for writing the finite-difference equations for the implicit method. It is important to keep in mind that (unlike in Table 4.4) this time the heat generation term may be present. The equation for the first node (adiabatic "boundary") is

$$\theta_1^{m+1} = \frac{\theta_1^m}{1 + 2\,Fo} + \frac{2\,Fo}{1 + 2\,Fo}\,\theta_2^{m+1} + \frac{Q\,\Delta\tau}{1 + 2\,Fo} \tag{16}$$

while the equations for the subsequent nodes (internal planes) are

$$\theta_i^{m+1} = \frac{1}{1 + 2\,Fo}\,\theta_i^m + \frac{Fo}{1 + 2\,Fo}\left(\theta_{i-1}^{m+1} + \theta_{i+1}^{m+1}\right) + \frac{Q\,\Delta\tau}{1 + 2\,Fo} \tag{17}$$

At time $\tau = 0$ (or m = 0), we initialize all the node temperatures in accordance with eqs. (9),

$$\theta_i^0 = \begin{cases} B, & 1 \le i \le N \\ \dfrac{B}{2}, & i = N + 1 \\ 0, & N + 1 < i \le M \end{cases} \tag{18}$$

where M >> N. We take M = 3N. Next, we increase the time by one step $\Delta\tau$, and sweep the conductor from left to right by using eq. (16) followed by eqs. (17). The latter does not have to be used all the way to i = M because the boundary condition (11) is met at smaller i's, i.e. closer to the normal (heated) zone. That location (i = K) is determined by the criterion

$$\theta_K^m < \varepsilon \tag{19}$$

where $\varepsilon \ll 1$. The following numerical example is based on $\varepsilon = 10^{-3}$. We increase the time again by $\Delta\tau$, and the sweep is repeated to a new (larger) K. The current time is

$$\tau = m\,\Delta\tau \tag{20}$$

The accuracy test was presented in the solution to Problem 4.31. The test showed that the choice of $\Delta\xi$ and $\Delta\tau$ loses its influence on the numerical results if $\Delta\tau = 0.001$ or smaller.

To investigate the growth of the normal zone, we monitor the time evolution of the temperature θ_1 in the centerplane (the hottest plane). This depends on time, the heat generation rate in the expanding or collapsing normal zone, and the severity of the initial temperature disturbance, i.e.

$$\theta_1 = \text{function}\,(\tau, Q, B) \tag{21}$$

If the disturbance "height" B is fixed, the θ_1 versus τ curve has the features shown in the sketch. The behavior can be of two types: a) θ_1 drops below 1, meaning that the

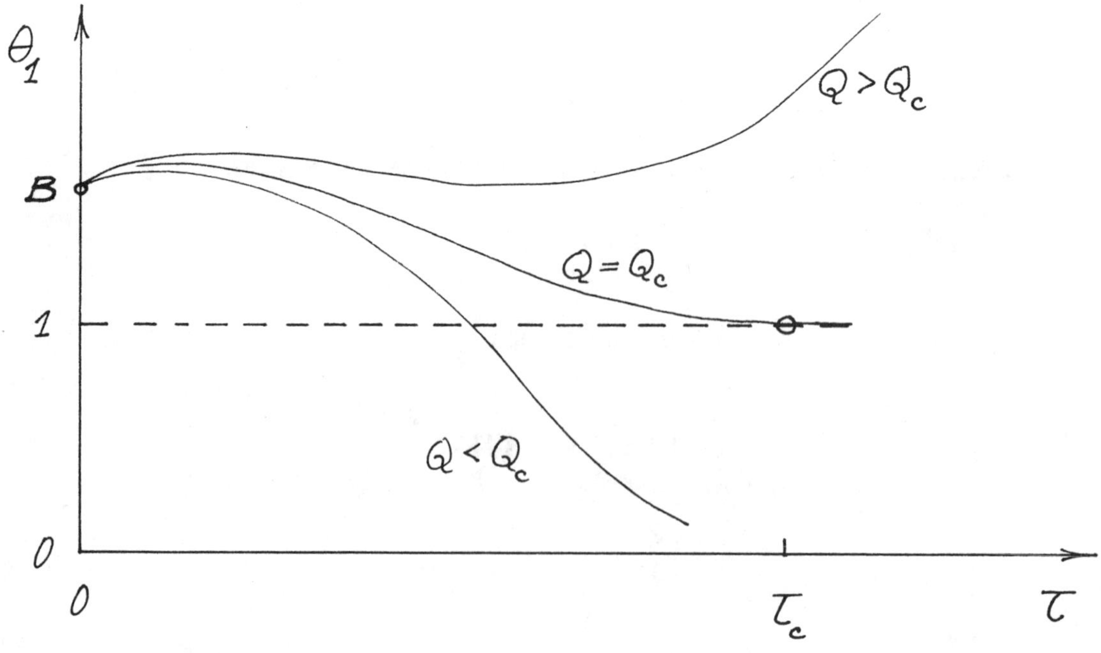

normal zone shrinks to zero, and the superconductor is stable, and b) θ_1 does not drop below 1, indicating that the disturbance (B,Q) is severe enough to cause instability. For any B, there exists a critical parameter $Q_c(B)$, below which the superconductor is stable,

$$\begin{aligned} &Q < Q_c && \text{stable} \\ &Q > Q_c && \text{unstable} \end{aligned} \tag{22}$$

There is also a critical time τ_c that marks the "last chance" that the conductor has to return to the superconducting state. The "critical" curve is tangent to $\theta_1 = 1$ at $\tau = \tau_c$, and can be found by varying Q.

The critical values in the attached table were obtained based on N = 5. They show that $\Delta\tau$ has a small effect if it is greater than 0.001. The instructor may ask the students to repeat this work for other values of B, so that the class may generate the important stability curve $Q_c(B)$. For this curve I expect the trend shown in the sketch. The stability curve separates the B-Q field (the field of operating conditions) into a region of stable disturbances, which is of interest to the designer, and a region of unstable disturbances, which must be avoided.

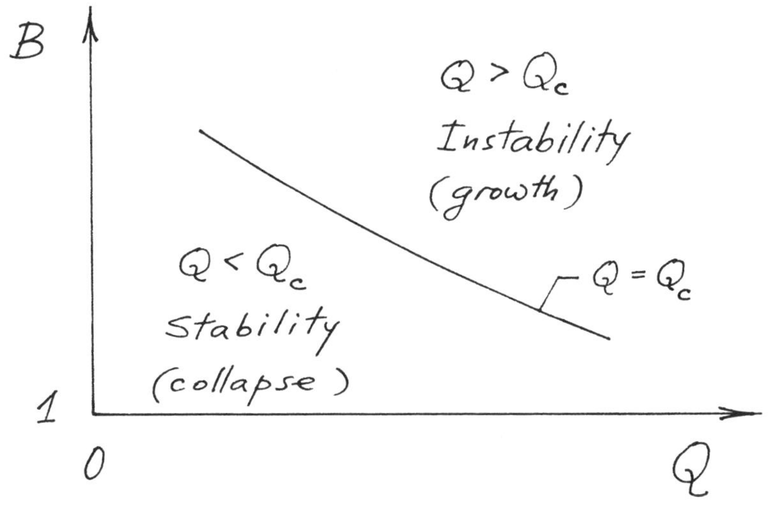

B	$\Delta\tau$	Q_c	τ_c
1.1	0.01	1.163	1.27
	0.001	1.155	1.00
	0.0001	1.154	0.96
1.2	......	......	......
⋮			

Chapter 5

EXTERNAL FORCED CONVECTION

Problem 5.1. After substituting the σ_{xx} expression (5.9) and the τ_{xy} expression (5.10) into eq. (5.8), the right side ("RS") of eq. (5.8) becomes [note the treatment of μ as a constant]:

$$RS = -\frac{\partial \sigma_{xx}}{\partial x} + \frac{\partial \tau_{xy}}{\partial y} + X =$$

$$= -\frac{\partial P}{\partial x} + 2\mu \frac{\partial^2 u}{\partial x^2} - \frac{2}{3}\mu \frac{\partial}{\partial x}\left(\frac{\partial u}{\partial x} + \frac{\partial v}{\partial y}\right) + \mu \frac{\partial}{\partial y}\left(\frac{\partial u}{\partial y} + \frac{\partial v}{\partial x}\right) + X \qquad \text{(a)}$$

The third term on the right side is zero, in accordance with the constant-ρ model, eq. (5.6); therefore RS reduces to

$$RS = -\frac{\partial P}{\partial x} + 2\mu \frac{\partial^2 u}{\partial x^2} + \mu \frac{\partial^2 u}{\partial y^2} + \mu \frac{\partial^2 v}{\partial y \partial x} + X \qquad \text{(b)}$$

The fourth term of this expression can be transformed using again the constant-ρ model, eq. (5.6):

$$\frac{\partial^2 v}{\partial y \partial x} = \frac{\partial}{\partial x}\left(\frac{\partial v}{\partial y}\right) = \frac{\partial}{\partial x}\left(-\frac{\partial u}{\partial x}\right) = -\frac{\partial^2 u}{\partial x^2} \qquad \text{(c)}$$

Substituting this new form into eq. (b) leads to eq. (5.11), the right side of which is

$$RS = -\frac{\partial P}{\partial x} + \mu\left(\frac{\partial^2 u}{\partial x^2} + \frac{\partial^2 u}{\partial y^2}\right) + X \qquad \text{(d)}$$

Problem 5.2. Let "LS" represent the left side of eq. (5.14). Eliminating e using eq. (5.16), and treating ρ as a constant, we obtain

$$LS = \rho\left(\frac{\partial i}{\partial t} + u\frac{\partial i}{\partial x} + v\frac{\partial i}{\partial y}\right) - \left(\frac{\partial P}{\partial t} + u\frac{\partial P}{\partial x} + v\frac{\partial P}{\partial y}\right) \qquad \text{(a)}$$

Each of the enthalpy derivatives can be written in terms of temperature and pressure derivatives using eq. (5.17):

$$\frac{\partial i}{\partial t} = c_P \frac{\partial T}{\partial t} + \frac{1 - \beta T}{\rho} \frac{\partial P}{\partial t} \qquad \text{(b)}$$

$$\frac{\partial i}{\partial x} = c_P \frac{\partial T}{\partial x} + \frac{1 - \beta T}{\rho} \frac{\partial P}{\partial x} \qquad \text{(c)}$$

$$\frac{\partial i}{\partial y} = c_P \frac{\partial T}{\partial y} + \frac{1 - \beta T}{\rho} \frac{\partial P}{\partial y} \qquad \text{(d)}$$

Substituting eqs. (b)-(d) into eq. (a) leads to

$$LS = \rho c_P \left(\frac{\partial T}{\partial t} + u \frac{\partial T}{\partial x} + v \frac{\partial T}{\partial y} \right) +$$

$$+ \left(\rho \frac{1 - \beta T}{\rho} - 1 \right) \left(\frac{\partial P}{\partial t} + u \frac{\partial T}{\partial x} + v \frac{\partial T}{\partial y} \right) \qquad \text{(e)}$$

which, after one more simplification, becomes the same as the expression listed in eq. (5.19).

Problem 5.3. By using the properties of water at 25°C and the data listed in the caption of Fig. 5.8, we calculate

$$Re_x = \frac{U_\infty x}{\nu} = \frac{0.6 \frac{cm}{s} 20 \ cm}{0.00894 \frac{cm^2}{s}} = 1340$$

$$Nu_x = 0.332 \ Pr^{1/3} \ Re_x^{1/2} \qquad \text{(5.79b)}$$

$$= 0.332 \ (6.21)^{1/3} \ (1340)^{1/2} = 22.34$$

$$Nu_x = 22.34 = \frac{q''_{w,x}}{\Delta T} \frac{x}{k}$$

$$q''_{w,x} = 22.34 \times 1K \ 0.6 \frac{W}{m \cdot K} \frac{1}{0.2m}$$

$$= 67 \frac{W}{m^2}$$

In order to calculate the x-averaged heat flux, it is useful to compare eq. (5.82b) with eq. (5.79b) and note that

$$\overline{Nu_x} = 2 \ Nu_x = 44.68$$

hence

$$\overline{q}''_{w,x} = 2\, q''_{w,x} = 134 \frac{W}{m^2}$$

Problem 5.4. a) We evaluate the water properties at the film temperature (20°C + 30°C)/2 = 25°C, and calculate in order

$$Re_x = \frac{U_\infty x}{\nu} = 0.6 \frac{cm}{s}\, 10\ cm \frac{s}{0.00894\ cm^2} = 671$$

$$\overline{Nu}_x = 0.664\, Pr^{1/3}\, Re_x^{1/2}$$

$$= 0.664\, (6.21)^{1/3}\, (671)^{1/2} = 31.62 = \frac{\overline{h}_x\, x}{k}$$

$$\overline{h}_x = 31.62 \frac{k}{x} = 31.62 \frac{0.6 \frac{W}{m \cdot K}}{0.1 m} = 189.7 \frac{W}{m^2\, K}$$

$$\overline{q}''_{w,x} = \overline{h}_x \left(T_\infty - T_w\right) = 189.7 \frac{W}{m^2\, K} (30\text{-}20)\ K$$

$$= 1897 \frac{W}{m^2}$$

$$q' = 2x\, \overline{q}''_{w,x} = 2 \times 0.1m\ 1897 \frac{W}{m^2} \cong 380 \frac{W}{m}$$

b) The calculation of the total drag force (F') experienced by the two-sided plate begins with eq. (5.55):

$$\overline{C}_{f,x} = 1.328\, Re_x^{-1/2} = 1.328\, (671)^{-1/2} = 0.0513$$

$$\overline{C}_{f,x} = \frac{\overline{\tau}_{w,x}}{\frac{1}{2} \rho\, U_\infty^2}$$

$$\overline{\tau}_{w,x} = \frac{1}{2} \rho\, U_\infty^2\, \overline{C}_{f,x}$$

$$= \frac{1}{2}\, 1 \frac{g}{cm^3} \left(0.6 \frac{cm}{s}\right)^2 0.0513 =$$

$$= 0.0092 \frac{g}{cm \cdot s^2} = 9.2 \times 10^{-4} \frac{N}{m^2}$$

$$F' = 2x \bar{\tau}_{w,x} = 2 \times 0.1m \; 9.2 \times 10^{-4} \frac{N}{m^2}$$

$$= 1.8 \times 10^{-4} \frac{N}{m}$$

Problem 5.5. a) The unknown temperature of the plate conductor is higher than the water temperature (20°C), therefore we begin with the assumption that the film temperature is

$$T_{film} = 25°C$$

By evaluating the water properties at 25°C, we calculate in order

$$Re_x = \frac{U_\infty x}{\nu} = 0.6 \frac{cm}{s} \; 10 \; cm \frac{s}{0.00894 \; cm^2} = 671$$

$$T_w(x) - T_\infty = \frac{q''_w x}{0.453 \; k \; Pr^{1/3} \; Re_x^{1/2}} \qquad \text{(see Table 5.4)}$$

$$= \frac{1000 \frac{W}{m^2} 0.1m}{0.453 \times 0.6 \frac{W}{m \cdot K} (6.21)^{1/3} (671)^{1/2}} = 7.73°C$$

$$T_w(x) = T_\infty + 7.73°C = 27.73°C$$

b) In order to calculate the x-averaged temperature of the plate, it is useful to note that the local temperature difference is proportional to $x^{1/2}$,

$$T_w(x) - T_\infty = C \, x^{1/2} \qquad (1)$$

We average this quantity over the swept length x,

$$\overline{T_w(x) - T_\infty} = \frac{1}{x} \int_0^x \left[T_w(\xi) - T_\infty\right] d\xi$$

$$= \frac{1}{x} \int_0^x C \, \xi^{1/2} \, d\xi = \frac{2}{3} C \, x^{1/2}$$

and, after looking again at eq. (1), note that

$$\overline{T_w(x) - T_\infty} = \frac{2}{3}\left[T_w(x) - T_\infty\right] = \frac{2}{3}\, 7.73°C = 5.2°C$$

Said another way, the x-averaged temperature difference is only 2/3 of the local temperature difference that occurs at x. The average plate temperature is therefore

$$\overline{T_w}(x) = T_\infty + 5.2°C = 25.2°C$$

The proper film temperature, (25.2°C + 20°C)/2 = 22.6°C is fairly close to the value assumed at the start, therefore there is no need to repeat the calculation by using T_{film} = 22.6°C.

Problem 5.6. a) We recognize, in order, the L-averaged shear stress, the total tangential force experienced by the wall, and the mechanical power spent on dragging the wall through the fluid:

$$\bar{\tau} = 0.664\, \rho\, U^2 \left(\frac{UL}{\nu}\right)^{-1/2}$$

$$F' = \bar{\tau} L$$

$$\mathbf{P} = F'U = 0.664\, \rho\, U^3 L \left(\frac{\nu}{UL}\right)^{1/2}$$

$$= 0.664\, \rho\, \nu^{1/2}\, U^{5/2}\, L^{1/2}$$

If $(\ )_c$ and $(\ )_h$ represent the "cold" and "hot" flow conditions, the dissipated power changes according to the ratio:

$$\frac{\mathbf{P}_h}{\mathbf{P}_c} = \frac{\rho_h}{\rho_c}\left(\frac{\nu_h}{\nu_c}\right)^{1/2} = \left(\frac{\rho_h\, \mu_h}{\rho_c\, \mu_c}\right)^{1/2}$$

In the cold case (no heating), T_c = 10°C and

$$\rho_c \cong 1\, \frac{g}{cm^3} \qquad \mu_c = 0.013\, \frac{g}{cm \cdot s}$$

In the hot case, the heat transfer from the 90°C wall to the 10°C water takes place across a boundary layer with the film temperature T_h = 50°C. The properties corresponding to T_h are

$$\rho_h \cong 1\, \frac{g}{cm^3} \qquad \mu_h = 0.00548\, \frac{g}{cm \cdot s}$$

The power ratio

$$\frac{P_h}{P_c} = \left(\frac{1}{1} \cdot \frac{0.00548}{0.013}\right)^{1/2} = 0.65$$

shows that the heating of the boundary layer decreases the dissipated power (as well as F' and $\bar{\tau}$) by 35 percent.

b) By retaining the observation that the water density is practically constant, we find that the power that has been saved by heating the boundary layer is

$$P_c - P_h = 0.664\, \rho\, U^{5/2}\, L^{1/2}\, \nu_c \left[1 - \left(\frac{\nu_h}{\nu_c}\right)^{1/2}\right] \tag{1}$$

The electric power spent on heating the wall is given sequentially by

$$\frac{\bar{h}L}{k} = \frac{\overline{q''}}{\Delta T}\frac{L}{k} = 0.664\, Pr^{1/3}\, Re_L^{1/2}$$

$$\overline{q''}L = 0.664\, k_h\, \Delta T\, Pr_h^{1/3} \left(\frac{UL}{\nu_h}\right)^{1/2} \tag{2}$$

in which the properties (k, Pr, ν) are evaluated at the film temperature T_h = 50°C,

$$k_h = 0.64 \frac{W}{m \cdot K}, \qquad Pr_h = 3.57$$

By dividing equations (1) and (2), we obtain (after some manipulation) a dimensionless measure of how effectively the heating of the wall has been converted into saved mechanical power:

$$\frac{P_c - P_h}{\overline{q''}L} = \frac{U^2}{c_h\, \Delta T} Pr_h^{2/3} \left(\frac{\nu_c}{\nu_h}\right)^{1/2} \left[1 - \left(\frac{\nu_h}{\nu_c}\right)^{1/2}\right]$$

In this expression we substitute c_h = 4.18 kJ/kg·K, ΔT = 90°C - 10°C = 80°C and $(\nu_h/\nu_c)^{1/2}$ = 0.65, and obtain

$$\frac{P_c - P_h}{\overline{q''}L} = \left(\frac{U}{516 \text{ m/s}}\right)^2$$

This ratio is appealing (greater than 1) only when U > 516 m/s. The length L, however, must be sufficiently small if the boundary layer is to remain laminar

$$\frac{UL}{\nu_h} < 5 \times 10^5$$

$$L < 5 \times 10^5 \frac{\nu_h}{U} < 5 \times 10^5 \frac{\nu_h}{516 \text{ m/s}}$$

$$< 5 \times 10^5 \ \frac{0.00554 \text{ cm}^2/\text{s}}{51600 \text{ cm/s}}$$

$$< 0.54 \text{ mm}$$

In conclusion, when L > 0.54 mm and the flow is laminar the electric power used to heat the wall is greater than the fluid-drag power saved.

Problem 5.7. The film temperature in this configuration is (40°C + 20°C) = 30°C. The calculation of the total force F proceeds in this order:

$$Re_x = \frac{U_\infty x}{\nu} \ 0.5 \frac{\text{m}}{\text{s}} \ 10\text{m} \frac{\text{s}}{0.16 \text{ cm}^2} = 3.1 \times 10^5$$

(laminar, just barely)

$$\overline{C}_{f,x} = 1.328 \ Re_x^{-1/2} = 0.00238$$

$$A = 10\text{m} \ 20\text{m} = 200 \text{ m}^2 \qquad \text{(roof area)}$$

$$F = A \ \overline{\tau}_{w,x} = A \ \overline{C}_{f,x} \frac{1}{2} \rho \ U_\infty^2$$

$$= 200 \text{ m}^2 \ 0.00238 \frac{1}{2} \ 1.165 \frac{\text{kg}}{\text{m}^3} \left(0.5 \frac{\text{m}}{\text{s}}\right)^2$$

$$= 0.069 \text{ N} \cong 0.016 \text{ lbf}$$

In order to calculate the total heat transfer rate q, we must first evaluate the average heat flux:

$$\overline{Nu}_x = 0.664 \ Pr^{1/3} \ Re_x^{1/2}$$

$$= 0.664 \ (0.72)^{1/3} \left(3.1 \times 10^5\right)^{1/2} = 331.4$$

$$\overline{h}_x = \overline{Nu}_x \frac{k}{x} = 331.4 \times 0.026 \frac{\text{W}}{\text{m} \cdot \text{K}} \ \frac{1}{10\text{m}}$$

$$= 0.862 \frac{\text{W}}{\text{m}^2 \text{ K}}$$

$$\overline{q}''_x = \overline{h}_x \left(T_w - T_\infty\right) = 0.862 \frac{W}{m^2\ K} (40 - 20)\ K$$

$$= 17.2 \frac{W}{m^2}$$

$$q = \overline{q}''_x\ A = 17.2 \frac{W}{m^2}\ 200\ m^2 = 3.4\ kW$$

Problem 5.8. a) The local heat flux varies as

$$q''_{w,x} = C\ x^n$$

in which n = - 1/2, and C is a constant. The L-averaged heat flux is therefore

$$\overline{q''_{w,L}} = \frac{1}{1+n} q''_{w,L} = 2q''_{w,L} = 2\ CL^{-1/2}$$

The position x where the local flux matches the L-averaged value is obtained by writing

$$Cx^{-1/2} = 2\ CL^{-1/2}$$

and this yields x = L/4. The attached figure confirms this finding, because it is drawn to scale.

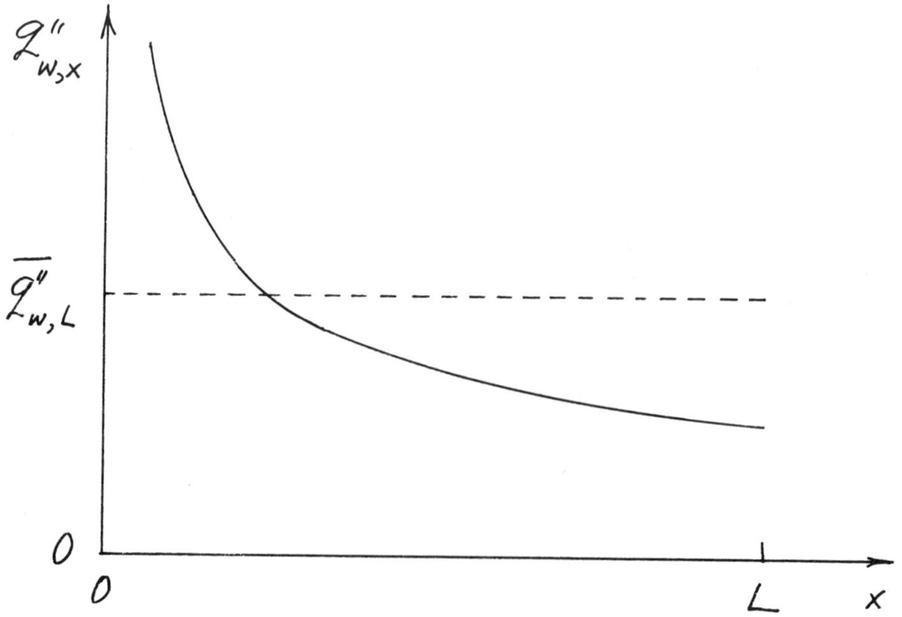

b) The relationship between the local heat flux in the middle and the L-averaged flux is obtained by first writing

$$q''_{w,L/2} = C\left(\frac{L}{2}\right)^{-1/2}$$

$$\overline{q}''_{w,L} = 2\ CL^{-1/2}$$

Dividing side by side, we conclude that the mid-point heat flux is about 30 percent smaller than the average flux:

$$\frac{q''_{w,L/2}}{\overline{q}''_{w,L}} = \frac{C\ 2^{1/2}\ L^{-1/2}}{2\ CL^{-1/2}} = \frac{1}{2^{1/2}} = 0.707$$

<u>Problem 5.9</u>. When the boundary layer is much thinner than the pipe diameter D, the wall friction force F and the total heat transfer rate q can be estimated based on the formulas for the flat wall:

$$\begin{aligned} F &= \pi\ DL\ \overline{\tau}_{w,L} = \pi\ DL\ \overline{C}_{f,L}\ \frac{1}{2}\ \rho\ U_\infty^2 \\ &= \pi\ DL\ 1.328\ Re_L^{-1/2}\ \frac{1}{2}\ \rho\ U_\infty^2 \\ &= 2.086\ \rho\ U_\infty^2\ DL\ Re_L^{-1/2} \end{aligned} \qquad (1)$$

$$\begin{aligned} q &= \pi\ DL\ \overline{q}''_{w,L} = \pi\ DL\ \Delta T\ \overline{h}_L \\ &= \pi\ DL\ \Delta T\ \frac{k}{L}\ 0.664\ Pr^{1/3}\ Re_L^{1/2} \\ &= 2.086\ k\ \Delta T\ D\ Pr^{1/3}\ Re_L^{1/2} \end{aligned} \qquad (2)$$

Dividing (1) and (2) side by side we obtain

$$\frac{q}{F} = Pr^{-2/3}\ \frac{c_P\ \Delta T}{U_\infty}$$

Problem 5.10. We evaluate the properties of water at the film temperature (20°C + 50°C)/2 = 35°C, and calculate in order:

$$Re_x = \frac{U_\infty x}{\nu} = 5 \frac{cm}{s} 100\ cm \frac{s}{0.00725\ cm^2} = 6.9 \times 10^4 \quad \text{(laminar)}$$

$$Nu_x = 0.332\ Pr^{1/3}\ Re_x^{1/2}$$

$$= 0.332\ (4.87)^{1/3} \left(6.9 \times 10^4\right)^{1/2} = 148$$

$$Nu_x = \frac{h_x\, x}{k}$$

$$h_x = 148 \frac{k}{x} = 148 \times 0.62 \frac{W}{m \cdot K} \frac{1}{1m} = 91.6 \frac{W}{m^2 K}$$

$$\overline{h}_x = 2\ h_x = 183.3 \frac{W}{m^2\ K}$$

$$\overline{q}''_{w,x} = \overline{h}_x \Delta T = 183.3 \frac{W}{m^2 K}\ 30\ K = 5500 \frac{W}{m^2}$$

Since the total area of the wall of length x = 1m is

$$A = 4 \times 20\ cm \times 1m = 0.8\ m^2$$

the total heat transfer rate is

$$q = \overline{q}''_{w,x}\ A = 4400\ W$$

The velocity boundary layer thickness at the same location (x)

$$\delta_{99} = 4.92 \times Re_x^{-1/2} =$$

$$= 4.92 \times 1m \left(6.9 \times 10^4\right)^{-1/2} = 1.9\ cm$$

is much smaller than the 20 cm-side of the duct cross-section. In conclusion, the use of the flat wall boundary layer results is justified.

Problem 5.11. i) Multiplying both terms of eq. (5.21) by u, we obtain

$$u\frac{\partial u}{\partial x} + u\frac{\partial v}{\partial y} = 0 \qquad (a)$$

This equation can be added together (side-by-side) with eq. (5.32), to obtain sequentially:

$$u\frac{\partial u}{\partial x} + v\frac{\partial u}{\partial y} + u\frac{\partial u}{\partial x} + u\frac{\partial v}{\partial y} = \nu\frac{\partial^2 u}{\partial y^2} \qquad (b)$$

$$\frac{\partial(u^2)}{\partial x} + \frac{\partial(uv)}{\partial y} = \nu\frac{\partial^2 u}{\partial y^2} \qquad (c)$$

ii) The analysis continues with the new momentum eq. (c). Integrating this equation from the wall (y = 0) to a distance at the outer edge of the boundary layer (y = δ), we obtain

$$\int_0^\delta \frac{\partial(u^2)}{\partial x}\,dy + (uv)_{y=\delta} - (uv)_{y=0} = \nu\left(\frac{\partial u}{\partial y}\right)_{y=\delta} - \nu\left(\frac{\partial u}{\partial y}\right)_{y=0} \qquad (d)$$

Counting from the left, the third and fourth terms are zero because of the wall conditions (no slip, impermeability) and the free-stream condition ($u = U_\infty$, constant). The second term,

$$(uv)_{y=\delta} = U_\infty (v)_{y=\delta} \qquad (e)$$

demands an estimate for the transversal velocity at the outer edge, $(v)_{y=\delta}$. This can be found by integrating eq. (5.21) across the boundary layer,

$$\int_0^\delta \frac{\partial u}{\partial x}\,dy + (v)_{y=\delta} - (v)_{y=0} = 0 \qquad (f)$$

in which $(v)_{y=0} = 0$. Combining eqs. (d) and (f) we conclude that

$$\int_0^\delta \frac{\partial}{\partial x}\left[u\,(U_\infty - u)\right]dy = \nu\left(\frac{\partial u}{\partial y}\right)_{y=0} \qquad (g)$$

The integral can be expanded using Leibniz' formula for differentiation,

$$\int_0^\delta \frac{\partial}{\partial x}[u(U_\infty - u)]\, dy = \frac{d}{dx}\int_0^\delta u(U_\infty - u)\, dy - \frac{d\delta}{dx}[u(U_\infty - u)]_{y=\delta}$$

$$= \frac{d}{dx}\int_0^\delta u(U_\infty - u)\, dy \qquad \text{(h)}$$

where it has been noted that $U_\infty - u = 0$ at $x = \delta$. Combining eqs. (g) and (h), we arrive at

$$\frac{d}{dx}\int_0^\delta u(U_\infty - u)\, dy = \nu \left(\frac{\partial u}{\partial y}\right)_{y=0} \qquad \text{(i)}$$

iii) Let the function $m(y/\delta)$ represent the dimensionless velocity profile, in other words, let

$$\frac{u}{U_\infty} = m\left(\frac{y}{\delta}\right) \qquad \text{(j)}$$

where $\delta = \delta(x)$ is the unknown of the problem. Soon we will assume a reasonable expression (shape) for the function $m(y/\delta)$. The substitution of the assumed profile (h) into the "integral momentum equation" (i) leads to the ordinary differential equation

$$C_1\, \delta \frac{d\delta}{dx} = \frac{\nu}{U_\infty} C_2 \qquad \text{(k)}$$

where C_1 and C_2 are two numerical constants that depend on the assumed shape m:

$$C_1 = \int_0^1 m(1 - m)\, d\left(\frac{y}{\delta}\right), \qquad C_2 = \left[\frac{dm}{d(y/\delta)}\right]_{y=0} \qquad \text{(l)}$$

Integrating eq. (k) we obtain the following general solution for boundary layer thickness and local skin-friction coefficient

$$\frac{\delta}{x} = a_1\, Re_x^{-1/2}, \quad \text{with} \quad a_1 = \left(\frac{2\, C_2}{C_1}\right)^{1/2} \qquad \text{(p)}$$

$$C_{f,x} = a_2\, Re_x^{-1/2}, \quad \text{with} \quad a_2 = (2\, C_2\, C_1)^{1/2} \qquad \text{(q)}$$

Writing $n = y/\delta$ for the dimensionless transversal coordinate, we can assume a certain shape function m (the left column of the table), and calculate the corresponding numerical coefficients for the integral solution (p)-(q):

Assumed profile		a_1	a_2
linear:	$m = n$	3.46	0.577
parabolic:	$m = \frac{n}{2}(3 - n^2)$	4.64	0.646
sine arc:	$m = \sin(\pi n/2)$	4.8	0.654

Problem 5.12. i) Multiplying eq. (5.21) by T, and adding it side-by-side to the energy eq. (5.62) we obtain, in order,

$$T\frac{\partial u}{\partial x} + T\frac{\partial v}{\partial y} + u\frac{\partial T}{\partial x} + v\frac{\partial T}{\partial y} = \alpha\frac{\partial^2 T}{\partial y^2} \tag{1}$$

$$\frac{\partial(uT)}{\partial x} + \frac{\partial(vT)}{\partial y} = \alpha\frac{\partial^2 T}{\partial y^2} \tag{2}$$

ii) Next, we integrate eq. (2) across the thermal boundary layer, and obtain

$$\int_0^{\delta_T} \frac{\partial(uT)}{\partial x}\,dy + (vT)_{\delta_T} - (vT)_0 = \alpha\left(\frac{\partial T}{\partial y}\right)_{\delta_T} - \alpha\left(\frac{\partial T}{\partial y}\right)_0 \tag{3}$$

in which the second and third terms (counted from the left) are zero, and

$$(vT)_{\delta_T} = v_{\delta_T} T_\infty \tag{4}$$

The outer-edge transversal velocity v_{δ_T} can be obtained by integrating the mass conservation eq. (5.21),

$$\int_0^{\delta_T} \frac{\partial u}{\partial x}\,dy + v_{\delta_T} - v_0 = 0 \tag{5}$$

where $v_0 = 0$ (impermeable wall). Together, eqs. (3)-(5) yield:

$$\int_0^{\delta_T} \frac{\partial}{\partial x}\left[u\,(T_\infty - T)\right] dy = \alpha\left(\frac{\partial T}{\partial y}\right)_0 \tag{6}$$

and, after invoking Leibniz' formula on the left side,

$$\frac{d}{dx}\int_0^{\delta_T} u\,(T_\infty - T)\,dy - \frac{d\,\delta_T}{dx}\left[u\,(T_\infty - T)\right]_{\delta_T} = \alpha\left(\frac{\partial T}{\partial y}\right)_0 \qquad (7)$$

The second term on the left side is zero, because $T = T_\infty$ at $y = \delta_T$. In conclusion, the integral energy equation reduces to

$$\frac{d}{dx}\int_0^{\delta_T} u\,(T_\infty - T)\,dy = \alpha\left(\frac{\partial T}{\partial y}\right)_0 \qquad (8)$$

iii) Assume the parabolic temperature and velocity profiles

$$\frac{T_w - T}{T_w - T_\infty} = \frac{p}{2}\,(3 - p^2), \qquad \text{with} \qquad p = \frac{y}{\delta_T} \qquad (9)$$

$$\frac{u}{U_\infty} = \frac{n}{2}\,(3 - n^2), \qquad \text{with} \qquad n = \frac{y}{\delta} \qquad (10)$$

and note that $\delta(x)$ is known from the statement of the preceding problem:

$$\delta = 4.64\ x\ Re_x^{-1/2} \qquad (11)$$

In high-Pr fluids, δ is greater than δ_T: therefore, in the δ_T-thin layer the velocity distribution is the same as the $n << 1$ limit of eq. (10):

$$\frac{u}{U_\infty} \cong \frac{3}{2}\,n = \frac{3y}{2\delta} = 0.323\ y\left(\frac{U_\infty}{\nu x}\right)^{1/2} \qquad (12)$$

The unknown in this problem is δ_T. About it we know the x dependence: in subsection 5.3.2. we learned that δ_T increases as $x^{1/2}$ regardless of the Pr range:

$$\delta_T = C\,x^{1/2} \qquad (13)$$

The problem reduces to figuring out the coefficient C. The analysis consists of substituting the profiles (9) and (12) in the integral energy eq. (8), and using $p = y/\delta_T$ to eliminate y, and eq. (13) to eliminate δ_T. What is left in the end is an equation for C, which yields

$$C = 4.53\ Pr^{-1/3}\left(\frac{\nu x}{U_\infty}\right)^{1/2} \qquad (14)$$

or, in view of eq. (13),

$$\delta_T = 4.53\, x\, Pr^{-1/3}\, Re_x^{-1/2} \qquad (15)$$

The local heat flux that corresponds to this result can be calculated using eqs. (9) and (15):

$$q''_{w,x} = -k\left(\frac{\partial T}{\partial y}\right)_0 = k\,\Delta T\,\frac{3}{2}\,\frac{1}{\delta_T}$$

$$= 0.331\,\frac{k\,\Delta T}{x}\,Pr^{1/3}\,Re_x^{1/2} \qquad (16)$$

The local Nusselt number is almost identical to Pohlhausen's solution, eq. (5.79b):

$$Nu_x = 0.331\, Pr^{1/3}\, Re_x^{1/2} \qquad (17)$$

Problem 5.13. In the $Pr \rightarrow 0$ limit, the flow is parallel and uniform in the thermal boundary layer. Therefore, substituting

$$u = U_\infty \qquad \text{and} \qquad v = 0$$

into the energy eq. (5.62), we obtain the energy equation listed at the top of the left column of the following table. The left column lists also the appropriate boundary conditions faced by the thermal boundary layer.

$\frac{U_\infty}{\alpha}\frac{\partial T}{\partial x} = \frac{\partial^2 T}{\partial y^2}$	$\frac{1}{\alpha}\frac{\partial T}{\partial t} = \frac{\partial^2 T}{\partial x^2}$	(4.22)
$T = T_\infty$ at $x = 0$	$T = T_i$ at $t = 0$	(4.23)
$T = T_w$ at $y = 0$	$T = T_\infty$ at $x = 0$	(4.24)
$T \rightarrow T_\infty$ as $y \rightarrow \infty$	$T \rightarrow T_i$ as $x \rightarrow \infty$	(4.25)
solution:	solution:	
????	$\frac{T - T_\infty}{T_i - T_\infty} = \text{erf}\left[\frac{x}{2\,(\alpha t)^{1/2}}\right]$	(4.42)

The right column of this table reviews the complete statement and solution of the unsteady conduction problem solved in section 4.3.1. The problem statements (the two columns) are analogous, therefore the solution marked (????) can be written down by inspecting the solution given in the right column, eq. (4.42). Note that as we shift from the right to the left in this table, the notation changes according to these rules:

$$y \leftarrow x$$

$$x \leftarrow t$$

$$\frac{\alpha}{U_\infty} \leftarrow \alpha$$

$$T_w \leftarrow T_\infty$$

$$T_\infty \leftarrow T_i$$

Applying this transformation to eq. (4.42), we obtain the T(x,y) solution for the convection problem listed on the left side of the table:

$$\frac{T - T_w}{T_\infty - T_w} = \text{erf}\left[\frac{y}{2\,(\alpha x/U_\infty)^{1/2}}\right]$$

The local wall heat flux that corresponds to this solution is

$$q''_{w,x} = -k\left(\frac{\partial T}{\partial y}\right)_{y=0} = k\,\Delta T\left(\frac{U_\infty}{\pi \alpha x}\right)^{1/2}$$

where $\Delta T = T_w - T_\infty$. This $q''_{w,x}$ result can be nondimensionalized to give birth to the local Nusselt number

$$Nu_x = \frac{q''_{w,x}}{\Delta T}\frac{x}{k} = \pi^{-1/2}\, Pe_x^{1/2}$$

where, as listed in eq. (5.79a), $\pi^{-1/2} = 0.564$.

Problem 5.14. The two Nu_x formulas (5.79a,b) can be written as one

$$Nu_x = C\, Re_x^{1/2} \qquad (1)$$

in which C is a function of Pr only:

$$C = 0.564\, Pr^{1/2}, \qquad (Pr \lesssim 0.5) \qquad (2)$$

$$C = 0.332\, Pr^{1/3}, \qquad (Pr \gtrsim 0.5) \qquad (3)$$

The average heat flux definition (5.80) states that

$$\overline{q''_{w,x}} = \frac{1}{x}\int_0^x q''_{w,\xi}\, d\xi$$

$$= \frac{1}{x}\int_0^x \frac{k\,\Delta T}{\xi}\, Nu_\xi\, d\xi \qquad (4)$$

in which ξ is a dummy variable, $0 \le \xi \le x$. Using eq. (1) in the integrand of eq. (4), we obtain

$$\overline{q''_{w,x}} = \frac{k\,\Delta T}{x}\int_0^x \frac{1}{\xi}\, C\, Re_\xi^{1/2}\, d\xi$$

$$= \frac{k\,\Delta T}{x}\, C \left(\frac{U_\infty}{\nu}\right)^{1/2} \int_0^x \frac{d\xi}{\xi^{1/2}}$$

$$= \frac{k\,\Delta T}{x}\, C \left(\frac{U_\infty}{\nu}\right)^{1/2} 2x^{1/2} \qquad (5)$$

The x-averaged Nusselt number defined in eq. (5.81) is therefore

$$\overline{Nu_x} = \frac{\overline{q''_{w,x}}}{\Delta T}\,\frac{x}{k} = 2C\, Re_x^{1/2} \qquad (6)$$

or, in view of eqs. (2,3),

$$\overline{Nu_x} = 1.128\ Pe_x^{1/2}, \qquad (Pr \lesssim 0.5) \qquad (7a)$$

$$\overline{Nu_x} = 0.664\, Pr^{1/3}\ Re_x^{1/2}, \qquad (Pr \gtrsim 0.5) \qquad (7b)$$

Problem 5.15. The narrow-strip configuration is a case of wall with non-uniform heat flux, in which

$$q''_w(\xi) = \begin{cases} 0, & 0 < \xi < x_1 \\ q'_w/\Delta x, & x_1 < \xi < x_1 + \Delta x \\ 0, & \xi > x_1 + \Delta x \end{cases}$$

Substituting this expression into the integrand of the last formula of Table 5.4 we obtain, in order:

$$T_w(x) - T_\infty = \frac{0.623}{k\, Pr^{1/3}\, Re_x^{1/2}} \left(\int_0^{x_1} + \int_{x_1}^{x_1+\Delta x} + \int_{x_1+\Delta x}^{x} \right)$$

$$= \frac{0.623}{k\, Pr^{1/3}\, Re_x^{1/2}} \int_{x_1}^{x_1+\Delta x} \left[1 - \left(\frac{\xi}{x}\right)^{3/4}\right] \frac{q'_w}{\Delta x}\, d\xi$$

$$= \frac{0.623}{k\, Pr^{1/3}\, Re_x^{1/2}} \left[1 - \left(\frac{x_1}{x}\right)^{3/4}\right]^{-2/3} q'_w$$

Problem 5.16. In the last (fourth) formula of Table 5.4 we set $q''_w(\xi) = q''_w$ (constant),

$$T_w(x) - T_\infty = \frac{0.623}{k\, Pr^{1/3}\, Re_x^{1/2}} \int_{\xi=0}^{x} \left[1 - \left(\frac{\xi}{x}\right)^{3/4}\right]^{-2/3} q''_w\, d\xi$$

$$= \frac{q''_w\, x}{k\, Pr^{1/3}\, Re_x^{1/2}}\, 0.623 \underbrace{\int_0^{x} \frac{dm}{\left(1 - m^{3/4}\right)^{2/3}}}_{B = 3.54}$$

and arrive at the formula for the uniform flux case (the third in Table 5.4):

$$T_w(x) - T_\infty = \frac{q''_w\, x}{0.453\, k\, Pr^{1/3}\, Re_x^{1/2}}$$

Problem 5.17. The flux q" covers only the front half of the plate length L. We use the last formula of Table 5.4

$$T_w(L) - T_\infty = \frac{0.623}{k\,Pr^{1/3}\,Re_x^{1/2}} \int_{\xi=0}^{L/2} \left[1 - \left(\frac{\xi}{L}\right)^{3/4}\right]^{-2/3} q''\,d\xi$$

$$= \frac{q''\,L}{k\,Pr^{1/3}\,Re_L^{1/2}}\; 0.623 \underbrace{\int_0^{1/2} \left(1 - m^{3/4}\right)^{-2/3} dm}_{C = 0.682} \qquad (1)$$

The alternative calculation is based on the third formula of Table 5.4, in which we assume that the uniform flux q"/2 is spread over the entire length L:

$$\left[T_w(L) - T_\infty\right]_{approx.} = \frac{\frac{q''}{2}\,L}{0.453\; k\,Pr^{1/3}\,Re_x^{1/2}} \qquad (2)$$

The relative goodness of this approximation can be seen by dividing eq. (2) by eq. (1):

$$\frac{\left[T_w(L) - T_\infty\right]_{approx.}}{T_w(L) - T_\infty} = \frac{\frac{1}{2 \times 0.453}}{0.623 \times 0.682} = 2.60$$

In conclusion, the approximation is not very good.

Problem 5.18. The relevant properties of water at 20°C and atmospheric pressure are

$$\rho = 0.997 \frac{g}{cm^3} \qquad Pr = 7.07$$

$$\nu = 0.01 \frac{cm^2}{s} \qquad k = 0.59 \frac{W}{mK}$$

The L-averaged shear stress $\bar{\tau}_{w,L}$ can be calculated in the following order,

$$\bar{\tau}_{w,L} = \left(\frac{1}{2} \rho U_\infty^2\right) \bar{C}_{f,L}$$

$$= \left(\frac{1}{2} \rho U_\infty^2\right) 1.328 \, Re_L^{-1/2} \qquad \text{(a)}$$

where the Reynolds number has a value that falls in the laminar range:

$$Re_L = \frac{U_\infty L}{\nu} = \frac{0.5 \text{ m}}{s} \frac{1 \text{ cm}}{0.01 \text{ cm}^2/s} = 5000 \qquad \text{(b)}$$

Combining (a) and (b) we obtain

$$\bar{\tau}_{w,L} = \frac{1}{2} 0.997 \frac{g}{cm^3} (0.5)^2 \frac{m^2}{s^2} 1.328 \, (5000)^{-1/2}$$

$$= 0.00234 \frac{g \cdot m^2}{cm^3 \, s^2} = 2340 \frac{N}{m^2} \qquad \text{(c)}$$

The L-averaged heat flux can be calculated based on a similar sequence:

$$\bar{q}''_{w,L} = \frac{k \Delta T}{L} \overline{Nu}_L$$

$$= \frac{k \Delta T}{L} 0.664 \, Pr^{1/3} \, Re_L^{1/2}$$

$$= 0.59 \frac{W}{m \, K} \frac{1 \text{ K}}{1 \text{ cm}} 0.664 \, (7.07)^{1/3} (5000)^{1/2}$$

$$= 53.17 \frac{W}{m \cdot cm} = 5317 \frac{W}{m^2} \qquad \text{(d)}$$

Problem 5.19. The $C_{f,x}$ and Nu_x formulas recommended in the text are

$$C_{f,x} = 0.664\ Re_x^{-1/2} \qquad (5.53)$$

$$Nu_x = 0.332\ Pr^{1/3}\ Re_x^{1/2} \qquad (5.79b)$$

Eliminating $Re_x^{1/2}$, we obtain

$$Nu_x = 0.332\ Pr^{1/3}\ \frac{0.664}{C_{f,x}}$$

$$= 0.332\ (17.8)^{1/3}\ \frac{0.664}{0.008} = 71.95$$

Equation (5.53) should be used to verify that the Reynolds number is indeed in the laminar range:

$$Re_x = \left(\frac{0.664}{C_{f,x}}\right)^2 = \left(\frac{0.664}{0.008}\right)^2 = 6889, \qquad \text{(laminar)}$$

Problem 5.20. The Prandtl number of air at 20°C is 0.72, therefore we can use the third case of Table 5.4 as an approximate estimate of the local Nusselt number along the strip:

$$\frac{q_w''}{T_w(L) - T_\infty}\frac{L}{k} = 0.453\ Pr^{1/3}\ Re_L^{1/2} \qquad (a)$$

In this equation q_w'' is the heat flux to one side of the strip, $q_w'' = (300/2)W/m^2 = 150\ W/m^2$, and x = L is the position of the trailing edge where the temperature sensor is located, $T_w(L) = 30°C$. Noting further that $k = 2.5 \times 10^{-4} W/cm\ K$, we use eq. (a) to calculate $Re_L^{1/2}$:

$$Re_L^{1/2} = \frac{q_w''}{T_w(L) - T_\infty}\frac{L}{k}\frac{1}{0.453\ Pr^{1/3}}$$

$$= \frac{150\ W/m^2}{(30 - 20)°C}\ \frac{2\ cm}{2.5 \times 10^{-4}\ W/cm\ K}\ \frac{1}{0.453\ (0.72)^{1/3}}$$

$$= 29.56$$

From this we deduce that $Re_L = 873.5 = U_\infty L/\nu$, in which $\nu = 0.15\ cm^2/s$. The U_∞ velocity of the free stream is therefore

$$U_\infty = 873.5 \frac{0.15\ \text{cm}^2/\text{s}}{2\ \text{cm}} = 0.66 \frac{\text{m}}{\text{s}}$$

Note that $Re_L = 873.5$ means also that the flow at x = L is laminar, i.e. that the use of eq. (a) is justified.

Problem 5.21. The instantaneous form of the mass conservation equation for incompressible flow is

$$\frac{\partial u}{\partial x} + \frac{\partial v}{\partial y} + \frac{\partial w}{\partial z} = 0 \qquad (1)$$

Decomposing the velocities according to eqs. (5.88)-(5.89), we obtain six terms on the left side of eq. (1):

$$\frac{\partial \bar{u}}{\partial x} + \frac{\partial u'}{\partial x} + \frac{\partial \bar{v}}{\partial y} + \frac{\partial v'}{\partial y} + \frac{\partial \bar{w}}{\partial z} + \frac{\partial w'}{\partial z} = 0 \qquad (2)$$

Next, we subject the entire left side of eq. (2) to the time-averaging operation (5.87):

$$\overline{\frac{\partial \bar{u}}{\partial x}} + \overline{\frac{\partial u'}{\partial x}} + \overline{\frac{\partial \bar{v}}{\partial y}} + \overline{\frac{\partial v'}{\partial y}} + \overline{\frac{\partial \bar{w}}{\partial z}} + \overline{\frac{\partial w'}{\partial z}} = 0 \qquad (3)$$

$$\downarrow \qquad \downarrow$$

$$\frac{\partial \bar{u}}{\partial x} \qquad \frac{\partial \overline{u'}}{\partial x} \qquad \text{etc.} \rightarrow$$

$$\downarrow$$

zero

The three terms of type $\overline{\partial u'/\partial x}$ are zero in accordance with the first of the rules listed in eq. (5.90). In conclusion, the time-averaged mass conservation equation reduces to

$$\frac{\partial \bar{u}}{\partial x} + \frac{\partial \bar{v}}{\partial y} + \frac{\partial \bar{w}}{\partial z} = 0 \qquad (5.91)$$

Problem 5.22. To the left side of the energy eq. (E) of Table 5.1 we add the mass conservation eq. (m) multiplied by T,

$$T\left(\frac{\partial u}{\partial x}+\frac{\partial v}{\partial y}+\frac{\partial w}{\partial z}\right)+$$

$$\frac{\partial T}{\partial t}+u\frac{\partial T}{\partial x}+v\frac{\partial T}{\partial y}+w\frac{\partial T}{\partial z}=\alpha\left(\frac{\partial^2 T}{\partial x^2}+\frac{\partial^2 T}{\partial y^2}+\frac{\partial^2 T}{\partial z^2}\right)$$

and the resulting (E) equation is

$$\frac{\partial T}{\partial t}+\frac{\partial (uT)}{\partial x}+\frac{\partial (vT)}{\partial y}+\frac{\partial (wT)}{\partial z}=\alpha\left(\frac{\partial^2 T}{\partial x^2}+\frac{\partial^2 T}{\partial y^2}+\frac{\partial^2 T}{\partial z^2}\right) \tag{1}$$

Next, we decompose each variable into a mean plus a fluctuating part (e.g. $T = \overline{T} + T'$), and then time-average each of the terms that appear on both sides of the equation:

$$\overline{\frac{\partial T}{\partial t}}=\overline{\frac{\partial}{\partial t}\left(\overline{T}+T'\right)}=\underbrace{\frac{\partial \overline{T}}{\partial t}}_{\text{zero}}+\underbrace{\frac{\partial \overline{T'}}{\partial t}}_{\text{zero}}=0$$

$$\overline{\frac{\partial (uT)}{\partial x}}=\overline{\frac{\partial}{\partial x}(\overline{u}+u')\left(\overline{T}+T'\right)}=$$

$$=\overline{\frac{\partial}{\partial x}\left(\overline{u}\,\overline{T}+u'\overline{T}+\overline{u}\,T'+u'T'\right)}$$

$$=\frac{\partial}{\partial x}\left(\overline{u}\,\overline{T}+\underset{\text{zero}}{\underset{\uparrow}{\overline{u'}}}\,\overline{T}+\overline{u}\,\underset{\text{zero}}{\underset{\uparrow}{\overline{T'}}}+\overline{u'\,T'}\right)$$

$$=\frac{\partial}{\partial x}\left(\overline{u}\,\overline{T}\right)+\frac{\partial}{\partial x}\left(\overline{u'\,T'}\right)$$

$$\overline{\frac{\partial (vT)}{\partial y}}=\cdots=\frac{\partial}{\partial y}\left(\overline{v}\,\overline{T}\right)+\frac{\partial}{\partial y}\left(\overline{v'\,T'}\right)$$

$$\overline{\frac{\partial (wT)}{\partial z}}=\cdots=\frac{\partial}{\partial z}\left(\overline{w}\,\overline{T}\right)+\frac{\partial}{\partial z}\left(\overline{w'\,T'}\right)$$

$$\overline{\frac{\partial^2 T}{\partial x^2}}=\frac{\partial^2 \overline{T}}{\partial x^2}$$

$$\overline{\frac{\partial^2 T}{\partial y^2}} = \frac{\partial^2 \overline{T}}{\partial y^2}$$

$$\overline{\frac{\partial^2 T}{\partial z^2}} = \frac{\partial^2 \overline{T}}{\partial z^2}$$

Substituting all these results into the energy eq. (1), we obtain

$$\frac{\partial(\overline{u}\,\overline{T})}{\partial x} + \frac{\partial(\overline{v}\,\overline{T})}{\partial y} + \frac{\partial(\overline{w}\,\overline{T})}{\partial z} = \alpha\left(\frac{\partial^2 \overline{T}}{\partial x^2} + \frac{\partial^2 \overline{T}}{\partial y^2} + \frac{\partial^2 \overline{T}}{\partial z^2}\right)$$

$$- \frac{\partial}{\partial x}\left(\overline{u'\,T'}\right) - \frac{\partial}{\partial y}\left(\overline{v'\,T'}\right) - \frac{\partial}{\partial z}\left(\overline{w'\,T'}\right)$$

The left side of this equation can be written alternatively as

$$\overline{u}\frac{\partial \overline{T}}{\partial x} + \overline{v}\frac{\partial \overline{T}}{\partial y} + \overline{w}\frac{\partial \overline{T}}{\partial z} + \overline{T}\underbrace{\left(\frac{\partial \overline{u}}{\partial x} + \frac{\partial \overline{v}}{\partial y} + \frac{\partial \overline{w}}{\partial z}\right)}_{\text{zero, cf. eq. (5.91)}}$$

Therefore, the final time-averaged energy equation is the form listed as eq. (5.95) in the text.

Problem 5.23. The constant-τ_{app} condition (5.115) means that

$$(\nu + \varepsilon_M)\frac{\partial \overline{u}}{\partial y} = \frac{\tau_{w,x}}{\rho} \tag{1}$$

In the viscous sublayer ($\nu >> \varepsilon_M$), we neglect ε_M on the left side, and integrate from $y = 0$, where $\overline{u} = 0$,

$$\overline{u} = \frac{\tau_{w,x}}{\rho}\frac{y}{\nu} \tag{2}$$

In terms of the dimensionless wall coordinates (5.117), this conclusion reads:

$$u^+ = y^+, \qquad (\nu >> \varepsilon_M) \tag{3}$$

In the fully turbulent sublayer ($\varepsilon_M >> \nu$), we neglect ν on the left side of eq. (1), and replace ε_M with the mixing-length model (5.113):

$$\kappa^2 y^2 \left(\frac{\partial \overline{u}}{\partial y}\right)^2 = \frac{\tau_{w,x}}{\rho} \tag{4}$$

This can be rewritten using the wall coordinates (5.117),

$$\kappa^2 (y^+)^2 \left(\frac{du^+}{dy^+}\right)^2 = 1 \tag{5}$$

which reduces to

$$\kappa\, y^+ \frac{du^+}{dy^+} = 1 \tag{6}$$

Integrating away from the outer edge of the viscous sublayer, $y^+ = y^+_{VSL}$, where $u^+ = y^+_{VSL}$ according to eq. (3), we obtain

$$u^+ = \frac{1}{\kappa} \ln y^+ + \underbrace{y^+_{VSL} - \frac{1}{\kappa} \ln y^+_{VSL}}_{B} \tag{7}$$

The constant identified as B shows that if experiments recommend the value $B \cong 5.5$ then the dimensionless thickness of the viscous sublayer is $y^+_{VSL} \cong 11.6$.

Problem 5.24. The analysis consists of solving the 2-equation system (5.119)-(5.120) for $\tau_{w,x}$ and δ. Starting with the momentum integral eq. (5.120), we note that in the integrand we need an expression for the $\overline{u}$ profile. This follows from eq. (5.119), which can be written for any y,

$$\frac{\overline{u}}{(\tau_{w,x}/\rho)^{1/2}} = 8.7 \left[\frac{y}{\nu}\left(\frac{\tau_{w,x}}{\rho}\right)^{1/2}\right]^{1/7} \tag{5.118}$$

and also for $y = \delta$:

$$\frac{U_\infty}{(\tau_{w,x}/\rho)^{1/2}} = 8.7 \left[\frac{\delta}{\nu}\left(\frac{\tau_{w,x}}{\rho}\right)^{1/2}\right]^{1/7} \tag{5.119}$$

Dividing these side by side we obtain

$$\frac{\overline{u}}{U_\infty} = \left(\frac{y}{\delta}\right)^{1/7} \tag{a}$$

Substituting this $\overline{u}$ profile in the integrand of eq. (5.120) leads to

$$\frac{d}{dx}\left[U_\infty^2\delta \underbrace{\int_0^1 n^{1/7}(1 - n^{1/7})\,dn}_{7/72}\right] = \frac{\tau_{w,x}}{\rho} \tag{b}$$

where the dummy variable n is shorthand for y/δ. Rearranged, eq. (b) states that

$$\frac{d\delta}{dx} = \frac{72}{7}\,\frac{\tau_{w,x}}{\rho U_\infty^2} \tag{c}$$

The second relation between δ and $\tau_{w,x}$ is eq. (5.119), which is equivalent to

$$\frac{\tau_{w,x}}{\rho} = \frac{U_\infty^{7/4}\ \nu^{1/4}}{(8.7)^{7/4}\ \delta^{1/4}} \tag{d}$$

Eliminating $\tau_{w,x}/\rho$ between eqs. (c) and (d),

$$\delta^{1/4}\frac{d\delta}{dx} = \left(\frac{\nu}{U_\infty}\right)^{1/4}\frac{72}{7\,(8.7)^{7/4}} \tag{e}$$

and integrating from x = 0, where – as an approximation – δ is zero, we obtain

$$\frac{\delta}{x} = 0.37\left(\frac{\nu}{U_\infty x}\right)^{1/5} \tag{5.122}$$

Finally, combining eq. (5.122) with eq. (d) produces a formula for the wall shear stress,

$$\frac{\tau_{w,x}}{\rho U_\infty^2} = (8.7)^{-7/4}\left(\frac{\nu}{U_\infty \delta}\right)^{1/4} = 0.0227\left(\frac{U_\infty \delta}{\nu}\right)^{-1/4}$$

$$= 0.029\left(\frac{U_\infty x}{\nu}\right)^{-1/5} \tag{5.121}$$

Problem 5.25. a) Starting with the local Stanton number definition (5.131), we write in order

$$St_x = \frac{h_x}{\rho c_P U_\infty} = \frac{h_x}{\rho c_P U_\infty} \frac{x}{k} \frac{k}{x}$$

$$= Nu_x \frac{\alpha}{U_\infty x} = Nu_x \frac{\nu}{U_\infty x} \frac{\alpha}{\nu}$$

$$= \frac{Nu_x}{Pe_x} = \frac{Nu_x}{Re_x\, Pr} \qquad (1)$$

b) According to eq. (5.53), the right side of the Colburn analogy (5.131) is

$$\frac{1}{2} C_{f,x} = 0.332\, Re_x^{-1/2} \qquad (2)$$

where $Re_x = U_\infty x/\nu$. The left side of the Colburn analogy can be estimated using eq. (1) and the Nu_x formula (5.79b), which holds for $Pr \gtrsim 0.5$ fluids,

$$St_x\, Pr^{2/3} = \frac{Nu_x}{Re_x\, Pr} Pr^{2/3}$$

$$= \frac{0.332\, Pr^{1/3}\, Re_x^{1/2}}{Re_x\, Pr} Pr^{2/3}$$

$$= 0.332\, Re_x^{-1/2} \qquad (3)$$

Equations (2) and (3) show that the Colburn analogy $St_x\, Pr^{2/3} = (1/2)C_{f,x}$ applies to the laminar section of the boundary layer, provided the fluid is such that $Pr \gtrsim 0.5$.

Problem 5.26. The first of the integrals appearing in eq. (5.132) is evaluated using eq. (5.79b):

$$\int_0^{x_{tr}} h_x\, dx = \int_0^{x_{tr}} 0.332\, Pr^{1/3}\, Re_x^{1/2} \frac{k}{x} dx$$

$$= 0.332\, Pr^{1/3} \left(\frac{U_\infty}{\nu}\right)^{1/2} k \int_0^{x_{tr}} \frac{dx}{x^{1/2}}$$

$$= 0.664\, Pr^{1/3} \left(\frac{U_\infty x_{tr}}{\nu}\right)^{1/2} k$$

The second integral requires the use of eqs. (5.131) and (5.121):

$$\int_{x_{tr}}^{L} h_x \, dx = \int_{x_{tr}}^{L} \rho c_P U_\infty Pr^{-2/3} \, 0.0296 \left(\frac{U_\infty x}{\nu}\right)^{-1/5} dx$$

$$= \rho c_P U_\infty Pr^{-2/3} \, 0.037 \left(\frac{U_\infty}{\nu}\right)^{-1/5} \left(L^{4/5} - x_{tr}^{4/5}\right)$$

$$= 0.037 \, Pr^{1/3} \, k \left(Re_L^{4/5} - Re_{x,tr}^{4/5}\right)$$

The average heat transfer coefficient is therefore

$$\overline{h}_L = \frac{1}{L}\left[0.664 \, Pr^{1/3} \, Re_{x,tr}^{1/5} \, k + 0.037 \, Pr^{1/3} \, k \left(Re_L^{4/5} - Re_{x,tr}^{4/5}\right)\right]$$

This result is equivalent to the $\overline{Nu_L}$ formula listed as eq. (5.133).

Problem 5.27. For air at 20°C, we will need the following properties:

$$\rho = 1.205 \frac{kg}{m^3} \qquad \nu = 0.15 \frac{cm^2}{s}$$

a) At x = 30 cm downstream from the leading edge the Reynolds number is

$$Re_x = \frac{U_\infty x}{\nu}$$

$$= 2.93 \frac{m}{s} \, 0.3m \frac{s}{0.15 \times 10^{-4} \, m^2} \cong 6 \times 10^4$$

This Re_x value is consistent with the transition range indicated in the text subsection 5.4.1.

b) When the terminal velocity of free-fall is reached, the weight of the ribbon (mg) equals the total shear force $2A \, \overline{\tau}_{w,L}$, where A = LW is the surface of one side of ribbon:

$$mg = 2\,LW\,\bar{\tau}_{w,L}$$

$$= 2\,LW\,\frac{1}{2}\,\rho\,U_\infty^2\,1.328\,Re_L^{-1/2}$$

$$U_\infty^{3/2} = \frac{m\,g}{LW\,\rho\,1.328}\left(\frac{L}{\nu}\right)^{1/2}$$

$$= \frac{0.00321\text{ kg }9.81\,\frac{m}{s^2}}{1.25m\ 0.114m\ 1.205\,\frac{kg}{m^3}\ 1.328}\left(\frac{1.25m}{0.15\times10^{-4}\,\frac{m^2}{s}}\right) = 39.86\left(\frac{m}{s}\right)^{3/2}$$

$$U_\infty = 11.7\,\frac{m}{s}$$

The velocity measured in the experiment (2.93 m/s, Fig. 5.11) is considerably smaller because the ribbon does not remain straight, and the flow is not laminar all over L.

Problem 5.28. The properties of 20°C air are

$$\rho = 1.205\,\frac{kg}{m^3} \qquad \nu = 0.15\,\frac{cm^2}{s}$$

When the terminal speed is reached, the weight of the ribbon is matched by the total longitudinal shear force

$$mg = 2A\,\bar{\tau}_{w,L} \qquad (1)$$

where A = WL and

$$\bar{\tau}_{w,L} = 0.037\,\rho\,U_\infty^2\,Re_L^{-1/5}$$

Equation (1) yields, in order,

$$mg = 2\,WL\,0.037\,\rho\,U_\infty^2\left(\frac{\nu}{U_\infty\,L}\right)^{1/5}$$

$$U_\infty^{9/5} = \frac{m\,g}{0.074\,WL\,\rho}\left(\frac{L}{\nu}\right)^{1/5}$$

$$= \frac{0.00321\text{ kg }9.81\,\frac{m}{s^2}}{0.074\times0.114m\times1.25m\ 1.205\,\frac{kg}{m^3}}\left(\frac{1.25m}{0.15\times10^{-4}\,\frac{m^2}{s}}\right)^{1/5} = 23.89\left(\frac{m}{s}\right)^{9/5}$$

$$U_\infty = 5.83\,\frac{m}{s}$$

The measured free-fall velocity is smaller (2.93 m/s) because in real life the ribbon does not remain straight. The current U_∞ estimate is closer to the measured value than the estimate based on the assumption that the flow is laminar (11.7 m/s), because the real flow is laminar over only one fourth of the length L.

Problem 5.29. a) The properties of water at 20°C are

$$\rho \cong 1 \frac{g}{cm^3} \qquad c_P = 4.18 \frac{kJ}{kg\ K} \qquad \nu = 0.01 \frac{cm^2}{s}$$

$$k = 0.59 \frac{W}{m \cdot K} \qquad Pr = 7.07$$

In order to calculate y,

$$y = y^+ \nu \left(\frac{\tau_{w,x}}{\rho} \right)^{-1/2}$$

we must first evaluate $\tau_{w,x}/\rho$. For this we use eq. (5.121):

$$\frac{\tau_{w,x}}{\rho} = 0.0296\ U_\infty^2\ Re_x^{-1/5}$$

in which

$$Re_x = \frac{U_\infty x}{\nu}$$

$$= 20 \frac{cm}{s} 600\ cm \frac{s}{0.01\ cm^2} = 1.2 \times 10^6 \qquad \text{(turbulent)}$$

The result is

$$\frac{\tau_{w,x}}{\rho} = 0.0018\ U_\infty^2$$

$$\left(\frac{\tau_{w,x}}{\rho} \right)^{1/2} = 0.0424 \times 20 \frac{cm}{s} = 0.849 \frac{cm}{s}$$

$$y = y^+ \nu \left(\frac{\tau_{w,x}}{\rho} \right)^{-1/2}$$

$$= 2.7 \times 0.01 \frac{cm^2}{s} \frac{s}{0.849\ cm} = 0.3\ mm$$

b) The boundary layer thickness can be evaluated based on eq. (5.122):

$$\delta = 0.37\ x\ Re_x^{-1/5}$$
$$= 0.37 \times 6m \left(1.2 \times 10^6\right)^{-1/5} = 13.5\ \text{cm}$$

In the laminar regime, the corresponding thickness would be the δ_{99} given by eq. (5.57):

$$\delta_{99} = 4.92\ x\ Re_x^{-1/2}$$
$$= 4.92 \times 6m \left(1.2 \times 10^6\right)^{-1/2} = 2.7\ \text{cm}$$

We see that the laminar boundary layer would have been much thinner (at x) than the real (turbulent) boundary layer.

c) Finally, for the x-averaged heat transfer coefficient we rely on eq. (5.134):

$$\overline{Nu_x} = 0.037\ Pr^{1/3} \left(Re_x^{4/5} - 23\ 550\right)$$
$$= 0.037\ (7.07)^{1/3} \left[\left(1.2 \times 10^6\right)^{4/5} - 23\ 550\right]$$
$$= 3512$$
$$\bar{h} = \overline{Nu_x} \frac{k}{x}$$
$$= 3512 \times 0.59 \frac{W}{m \cdot K} \frac{1}{6m} = 345 \frac{W}{m^2\ K}$$

Problem 5.30. a) We begin by recognizing the wall averaged shear stress, the total tangential force experienced by the wall, and the mechanical power spent on dragging the flat surface through the fluid:

$$\bar{\tau} = 0.037\ \rho\ U^2\ Re_L^{-1/5}$$

$$F' = \bar{\tau}\ L$$

$$\mathbf{P} = F'U = 0.037\ \rho\ U^3\ L \left(\frac{\nu}{UL}\right)^{1/5}$$

If $(\ )_c$ and $(\ )_h$ represent the cold-wall and hot-wall conditions, the dissipated power changes according to the ratio

$$\frac{\mathbf{P}_h}{\mathbf{P}_c} = \frac{\rho_h}{\rho_c}\left(\frac{\nu_h}{\nu_c}\right)^{1/5}$$

b) When the wall is heated, the water film temperature is (90°C + 10°C)/2 = 50°C, with the corresponding properties

$$\rho_h \cong 1\ \frac{g}{cm^3} \qquad\qquad \mu_h = 0.00548\ \frac{g}{cm \cdot s}$$

The ratio

$$\frac{\mathbf{P}_h}{\mathbf{P}_c} = \frac{1}{1}\left(\frac{0.00548}{0.013}\right)^{1/5} = 0.84$$

shows that by heating the wall to 90°C we can reduce the drag power by 16 percent.

c) The power that has been saved by heating the water boundary layer is

$$\mathbf{P}_c - \mathbf{P}_h = 0.037\ \rho\ U^{14/5}\ L^{4/5}\ \nu_c^{1/5}\left[1 - \left(\frac{\nu_h}{\nu_c}\right)^{1/5}\right] \qquad (1)$$

The electric power spent on heating the wall is given sequentially by

$$\frac{\bar{h}\,L}{k_h} = 0.037\ Pr_h^{1/3}\left(\frac{UL}{\nu_h}\right)^{4/5}$$

$$\overline{q''}\ L = 0.037\ k_h\ \Delta T\ Pr_h^{1/3}\left(\frac{UL}{\nu_h}\right)^{4/5} \qquad (2)$$

in which (k, Pr, ν) are also evaluated at the 50°C film temperature,

$$k_h = 0.64 \frac{W}{m \cdot K} \qquad \qquad Pr_h = 3.57$$

By dividing eqs. (1) and (2) we obtain a dimensionless measure of how effectively the heating of the wall has been converted into **P** savings:

$$\frac{\mathbf{P}_c - \mathbf{P}_h}{\overline{q''} L} = \frac{U^2}{c \Delta T} Pr_h^{2/3} \left(\frac{\nu_c}{\nu_h}\right)^{1/5} \left[1 - \left(\frac{\nu_h}{\nu_c}\right)^{1/5}\right]$$

In this expression we substitute c = 4.18 kJ/kg·K, ΔT = 90°C - 10°C = 80°C and $(\nu_h/\nu_c)^{1/5} = 0.84$, and obtain

$$\frac{\mathbf{P}_c - \mathbf{P}_h}{\overline{q''} L} = \left(\frac{U}{867 \text{ m/s}}\right)^2$$

This shows that the drag power savings are much smaller than the heating power investment when the ship speed is of order 10 m/s.

Problem 5.31. The relevant properties of air at the film temperature (20°C) are

$$\rho = 1.205 \times 10^{-3} \frac{g}{cm^3}, \qquad \qquad \nu = 0.15 \frac{cm^2}{s}$$

When the iceberg drifts steadily, the sea water drag over the bottom surface matches the air drag over the top surface. Therefore, we can write in order

$$F_{D,air} = F_{D,water}$$

$$(\bar{\tau}_{w,L} A)_a = (\bar{\tau}_{w,L} A)_w$$

$$0.037 \rho_a U_a^2 \left(\frac{\nu_a}{U_a L}\right)^{1/5} = 0.037 \rho_w U_w^2 \left(\frac{\nu_w}{U_w L}\right)^{1/5}$$

and the last expression yields

$$\frac{U_{air}}{U_{water}} = \left(\frac{\rho_w}{\rho_a}\right)^{5/9} \left(\frac{\nu_w}{\nu_a}\right)^{1/9}$$

$$= \left[\frac{1 \text{ kg}}{(10 \text{ cm})^3} \frac{cm^3}{1.205 \times 10^{-3} g}\right]^{5/9} \left(\frac{0.015 \text{ cm}^2/s}{0.15 \text{ cm}^2/s}\right)^{1/9}$$

$$= 32.4$$

and since U_{water} = 10 cm/s, we conclude that the wind velocity is

$$U_{air} = 32.4 \times 10 \frac{cm}{s} = 3.24 \frac{m}{s} = 11.7 \frac{km}{h}$$

Problem 5.32. The air properties are evaluated at the film temperature,

$$\nu = 0.15 \frac{cm^2}{s}, \qquad k = 2.5 \times 10^{-4} \frac{W}{cm\ K}, \qquad Pr = 0.72$$

The air boundary layer is turbulent,

$$Re_L = 3.24 \frac{m}{s} 100\ m \frac{s}{0.15\ cm^2} \cong (2.16)\ 10^7$$

therefore, the $\overline{Nu_L}$ formula is eq. (5.134):

$$\overline{Nu_L} \cong 0.037\ Pr^{1/3}\ Re_L^{4/5}$$

$$= 0.037\,(0.72)^{1/2}\left(2.16 \times 10^7\right)^{4/5}$$

$$= 2.44 \times 10^4$$

The corresponding L-averaged heat flux into the top surface of the iceberg is

$$\overline{q''_L} = \overline{h_L}\ \Delta T = \overline{Nu_L} \frac{k}{L} \Delta T$$

$$= 2.44 \times 10^4\ 2.5 \times 10^{-4} \frac{W}{cm\ K} \frac{1}{100\ m}\ 40°C$$

$$= 244 \frac{W}{m^2}$$

The melting rate associated with this heat flux is

$$\frac{dH}{dt} = \frac{\overline{q''_L}}{\rho\ h_{sf}} = 244 \frac{W}{m^2} \frac{(0.1\ m)^3}{1\ kg} \frac{kg}{333.4\ kJ}$$

$$= 0.73 \times 10^{-6} \frac{m}{s} = 2.6 \frac{mm}{h}$$

Problem 5.33. The relevant properties of air at 0°C are

$$\nu = 0.132 \frac{cm^2}{s}, \qquad \alpha = 0.184 \frac{cm^2}{s}, \qquad Pr = 0.72$$

$$\rho c_P = \frac{k}{\alpha} = 2.4 \times 10^{-4} \frac{W}{cm\,K} \frac{1}{0.184\ cm^2/s} = 0.0013 \frac{J}{cm^3 K}$$

In order to arrive at the heat transfer coefficient for the external surface of the window, we calculate in order:

$$Re_x = \frac{U_\infty x}{\nu} = 15 \frac{m}{s} 60\ m \frac{1}{0.132\ cm^2/s}$$

$$= 6.8 \times 10^7$$

$$\frac{1}{2} C_{f,x} = 0.0296\ Re_x^{-1/5} = 8.03 \times 10^{-4} \qquad (5.121)$$

$$St_x = \frac{1}{2} C_{f,x}\ Pr^{-2/3} = 8.03 \times 10^{-4} (0.72)^{-2/3} \qquad (5.131)$$

$$\cong 0.001$$

$$h_x = St_x\ \rho\ c_P U_\infty = 0.001 \times 0.0013 \frac{J}{cm^3 K} 15 \frac{m}{s}$$

$$= 19.6 \frac{W}{m^2 K}$$

This h_x value is a lower bound for the actual heat transfer coefficient because

i) equation (5.121) underpredicts the skin friction coefficient at large Reynolds numbers,

ii) The external surface of the building is not "smooth", as was assumed in eq. (5.121), and

iii) the free stream U_∞ is not completely smooth, i.e. without eddies.

Problem 5.34. Using the properties of air at 10°C,

$$\nu = 0.141 \frac{cm^2}{s} \qquad k = 0.025 \frac{W}{m \cdot K} \qquad Pr = 0.72$$

we calculate first the Reynolds number

$$Re_D = \frac{UD}{\nu} = 4 \frac{cm}{s}\, 21 \times 10^{-6} m \frac{1}{0.141} \frac{s}{cm^2}$$

$$= 0.06$$

For such a low Reynolds number, Churchill and Bernstein's correlation (5.135) yields, in order,

$$\overline{Nu}_D = \frac{\overline{h}_D D}{k} = 0.3 + \frac{0.62\, Re_D^{1/2}\, Pr^{1/3}}{\left[1 + \left(\frac{0.4}{Pr}\right)^{2/3}\right]^{1/4}} \left[1 + \left(\frac{Re_D}{282000}\right)^{5/8}\right]^{4/5}$$

$$= 0.3 + 0.12\,[1 + (\sim 0)]$$

$$= 0.42$$

and

$$\overline{h}_D = 0.42 \frac{k}{D} = 0.42 \frac{0.025\, W}{m \cdot K} \frac{1}{21 \times 10^{-6} m}$$

$$= 500 \frac{W}{m^2 K}$$

Problem 5.35. a) The properties of 10°C water are

$$\rho \cong 1 \frac{g}{cm^3} \qquad \nu = 0.013 \frac{cm^2}{s}$$

$$k = 0.58 \frac{W}{m \cdot K} \qquad Pr = 9.45$$

We calculate the drag force by using Fig. 5.21:

$$Re_D = \frac{U_\infty D}{\nu} = 1 \frac{m}{s} 0.15m \frac{s}{0.013 \times 10^{-4} \, m^2}$$

$$= 1.15 \times 10^5$$

$C_D \cong 1.3$ (Fig. 5.21)

$A = LD = 0.75m \; 0.15m$

$$F_D = C_D A \frac{1}{2} \rho U_\infty^2$$

$$= 1.3 \times 0.75m \; 0.15m \frac{1}{2} 1 \frac{10^{-3} \, kg}{10^{-6} \, m^3} \left(1 \frac{m}{s}\right)^2 = 73 \text{ N}$$

b) How large is a force of 73 N? The weight (<u>in air</u>) of the same portion of the leg ($L \times D$, the portion that would be immersed in the river) is

$$W = \rho_{meat} \frac{\pi}{4} D^2 L g$$

$$\cong 1060 \frac{kg}{m^3} \frac{\pi}{4} (0.15m)^2 \, 0.75m \; 9.81 \frac{m}{s^2}$$

$$\cong 138 \text{ N}$$

The drag force is about half the weight of the immersed portion of the leg.

c) The heat transfer coefficient is furnished by eq. (5.135), in which we substitute $Re_D = 1.15 \times 10^5$ and $Pr = 9.45$:

$$\overline{Nu}_D = 0.3 + 380 \, (1 + 0.571)^{4/5} = 545.7$$

$$\bar{h} = \overline{Nu}_D \frac{k}{D} = 545.7 \times 0.58 \frac{W}{m \cdot K} \frac{1}{0.15m}$$

$$= 2110 \frac{W}{m^2 \, K}$$

d) $q = \bar{h} A_{lateral} \, \Delta T$

$$= \bar{h} \, \pi D L \, \Delta T = 2110 \frac{W}{m^2 \, K} \pi \; 0.15m \; 0.75m \; 1K$$

$$= 746 \text{ W}$$

Problem 5.36. a) The properties of air at 20°C are

$$\rho = 1.205 \frac{kg}{m^3} \qquad \nu = 0.15 \frac{cm^2}{s}$$

The drag force can be estimated based on Fig. 5.21:

$$Re_D = \frac{U_\infty D}{\nu} \doteq 35.76 \frac{m}{s} 0.074m \frac{s}{0.15 \times 10^{-4} m^2}$$

$$= 1.76 \times 10^5$$

$$C_D \cong 0.42 \quad \text{(Fig. 5.21)}$$

$$F_D = C_D A \frac{1}{2} \rho U_\infty^2 \quad \left(\text{where } A = \frac{\pi D^2}{4}\right)$$

$$= 0.42 \frac{\pi}{4} (0.074m)^2 \frac{1}{2} 1.205 \frac{kg}{m^3} \left(35.76 \frac{m}{s}\right)^2$$

$$= 1.39 \text{ N}$$

b) The weight of the baseball is comparable to the drag force:

$$mg = 0.145 \text{ kg } 9.81 \frac{m}{s^2} = 1.42 \text{ N}$$

c) Between the pitcher's mound and the catcher's mitt, the kinetic energy of the ball [(1/2) mU^2] drops by an amount equal to the work done against the atmosphere ($F_D \cdot x$). The first law of thermodynamics for the process 1 (pitcher) → 2 (catcher) is

$$- F_D \cdot x = \frac{1}{2} m U_2^2 - \frac{1}{2} m U_1^2$$

$$U_1^2 - U_2^2 = \frac{2}{m} F_D x$$

$$= \frac{2}{0.145 \text{ kg}} 1.39 \text{ N } 18.5\text{m} = 354.7 \left(\frac{m}{s}\right)^2$$

$$U_2^2 = \left(35.76 \frac{m}{s}\right)^2 - 354.7 \left(\frac{m}{s}\right)^2 = 924.1 \left(\frac{m}{s}\right)^2$$

$$U_2 = 30.4 \frac{m}{s}$$

The final horizontal velocity is 85 percent of the initial value.

Problem 5.37. a) In eq. (5.139) we evaluate all the air properties at $T_\infty = 20°C$, except $\mu_w = \mu(T_w = 30°C)$:

$$k_\infty = 0.025 \frac{W}{m \cdot K} \qquad \mu_\infty = 1.81 \times 10^{-5} \frac{kg}{s \cdot m}$$

$$\nu_\infty = 0.15 \frac{cm^2}{s} \qquad \mu_w = 1.86 \times 10^{-5} \frac{kg}{s \cdot m}$$

$$Pr_\infty = 0.72$$

The Reynolds number corresponds to the upper end of the Re_D range in which eq. (5.139) is valid,

$$Re_D = \frac{U_\infty D}{\nu_\infty} = 22.35 \frac{m}{s} 0.07m \frac{s}{0.15 \times 10^{-4} \, m^2}$$

$$= 1.043 \times 10^5$$

therefore we obtain, in order,

$$\overline{Nu}_D = 2 + \left(0.4 \, Re_D^{1/2} + 0.06 \, Re_D^{2/3}\right) Pr_\infty^{0.4} \left(\frac{\mu_\infty}{\mu_w}\right)^{1/4}$$

$$= 2 + 262.1 \, (0.72)^{0.4} \left(\frac{1.81}{1.86}\right)^{1/4}$$

$$= 230.3$$

$$\bar{h} = \overline{Nu}_D \frac{k_\infty}{D} = 230.3 \times 0.025 \frac{W}{m \cdot K} \frac{1}{0.07m}$$

$$= 82.25 \frac{W}{m^2 \, K}$$

$$A = \pi D^2 = \pi \, (0.07m)^2 = 0.0154 \, m^2$$

$$q = \bar{h} A \left(T_w - T_\infty\right) = 82.25 \frac{W}{m^2 \, K} 0.0154 \, m^2 \, (30 - 20) \, K$$

$$= 12.7 \, W$$

The time of travel and the total heat transfer (ball → air) are

$$t = \frac{x}{U} = \frac{18.5\text{m}}{22.35\text{ m/s}} = 0.83\text{s}$$

$$Q = qt = 12.7\text{ W } 0.83\text{s} = 10.5\text{ J}$$

b) The depth to which the cooling effect penetrates into the leather cover is

$$\delta \sim (\alpha_{leather}\, t)^{1/2} \sim \left(0.001\ \frac{\text{cm}^2}{\text{s}}\ 0.83\text{s}\right)^{1/2}$$

$$\sim 0.29\text{ mm}$$

The temperature drop that would be experienced by the 0.29 mm skin of the ball is

$$\Delta T_w = \frac{Q}{(mc)_{skin}} = \frac{Q}{(\rho c)_{leather}\, A\delta}$$

$$\cong \frac{10.5\text{ J}}{860\ \frac{\text{kg}}{\text{m}^3}\ 1.5\ \frac{\text{kJ}}{\text{kg}\cdot\text{K}}\ 0.0154\text{ m}^2\ 0.00029\text{m}} \cong 1.8°\text{C}$$

This temperature drop is small compared with the ball-air temperature difference $T_w - T_\infty = 10°C$, therefore the assumption that T_w is constant between the pitcher and the catcher is reasonable.

Another way of reaching the same conclusion is provided by the right side of Fig. 4.4, where the abscissa parameter has a value less than 1:

$$\frac{h}{k_{leather}}(\alpha_{leather}\, t)^{1/2} = \frac{h}{k_{leather}}\,\delta$$

$$\sim 82.25\ \frac{\text{W}}{\text{m}^2\text{ K}}\ \frac{\text{m}\cdot\text{K}}{0.135\text{ W}}\ 0.29 \times 10^{-3}\text{m} \sim 0.18$$

This means that the ordinate parameter is approximately:

$$\frac{T_w(t) - T_\infty}{T_w(0) - T_\infty} \sim 0.825$$

or

$$T_w(t) - T_\infty \sim 8.25°\text{C}$$

and

$$\Delta T_w = T_w(0) - T_w(t) \sim 1.75°\text{C}$$

Problem 5.38. The air properties that are required by a calculation based on eq. (5.139) are

$$\mu_w = 2 \times 10^{-5} \frac{kg}{s \cdot m} \qquad \nu_\infty = 0.141 \frac{cm^2}{s}$$

$$\mu_\infty = 1.76 \times 10^{-5} \frac{kg}{s \cdot m} \qquad k_\infty = 0.025 \frac{W}{m \cdot K}$$

$$Pr_\infty = 0.72$$

We begin with the Reynolds number,

$$Re_D = \frac{U_\infty D}{\nu_\infty} = 2 \frac{m}{s} 0.06m \frac{s}{0.141 \times 10^{-4}\ m^2}$$

$$= 8511$$

which falls in the range where eq. (5.139) is valid:

$$\overline{Nu_D} = \ldots = 2 + 61.91\ (0.72)^{0.4} \left(\frac{1.76}{2}\right)^{1/4} = 54.58$$

$$\bar{h} = \overline{Nu_D} \frac{k_\infty}{D} = 54.58 \times 0.025 \frac{W}{m \cdot K} \frac{1}{0.06m}$$

$$= 22.74 \frac{W}{m^2\ K}$$

$$A = \pi D^2 = \pi (0.06m)^2 = 0.0113\ m^2$$

$$q = \bar{h} A \left(T_w - T_\infty\right)$$

$$= 22.74 \frac{W}{m^2\ K} 0.0113\ m^2\ (60 - 10)\ K = 12.85\ W$$

Problem 5.39. In the pure-conduction limit $Re_D = 0$, the total heat transfer rate from the sphere to the surrounding (motionless) fluid is given by eq. (2.39), in which $r_o \to \infty$:

$$q = 4\pi k\ r_i \left(T_w - T_\infty\right)$$

$$= 4\pi k \frac{D}{2} \left(T_w - T_\infty\right) \qquad (1)$$

Recalling the overall Nusselt number definition,

$$\overline{Nu}_D = \frac{\bar{h}\,D}{k} = \frac{\bar{q}''_w}{\Delta T}\frac{D}{k} = \frac{q/\pi D^2}{\Delta T}\frac{D}{k}$$

the q formula (1) is equivalent to :

$$\overline{Nu}_D = \frac{2\pi k\,D\Delta T}{\pi D^2}\frac{D}{\Delta T\,k} = 2$$

The equivalent pure-conduction limit in eq. (5.142) is represented by the constant $\overline{Nu}^0_{\underline{L}}$. The value of this constant can be deduced from $\overline{Nu}_D = 2$, by first evaluating the new length scale

$$\underline{L} = A^{1/2} = \left(\pi D^2\right)^{1/2} = \pi^{1/2}D$$

and then writing

$$\overline{Nu}^0_{\underline{L}} = \frac{\bar{h}\,\underline{L}}{k} = \frac{\bar{h}\,D}{k}\frac{\underline{L}}{D} = \overline{Nu}_D\frac{\underline{L}}{D}$$

$$= 2\,\frac{\pi^{1/2}D}{D} = 2\pi^{1/2} = 3.545$$

Problem 5.40. When the glass bead reaches its terminal speed, the weight of the bead (mg) is balanced by the drag force (F_D). Starting with eq. (5.138), this balance provides a relationship between the drag coefficient C_D and the unknown speed U_∞,

$$F_D = C_D A\frac{1}{2}\rho_a U_\infty^2 = mg$$

$$C_D\frac{\pi}{4}D^2\frac{1}{2}\rho_a U_\infty^2 = \rho\frac{4\pi}{3}\left(\frac{D}{2}\right)^3 g$$

$$C_D = \frac{\rho}{\rho_a}\frac{4}{3}\frac{gD}{U^2}$$

in which we substitute

$\rho = 2800\ kg/m^3$	(density of glass)
$\rho_a = 1.205\ kg/m^3$	(density of air at 20°C)
$g = 9.81\ m/s^2$	
$D = 0.0005m$	

The end result is

$$C_D = 15.2 \left(\frac{m/s}{U_\infty}\right)^2 \qquad (1)$$

A second relation involving the unknown U_∞ is the definition of Re_D on the abscissa of Fig. 5.21,

$$Re_D = \frac{U_\infty D}{\nu_a}$$

in which $\nu_a = 0.15\ cm^2/s$ is the kinematic viscosity of air at 20°C. Numerically, the Re_D definition reduces to

$$Re_D = 33.3 \frac{U_\infty}{m/s} \qquad (2)$$

Now we must find the proper U_∞ value so that the C_D and Re_D values calculated with eqs. (1) and (2) represent a point on the "sphere" curve in Fig. 5.21. We do this by trial and error, proceeding from left to right in the following table:

pick U_∞	calculate C_D with eq. (1)	read Re_D off Fig. 5.21	calculate U_∞ with eq. (2)
3.5 m/s	1.24	85	2.44 m/s
3.8 m/s	1.05	123	3.7 m/s
3.9 m/s	1	138	4.14 m/s
4 m/s	0.95	150	4.5 m/s

The correct U_∞ value is located between the second and third guesses, more exactly at

$$U_\infty \cong 3.8\ \frac{m}{s}, \quad \text{for which} \quad Re_D \cong 127 \qquad (2)$$

In vacuum, the bead velocity would be increasing as gt. The approximate time when the terminal speed is reached (i.e. when the air drag effect becomes important) is when gt is comparable with the just calculated U_∞,

$$gt \sim U_\infty$$

$$t \sim \frac{U_\infty}{g} = \frac{3.8\ m/s}{9.8\ m/s^2} \cong 0.4\ s$$

The bead falls for a total of at least 10m/(3.8 m/s) = 2.6s, therefore it is reasonable to assume that during most of this travel its speed has the terminal value U_∞.

We can now evaluate the average heat transfer coefficient using eq. (5.139), in which we substitute $Re_D = 127$ and

$\mu_w = 3.58 \times 10^{-5}$ kg/s·m	(air viscosity at 500°C)
$\mu_\infty = 1.81 \times 10^{-5}$ kg/s·m	(air viscosity at 20°C)
$Pr = 0.72$	(air Prandtl number at 20°C)

In the end we obtain $\overline{Nu_D} \cong 6.47$, which means that

$$h = \frac{k_a}{D} Nu = \frac{0.025\ W}{m \cdot K} \frac{6.47}{0.0005m} = 324 \frac{W}{m^2K}$$

Treating the glass bead as a lumped capacitance of temperature T, we write the first law for the bead as a closed thermodynamic system,

$$\rho V c \frac{d\left(T - T_\infty\right)}{dt} = - hA_s\left(T - T_\infty\right)$$

and recognize that $V = (4\pi/3)(D/2)^3$, and $A_s = \pi D^2$. Integrating this equation from $t = 0$, where $T = T_i = 500°C$, we obtain

$$\frac{T - T_\infty}{T_i - T_\infty} = \exp\left(- \frac{6\ ht}{\rho c\ D}\right)$$

in which t = 2.6s and, for glass, c = 0.8 kJ/kg·K. Numerically, this yields in order

$$\frac{T - T_\infty}{T_i - T_\infty} = \exp\left(- 4.51\right) = 0.011$$

$$T = 20°C + 0.011\ (500\text{-}20)°C$$

$$= 25.3°C.$$

In conclusion, by the time it falls to the ground, the glass bead has almost the same temperature as the surrounding air.

Problem 5.41. a) The air properties are evaluated at the bulk temperature of 300°C,

$$k = 0.045\ \frac{W}{m\cdot K} \qquad Pr = 0.68$$

$$\nu = 0.481\ \frac{cm^2}{s} \qquad Pr_w = 0.72 \text{ (air at 30°C)}$$

Figure 5.26 shows that $C_n \cong 1$ for an array with 20 rows. For the aligned array we calculate,

$$U_{max} = U_\infty \frac{X_t}{X_t - D} = 2\ \frac{m}{s}\ \frac{7}{7-4} = 4.67\ \frac{m}{s} \qquad (5.146)$$

$$Re_D = U_{max}\frac{D}{\nu} = 4.67 \times 100\ \frac{cm}{s}\ \frac{4\ cm}{0.481\ cm^2/s} = 3881 \qquad (5.145)$$

$$\overline{Nu}_D = 0.27\ C_n\ Re_D^{0.63}\ Pr^{0.36}\left(\frac{Pr}{Pr_w}\right)^{1/4} \qquad (5.144)$$

$$= 0.27 \times 1 \times (3881)^{0.63}\ (0.68)^{0.36}\left(\frac{0.68}{0.72}\right)^{1/4}$$

$$= 42.26$$

$$\bar{h} = \frac{k}{D}\overline{Nu}_D$$

$$= 0.045\ \frac{W}{m\cdot K}\ \frac{42.26}{0.04 m} = 47.54\ \frac{W}{m^2 K}$$

b) When the cylinders are in a staggered array, we must decide which space (flow area) is narrower, between two cylinders in the same row

$$X_t - D = 3\ cm$$

or between two cylinders aligned with the direction "a" in Fig. 5.24 (right),

$$\left[X_l^2 + \left(\frac{1}{2}X_t\right)^2\right]^{1/2} - D = 3.83\ cm$$

The space between two cylinders in the same row is the narrower; this means that U_{max} and Re_D are the same as in the first part of the problem. By using eq. (5.147) we find that the heat transfer coefficient in the staggered array is approximately 30 percent larger than in the aligned array,

$$\overline{Nu_D} = 0.35\,(3881)^{0.6}\,(0.68)^{0.36}\left(\frac{0.68}{0.72}\right)^{1/4}\left(\frac{7\text{ cm}}{7\text{ cm}}\right)^{0.2}$$

$$= 54.78$$

$$\bar{h} = \frac{k}{D}\,\overline{Nu_D} = 61.63\,\frac{W}{m^2\,K}$$

Problem 5.42. Substituting the Gaussian profile (5.148) in the integrand on the left side of eq. (5.150), we obtain

$$2\pi\int_0^\infty \rho\bar{u}_c^2\exp\left[-2\left(\frac{r}{b}\right)^2\right] r\,dr = 2\pi\rho\bar{u}_c^2\frac{b^2}{2}\underbrace{\int_0^\infty e^{-m^2}m\,dm}_{1/2}$$

$$= \frac{\pi}{2}\rho\bar{u}_c^2\,b^2$$

Setting this result equal to the right side of eq. (5.150), we obtain the formula for the time-averaged axial velocity,

$$\frac{\pi}{2}\rho\bar{u}_c^2\,b^2 = \rho U_0^2\frac{\pi}{4}D_0^2$$

therefore

$$\bar{u}_c \cong \frac{U_0\,D_0}{2^{1/2}\,0.107x} = 6.61\,\frac{U_0\,D_0}{x}$$

This result shows that the centerline velocity is equal to the mean nozzle velocity at the distance $x \cong 6.61\,D_0$ downstream from the nozzle, which marks the beginning of the linear-growth region to which eqs. (5.148)-(5.153) apply.

Problem 5.43. Using the $\bar{u}$ and $(\bar{T} - T_\infty)$ profiles (5.148)-(5.149) on the left side of eq. (5.152), we obtain

$$2\pi \int_0^\infty \rho c_P \bar{u} (\bar{T} - T_\infty) r \, dr = 2\pi \rho c_P \bar{u}_c (\bar{T}_c - T_\infty) a^2 \underbrace{\int_0^\infty e^{-m^2} m \, dm}_{1/2}$$

where a^2 is shorthand for

$$a^2 = \frac{b^2 b_T^2}{b^2 + b_T^2} \cong \frac{0.0161 \, x^2}{2.409} \quad \text{[using } b(x) \text{ and } b_T(x) \text{ of eqs. (5.148)-(5.149)]}$$

In conclusion, the energy conservation theorem (5.152) reads

$$\pi \rho c_P \bar{u}_c (\bar{T}_c - T_\infty) a^2 = \rho c_P U_0 (T_0 - T_\infty) \frac{\pi}{4} D_0^2$$

or, after substituting the a(x) expression defined above, and the $\bar{u}_c(x)$ expression (5.151),

$$\bar{T}_c - T_\infty \cong 5.65 \frac{(T_0 - T_\infty) D_0}{x}$$

Problem 5.44. Multiplying the mass-conservation equation by $\bar{u}$, and adding it to the left side of the momentum equation, we obtain:

$$\frac{\partial \bar{u}^2}{\partial x} + \frac{1}{r} \frac{\partial}{\partial r} (r \bar{u} \bar{v}) = \frac{1}{r} \frac{\partial}{\partial r} \left[r (\nu + \varepsilon_M) \frac{\partial \bar{u}}{\partial r} \right]$$

Next, we integrate this equation term by term over a disc the radius of which is sufficiently larger than the characteristic radius of the turbulent jet,

$$\int_{\theta = 0}^{2\pi} \int_{r = 0}^{\infty} (\quad) \, r \, dr \, d\theta$$

The result is:

$$2\pi \int_0^\infty \frac{\partial \overline{u}^2}{\partial x} r\, dr + \underbrace{2\pi (r\, \overline{u}\, \overline{v})_\infty}_{\substack{\text{zero} \\ (\overline{u}=0)}} - \underbrace{2\pi (r\, \overline{u}\, \overline{v})_0}_{\substack{\text{zero} \\ (r,\, \overline{v}=0)}} =$$

$$= \underbrace{2\pi \left[r (\nu + \varepsilon_M) \frac{\partial \overline{u}}{\partial r} \right]_\infty}_{\substack{\text{zero} \\ \left(\frac{\partial \overline{u}}{\partial r} = 0\right)}} - \underbrace{2\pi \left[r (\nu + \varepsilon_M) \frac{\partial \overline{u}}{\partial r} \right]_0}_{\substack{\text{zero} \\ \left(r,\, \frac{\partial \overline{u}}{\partial r} = 0\right)}}$$

In conclusion, we are left with

$$2\pi \int_0^\infty \frac{\partial \overline{u}^2}{\partial x} r\, dr = 0$$

which is analogous to

$$\frac{d}{dx} \int_0^\infty \overline{u}^2\, r\, dr = 0$$

or

$$\int_0^\infty \overline{u}^2\, r\, dr = \text{constant} \tag{5.150}$$

This theorem applies also to a laminar round jet: the only changes that would take place in the preceding analysis would be $\overline{u} \to u$, and $\varepsilon_M \to 0$, so that instead of eq. (5.150) we obtain

$$\int_0^\infty u^2\, r\, dr = \text{constant}$$

Problem 5.45. Multiplying the mass conservation equation by $\overline{T}$, and adding it to the energy eq. (E) we obtain

$$\frac{\partial}{\partial x}\left(\overline{u}\,\overline{T}\right)+\frac{1}{r}\frac{\partial}{\partial r}\left(r\,\overline{v}\,\overline{T}\right)=\frac{1}{r}\frac{\partial}{\partial r}\left[r\,(\alpha+\varepsilon_H)\frac{\partial\overline{T}}{\partial r}\right]$$

Next, we integrate each term in a constant-x plane,

$$2\pi\int_0^\infty \frac{\partial}{\partial x}\left(\overline{u}\,\overline{T}\right) r\,dr+\underbrace{2\pi\left(r\,\overline{v}\,\overline{T}\right)_\infty}_{2\pi\,T_\infty(r\,\overline{v})_\infty}-\underbrace{2\pi\left(r\,\overline{v}\,\overline{T}\right)_0}_{\substack{\text{zero}\\(r=0)}}=$$

$$=\underbrace{2\pi\left[r\,(\alpha+\varepsilon_H)\frac{\partial\overline{T}}{\partial r}\right]_\infty}_{\substack{\text{zero}\\ \left(\frac{\partial\overline{T}}{\partial r}=0\right)}}-\underbrace{2\pi\left[r\,(\alpha+\varepsilon_H)\frac{\partial\overline{T}}{\partial r}\right]_0}_{\substack{\text{zero}\\ \left(r,\,\frac{\partial\overline{T}}{\partial r}\right)=0}}$$

In conclusion, only two terms survive:

$$\int_0^\infty \frac{\partial}{\partial x}\left(\overline{u}\,\overline{T}\right) r\,dr+T_\infty(r\,\overline{v})_\infty=0 \qquad (1)$$

The unknown $(r\,\overline{v})_\infty$ can be determined by integrating the mass conservation eq. (m) in the same constant-x plane,

$$2\pi\int_0^\infty \frac{\partial\overline{u}}{\partial x}\,r\,dr+2\pi\,(r\,\overline{v})_\infty-\underbrace{2\pi\,(r\,\overline{v})_0}_{\text{zero}}=0$$

therefore

$$(r\,\overline{v})_\infty=-\int_0^\infty \frac{\partial\overline{u}}{\partial x}\,r\,dr \qquad (2)$$

Combining eqs. (1) and (2), we obtain

$$\int_0^\infty \frac{\partial}{\partial x}\left[\bar{u}\left(\bar{T} - T_\infty\right)\right] r\,dr = 0$$

in other words,

$$\frac{d}{dx}\int_0^\infty \bar{u}\left(\bar{T} - T_\infty\right) r\,dr = 0$$

and, finally,

$$\int_0^\infty \bar{u}\left(\bar{T} - T_\infty\right) r\,dr = \text{constant} \tag{5.152}$$

The value of the constant on the right side can be identified by evaluating the integral in the plane of the nozzle. Note that eq. (5.152) holds for a laminar round jet as well, where it would be written as

$$\int_0^\infty u\left(T - T_\infty\right) r\,dr = \text{constant}$$

Project 5.1. a) Assume that T_0 (unknown) is the uniform temperature of the interface between the two bodies in rolling contact. Consider first the temperature distribution in body no. 1. It is a "thermal" boundary layer through which body no. 1 "flows" with the uniform speed U:

$$U \frac{\partial T}{\partial x} = \alpha_1 \frac{\partial^2 T}{\partial y^2} \tag{1}$$

$$T = T_0 \quad \text{at} \quad y = 0 \tag{2}$$

$$T \to T_1 \quad \text{as} \quad y \to \infty \tag{3}$$

The problem (1)-(3) is identical to that of the temperature distribution in a thermal boundary layer in the limit $Pr \to 0$. As shown in the top half of Fig. 5.7., when Pr is very small the velocity thickness δ is much smaller than δ_T. This means that in the $Pr \to 0$ limit the fluid flows with a uniform velocity (U_∞ in Fig. 5.7) through the thermal boundary layer region. The uniform longitudinal velocity (U) is a noteworthy feature of eq. (1), therefore, for the local heat flux we can simply use eq. (5.79a):

$$\frac{q''_x}{T_1 - T_0} \frac{x}{k_1} = 0.564 \left(\frac{Ux}{\alpha_1}\right)^{1/2} \tag{4}$$

A similar analysis of the thermal boundary layer that develops inside body no. 2 leads to an alternative expression for the same local heat flux q''_x:

$$\frac{q''_x}{T_0 - T_2} \frac{x}{k_2} = 0.564 \left(\frac{Ux}{\alpha_2}\right)^{1/2} \tag{5}$$

b) By dividing eqs. (4) and (5) we obtain an expression for the interface temperature

$$T_0 = \frac{r}{1+r} T_1 + \frac{1}{1+r} T_2 \tag{6}$$

in which r is the dimensionless group

$$r = \frac{(\rho c k)_1^{1/2}}{(\rho c k)_2^{1/2}} \tag{7}$$

c) The local heat flux can now be calculated by eliminating T_0 between eqs. (4) and (6), or between eqs. (5) and (6):

$$q_x'' = \frac{0.564}{1 + r} k_1 (T_1 - T_2) \left(\frac{U}{\alpha_1}\right)^{1/2} x^{-1/2} \tag{8}$$

The L-averaged heat flux that corresponds to this result is

$$\bar{q}'' = \frac{1}{L} \int_0^L q_x'' \, dx$$

$$= \frac{1.128}{1 + r} k_1 (T_1 - T_2) \left(\frac{U}{\alpha_1 L}\right)^{1/2} \tag{9}$$

d) The preceding results apply as long as the two boundary layer regions are sufficiently slender. According to eq. (5.69), in the small-Pr limit the slenderness ratio (δ_T/L) is of order $Pe_L^{-1/2}$. There are two slenderness criteria to meet, because there are two boundary layers:

$$\left(\frac{UL}{\alpha_1}\right)^{-1/2} << 1 \tag{10}$$

$$\left(\frac{UL}{\alpha_2}\right)^{-1/2} << 1 \tag{11}$$

These conditions can be rewritten as

$$U >> \frac{\alpha_1}{L} \tag{12}$$

$$U >> \frac{\alpha_2}{L} \tag{13}$$

therefore, the peripheral velocity must be higher than α_m/L, where α_m is the greater of the two thermal diffusivities (α_1, α_2).

Project 5.2. The objective of this problem is to show that if your job is to maximize a "complicated" function (q_B), then a good idea is to perform the analysis in dimensionless form. You have to maximize

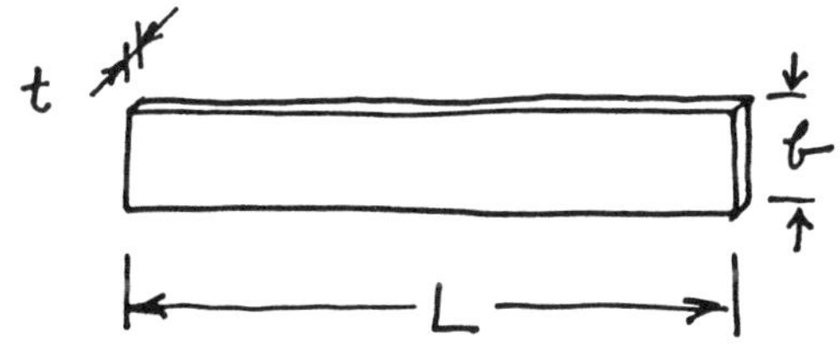

$$q_B = \left(T_B - T_\infty\right)(hpkA)^{1/2} \tanh\left[L\left(\frac{hp}{kA}\right)^{1/2}\right]$$

subject to the following

$V = bLt$, constant ; $p = 2b$

$t =$ constant , $A = bt$

$$h = \frac{k_f}{b} 0.664 \, Pr^{1/3} \left(\frac{U_\infty b}{\nu}\right)^{1/2}$$

There is only one degree of freedom in this design problem, b or L; choosing b, the objective function q_B can then be arranged in dimensionless form as

$$\frac{q_B}{kt\left(T_B - T_\infty\right)} = \left[\underbrace{1.328 \, Pr^{1/3} \frac{k_f}{k} Re_t^{1/2}}_{C} \left(\frac{b}{t}\right)^{3/2}\right]^{1/2} \tanh\left\{\frac{V}{t^3}\left[C\left(\frac{b}{t}\right)^{-5/4}\right]^{1/2}\right\}$$

where $Re_t = U_\infty t/\nu$ is a known constant. Setting

$$m^{5/4} = \frac{V}{t^3} C^{1/2}, \text{ constant,}$$

we rewrite the objective function as

$$\frac{q_B}{kt\left(T_B - T_\infty\right)} = \underbrace{C^{1/2} m^{3/4}}_{\text{constant}} \left(\frac{b}{mt}\right)^{3/4} \tanh\left[\left(\frac{b}{mt}\right)^{-5/4}\right]$$

Thus, b influences q_B in the same way that ξ influences $\xi^{3/4} \tanh(\xi^{-5/4})$, where $\xi = \frac{b}{mt}$. As shown in the graph, this function has a maximum at

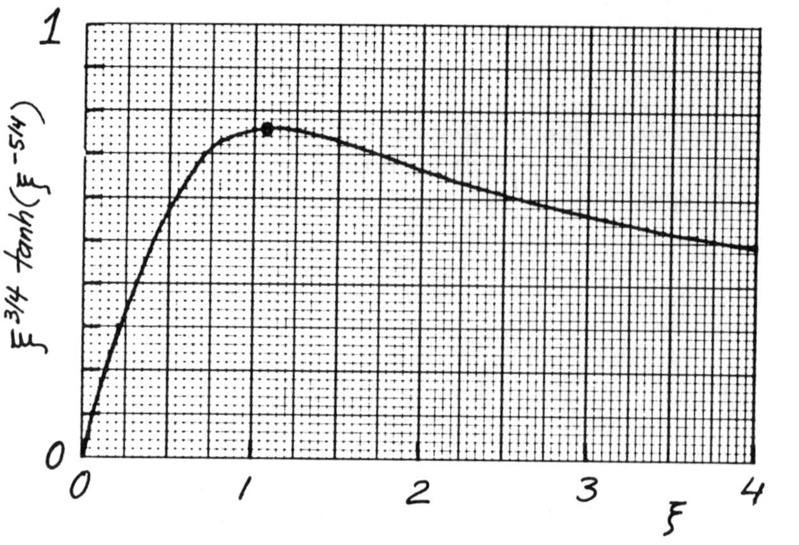

$\xi_{opt} = 1.071,$

which means that

$$\frac{b_{opt}}{t} = 1.071\ m$$

or, recalling the definitions of m and C,

$$\frac{b_{opt}}{t} = 1.071 \left(\frac{V}{t^3}\right)^{4/5} \left(1.328\ Pr^{1/3}\ \frac{k_f}{k}\ Re_t^{1/2}\right)^{2/5} .$$

Project 5.3. Correlations of type (5.135, cylinder), and (5.139, sphere) show that the overall Nusselt number generally increases with the Reynolds number,

$$\frac{h\ D}{k_{fluid}} \sim \left(\frac{U_\infty D}{\nu}\right)^n \tag{1}$$

In a sufficiently narrow Re_D range, the exponent n is a constant less than 1 (more in the vicinity of 0.5). We conclude that h varies as D^{n-1}, or, since $D = 2r_o$,

$$\frac{h}{h_*} = \left(\frac{r_o}{r_i}\right)^{n-1} \tag{2}$$

where h_* is the "reference" heat transfer coefficient evaluated with the appropriate correlation by setting $r_o = r_i$. The same correlation can be used [by comparison with eq. (1)] to evaluate the proper value of n.

i) The overall thermal resistance of the cylindrical shell with r_o-dependent h on the fluid side is (L = cylinder length, k = shell conductivity),

$$R_t = \frac{\ln \frac{r_o}{r_i}}{2\pi k\ L} + \frac{1}{2\pi\ r_o L\ h_*\ (r_o/r_i)^{n-1}} \tag{3}$$

It is more convenient to express R_t in terms of the "dimensionless" outer radius

$$\rho = \frac{r_o}{r_i} \tag{4}$$

i.e. to write

$$R_t = \frac{\ln \rho}{2\pi k\, L} + \frac{1}{2\pi\, r_i L\, h_*\, \rho^n} \tag{5}$$

and to solve $\partial R_t/\partial \rho = 0$ for the critical ratio $\rho_c = r_{o,c}/r_i$:

$$\frac{1}{2\pi k\, L}\,\frac{1}{\rho_c} + \frac{1}{2\pi\, r_i L\, h_*}\,(-\,n)\,\rho_c^{-n-1} = 0 \tag{6}$$

The answer is

$$\rho_c = \frac{r_{o,c}}{r_i} = \left(\frac{n\,k}{r_i\, h_*}\right)^{1/n} \tag{7}$$

As a check, we consider the limiting case of constant h (i.e. n = 1), and note that eq. (7) leads immediately to the classical result (2.44) in the text.

ii) Numerical example:

$T_i = 100°C$	Air properties at
$T_\infty = 20°C$, air	$(100°C + 20°C)/2 = 60°C$
$r_i = 4$ cm	Pr = 0.7
$U_\infty = 2\,\frac{m}{s}$	$\nu = 0.188\,\frac{cm^2}{s}$
$k = 0.16\,\frac{W}{m \cdot K}$, polystyrene	$k_a = 0.028\,\frac{W}{m \cdot K}$, air

We calculate the heat transfer coefficient h_* based on $D = 2\, r_i$,

$$Re_D = \frac{U_\infty D}{\nu} = \frac{2\,\frac{m}{s}\,0.08m}{0.118 \times 10^{-4}\,\frac{m^2}{s}} = 8511$$

$Nu_D = 48.81$, using eq. (5.135) with $Re_D = 8511$ and Pr = 0.7

$$\frac{h_*\, D}{k_a} = 48.81$$

$$h_* = 48.81 \frac{0.028 \frac{W}{m \cdot K}}{0.08 m} = 17.1 \frac{W}{m^2 K}$$

The classical (constant-h) critical radius is

$$r_{o,c,classical} = \frac{k}{h_*} = \frac{0.16 \frac{W}{m \cdot K}}{17.1 \frac{W}{m^2 K}} = 0.94 \text{ cm}$$

The exponent n of eq. (1) can be evaluated by replacing (approximating) eq. (5.135) with eq. (1), i.e. with

$$Nu_D \sim Re_D^n \tag{8}$$

The correct correlation, eq. (5.135), furnished one (Nu_D, Re_D) point already,

$$Nu_D = 48.81 \qquad \text{for} \qquad Re_D = 8511 \tag{9}$$

We generate a second point by using, say, $Re_D = 20000$ (i.e. a Re_D value close to 8511) in eq. (5.135), and obtain

$$Nu_D = 78.87 \qquad \text{for} \qquad Re_D = 20\,000 \tag{10}$$

By substituting (9) and (10) in proportionalities of type (8), and dividing the resulting proportionalities side by side we obtain the equation

$$\frac{78.87}{48.81} = \left(\frac{20\,000}{8511}\right)^n$$

which yields n = 0.562. We can now substitute the needed numerical values in eq. (7), to calculate the "exact" critical radius

$$\frac{r_{o,c}}{r_i} = \left(\frac{0.562 \times 0.16 \frac{W}{m \cdot K}}{0.04 m \; 17.1 \frac{W}{m \cdot K}} \right)^{1/0.562} = 0.027$$

$$r_{o,c} = 0.027 \times 0.04 m = 0.11 \text{ cm}$$

In conclusion, the classical critical radius is roughly nine times greater than the correct critical radius.

iii) The total thermal resistance of a spherical shell with r_o-dependent h on the fluid side is, cf. the method of part (i),

$$R_t = \frac{1}{4\pi k\, r_i}\left(1 - \frac{1}{\rho}\right) + \frac{1}{4\pi\, r_i^2\, h_*\, \rho^{n+1}}$$

$$\frac{\partial R_t}{\partial \rho} = \frac{1}{4\pi k\, r_i}\,\frac{1}{\rho_c^2} - \frac{(n+1)\,\rho_c^{-n-2}}{4\pi\, r_i^2\, h_*} = 0$$

$$\rho_c = \frac{r_{o,c}}{r_i} = \left[\frac{(n+1)\,k}{r_i\, h_*}\right]^{1/n} \qquad (11)$$

Equation (11) reduces to the classical result (2.47) when h is constant, i.e. when n = 1.

Chapter 6

INTERNAL FORCED CONVECTION

Problem 6.1. The second of the equations (6.14) can be written as

$$\frac{1}{r}\frac{d}{dr}\left(r\frac{du}{dr}\right) = c_1 \qquad (1)$$

which, integrated once, yields

$$r\frac{du}{dr} = c_1\frac{r^2}{2} + c_2 \qquad (2)$$

Dividing by r, we obtain

$$\frac{du}{dr} = c_1\frac{r}{2} + \frac{c_2}{r} \qquad (3)$$

and, in view of eq. (6.16b), $c_2 = 0$. Integrating eq. (3) one more time,

$$u = c_1\frac{r^2}{4} + c_3 \qquad (4)$$

and invoking the boundary condition (6.16a), we conclude that

$$u = a\,(r^2 - r_0^2) \qquad (5)$$

The constant a is related to the average velocity U. Substituting eq. (5) into eq. (6.1), we obtain

$$U = \frac{1}{\pi r_0^2}\int_0^{2\pi} d\theta \int_0^{r_0} u\, r\, dr$$

$$= a\, r_0^2 \int_0^1 (m - 1)\, dm = -\frac{a\, r_0^2}{2} \qquad (6)$$

Combined, eqs. (5) and (6) state that

$$u = 2U\left[1 - \left(\frac{r}{r_0}\right)^2\right] \qquad (7)$$

The relationship between U and longitudinal pressure gradient is obtained by substituting eq. (7) into the first of eqs. (6.14) in the text:

$$\frac{dP}{dx} = \mu\left[-\frac{4U}{r_0^2} + \frac{1}{r}\left(-\frac{4r\,U}{r_0^2}\right)\right] \qquad (8)$$

$$= -\mu\frac{8U}{r_0^2}$$

Rearranged, this conclusion becomes the same as eq. (6.18) in the text.

Problem 6.2. Integrated twice, the second of eqs. (6.20) yields

$$u = c_1 y^2 + c_2 y + c_3 \qquad (1)$$

The mid-plane symmetry condition

$$\frac{du}{dy} = 0 \quad \text{at} \quad y = 0 \qquad (2)$$

requires that $c_2 = 0$. The no-slip condition at the upper wall,

$$u = 0 \quad \text{at} \quad y = \frac{D}{2} \qquad (3)$$

requires further that $c_3 = -c_1(D/2)^2$. In the end, the u(y) solution reduces to

$$u = c_1\left[y^2 - \left(\frac{D}{2}\right)^2\right] \qquad (4)$$

Next, we note that the average velocity definition (6.1) relates c_1 to U:

$$U = \frac{1}{D}\int_{-D/2}^{D/2} u\,dy = \frac{1}{D}2\int_0^{D/2} u\,dy$$

$$= c_1\left(\frac{D}{2}\right)^2\int_0^1 (m^2 - 1)\,dm = -\frac{c_1 D^2}{6} \qquad (5)$$

Eliminating c_1 between eqs. (4) and (5) we obtain

$$u = \frac{3}{2}U\left[1 - \left(\frac{y}{D/2}\right)^2\right] \qquad (6)$$

The proportionality between U and pressure gradient is proven by substituting the solution (6) into the first of eqs. (6.20),

$$\frac{dP}{dx} = \mu\, 2c_1 = 2\mu\left(-\frac{6U}{D^2}\right) \tag{7}$$

This is the same as writing

$$U = \frac{D^2}{12\,\mu}\left(-\frac{dP}{dx}\right) \tag{6.22}$$

Problem 6.3. From a set of mathematical tables we learn that the hexagonal cross-section is characterized by

$$A = 2.598\, a^2 \qquad\qquad p = 6a$$

where a is the side of the regular hexagon. Now we can calculate in order

$$D_h = \frac{4 \times 2.598\, a^2}{6a} = 1.732\, a$$

$$B = \frac{\pi\, D_h^2 / 4}{A} = 0.907$$

$$C = 16 \exp\,(0.294\, B^2 + 0.068B - 0.318) \tag{6.30}$$

$$= 15.77$$

This C value is 4.7 percent higher than the 15.065 value recommended by Table 6.2.

Problem 6.4. We evaluate the water properties at 5°C, and begin with calculating the Reynolds number:

$$D_h = 2D = 4 \text{ cm}$$

$$Re_{D_h} = \frac{U\, D_h}{\nu}$$

$$= 3.2 \frac{\text{cm}}{\text{s}}\, 4 \text{ cm} \frac{\text{s}}{0.01514 \text{ cm}^2} = 845 \quad \text{(laminar)}$$

$$f = \frac{24}{Re_{D_h}} = \frac{24}{845} = 0.0284$$

$$\frac{\Delta P}{L} = f \frac{4}{D_h} \frac{1}{2} \rho U^2$$

$$= 0.0284 \frac{4}{4\ \text{cm}} \frac{1}{2} 1 \frac{\text{g}}{\text{cm}^3} \left(3.2 \frac{\text{cm}}{\text{s}}\right)^2$$

$$= 0.145 \frac{\text{g}}{\text{cm}^2 \text{s}^2} = 1.45 \frac{\text{N/m}^2}{\text{m}}$$

$$X \cong 0.05\ D_h\ Re_{D_h} = 0.05 \times 4\ \text{cm}\ 845 = 169\ \text{cm}$$

The length/spacing ratio of the flow region photographed in Fig. 6.4 is 9.3/1. The length of that region is therefore 9.3D = 18.6 cm, and represents roughly 10 percent of the calculated entrance length X.

Problem 6.5. It is advisable to evaluate the fluid (water) properties at the average mean temperature

$$T_m = \frac{1}{2}(T_{in} + T_{out})$$

in which T_{in} = 30°C, and T_{out} is unknown. The outlet temperature will be between 30°C and 10°C (the wall), therefore, as a start, we write

$$T_m \cong \frac{1}{2}(30 + 10)°\text{C} = 20°\text{C}$$

The needed properties are then evaluated at 20°C:

$$k = 0.59 \frac{\text{W}}{\text{m}\cdot\text{K}}, \qquad \rho \cong 1 \frac{\text{g}}{\text{cm}^3}, \qquad c_P = 4.18 \frac{\text{J}}{\text{gK}}$$

The flow is laminar (see Example 6.1). The outlet temperature is provided by eq. (6.55), for which we calculate in order

$$Nu_D = \frac{hD}{k} = 3.66$$

$$h = 3.66 \times 0.59 \frac{\text{W}}{\text{m}\cdot\text{K}} \frac{1}{2.7\ \text{cm}} = 80 \frac{\text{W}}{\text{m}^2\text{K}}$$

$$\frac{T_w - T_{out}}{T_w - T_{in}} = \exp\left(- \frac{2hL}{r_o \rho c_P U}\right) \tag{6.55}$$

$$= \exp\left(- \frac{2 \times 80 \frac{W}{m \cdot K} 10m}{1.35\ cm\ 1 \frac{g}{cm^3} 4.18 \frac{J}{gK} 6 \frac{cm}{s}}\right)$$

$$= \exp(- 0.473) = 0.623$$

$$T_{out} = T_w - 0.623 (T_w - T_{in})$$

$$= 10°C - 0.623 (10 - 30)°C = 22.5°C$$

It is useful to see what would have been the proper temperature for evaluating the water properties,

$$T_m = \frac{1}{2}(T_{in} + T_{out}) = \frac{30 + 22.5}{2}°C = 26.2°C$$

This mean temperature is only 6.2°C higher than the one assumed at the start. Fortunately, we do not have to repeat the calculations (for T_m = 26.2°C) because in the range 20°C - 26.2°C the needed properties (k, ρ, c_P) are practically constant.

<u>Problem 6.6</u>. The relevant properties of water at 50°C are

$$k = 0.64 \frac{W}{m \cdot K} \qquad \rho = 0.988 \frac{g}{cm^3}$$

$$\nu = 0.00554 \frac{cm^2}{s} \qquad c_P = 4.18 \frac{kJ}{kg \cdot K}$$

We begin with the calculation of the Reynolds number, in order to be sure that the flow is laminar:

$$D_h = 2D = 2\ cm$$

$$Re_{D_h} = \frac{U D_h}{\nu} = 3.2 \frac{cm}{s} 2\ cm \frac{s}{0.00554\ cm^2}$$

$$= 1155 \qquad \text{(laminar)}$$

$$Nu = \frac{h D_h}{k} = 4.364 \qquad \text{(fully developed)}$$

$$h = Nu \frac{k}{D_h} = 4.364 \frac{0.64 \frac{W}{m \cdot K}}{0.02 m} \cong 140 \frac{W}{m^2 K}$$

$$h = \frac{q''_w}{T_w - T_m}$$

$$q''_w = \frac{1}{2} q''_{one\ blade} = 800 \frac{W}{m^2}$$

$$T_w - T_m = \frac{q''_w}{h} = 800 \frac{W}{m^2} \frac{m^2 K}{140\ W} = 5.73°C$$

$$\frac{d T_m}{dx} = \frac{p}{A} \frac{q''_w}{\rho c_p U} = \frac{2W}{WD} \frac{q''_w}{\rho c_p U}$$

$$= \frac{2}{0.01 m} \frac{800 \frac{W}{m^2}}{0.988 \frac{g}{cm^3} 4.18 \frac{J}{g \cdot K} 3.2 \frac{cm}{s}}$$

$$= 1.21 \frac{°C}{m}$$

For the thermal entrance length, we use eq. (6.32):

$$X_T \cong 0.05\ Re_{D_h}\ Pr\ D_h$$

$$\cong 0.05 \times 1155 \times 3.57 \times 2\ cm = 4.1 m$$

In conclusion, the length of the parallel-plate channel must be considerably larger than 4m if the above calculations are to be valid.

Problem 6.7. a) The hydrostatic pressure distributions $P_c(y)$ and $P_h(y)$ must cross at $y = H/2$ so that the height-averaged pressure is the same on both sides of the door,

$$P_h\left(\frac{H}{2}\right) = P_c\left(\frac{H}{2}\right)$$

The pressure difference that drives the air leak through the bottom gap is

$$\Delta P = P_c(0) - P_h(0) = \rho_c g \frac{H}{2} - \rho_h g \frac{H}{2}$$

$$= \Delta\rho \cdot g \cdot \frac{H}{2}, \quad \text{where } \Delta\rho = \rho_c - \rho_h \qquad (1)$$

The gap is a parallel-plate channel (D thin, W wide, L long), therefore

$$U = \frac{D^2}{12\mu}\frac{\Delta P}{L}$$

or

$$\Delta P = U \frac{12\ \mu L}{D^2}\frac{\rho\ DW}{\rho\ DW} = \dot{m}\frac{12\ \nu L}{D^3 W} \qquad (2)$$

Combining eqs. (1) and (2) we find that

$$\dot{m} = \frac{\Delta\rho\ g\ D^3\ WH}{24\ \nu L} \qquad (3)$$

The warm chamber (T_h) loses energy by convection, because of the $\dot{m}$ counterflow, warm over the top of the door, and cold under the bottom:

$$q = \dot{m}\ i_{top} - \dot{m}\ i_{bottom} \qquad (i = \text{specific enthalpy})$$

$$= \dot{m}\ c_P (T_h - T_c) \qquad (4)$$

b) For the numerical part of the problem we have

$$T_c = 10°C \qquad T_h = 30°C$$

$$\rho_c = 1.247 \frac{kg}{m^3} \qquad \rho_h = 1.165 \frac{kg}{m^3}$$

$$H = 2.2m, \quad D = 0.5mm, \quad W = 1.5m, \quad L = 5\ cm$$

We evaluate the other air flow properties (ν, c_P, ρ) at the "representative" temperature of (10°C + 30°C)/2 = 20°C, so that we perform the flow calculations only once (instead of doing them for each air gap separately). We obtain in order

$$\Delta\rho = \rho_c - \rho_h = 0.082 \frac{kg}{m^3}$$

$$\Delta P = \Delta\rho \cdot g \cdot \frac{H}{2} = 0.082 \frac{kg}{m^3} 9.81 \frac{m}{s^2} \frac{2.2m}{2}$$

$$= 0.885 \frac{kg}{s^2 m} = 0.885 \frac{N}{m^2}$$

$$\dot{m} = \Delta P \frac{D^3 W}{12 \nu L}$$

$$= 0.885 \frac{kg}{s^2 m} \frac{(0.0005m)^3 \, 1.5m}{12 \times 0.15 \frac{(0.01m)^2}{s} 0.05m}$$

$$= 1.843 \times 10^{-5} \frac{kg}{s}$$

$$q = \dot{m} \, c_P (T_h - T_c) = 1.843 \times 10^{-5} \frac{kg}{s} 1.006 \frac{kJ}{kg\,K} 20°C$$

$$= 0.37W$$

It can be verified that the air flow in the gap is laminar ($Re_{D_h} \cong 1.4$), and that it is also fully developed over most of the gap length L. The important conclusion made visible by eqs. (3) and (4) is that the air leak m and the heat leak q are proportional to the gap spacing cubed (D^3). For example, if D is 2mm instead of 0.5mm, the heat leak q jumps to 24 watts.

Problem 6.8. For lack of a more accurate picture of the temperature distribution in the fluid, we evaluate the water properties at the average bulk (mean) temperature $(T_{in} + T_{out})/2 = (10°C + 20°C)/2 = 15°C$. The properties that will be used are:

$$\nu = 0.01138 \frac{cm^2}{s} \qquad \rho \cong 1 \frac{g}{cm^3}$$

$$Pr = 8.13 \qquad c_P = 4.186 \frac{J}{g{\cdot}K}$$

$$k = 0.59 \frac{W}{m{\cdot}K}$$

In order to find out if $x = 2.86m$ belongs to the thermally fully developed region, we calculate the thermal entrance length:

$$Re_D = \frac{UD}{\nu} = 10 \frac{cm}{s} 2\ cm \frac{s}{0.01138\ cm^2}$$

$$= 1757.5 \quad \text{(laminar)}$$

$$X_T \cong 0.05\ D\ Re_D\ Pr$$

$$\cong 0.05 \times 0.02m\ 1757.5 \times 8.13 = 14.3m$$

We find that $x < X_T$, however x is not "much smaller" than X_T. This means that $x = 2.86$ m belongs to the thermal entrance region, not too far from the start of the fully developed region.

Nu_D can be calculated in four different ways:

1. Graphically, by reading Fig. 6.8:

$$\frac{x/D}{Re_D\ Pr} = \frac{2.86m/0.02m}{1757.5 \times 8.13} = 0.01 \quad \text{(abscissa)}$$

$$Nu_D \cong 6.1 \quad \text{(ordinate)}$$

2. By assuming that the temperature profile is fully developed:

$$Nu_D = 4.364$$

3. By assuming that the thermal boundary layer (which begins at $x = 0$) extends all the way to $x = 2.86$m:

$$Nu_x = 0.332\, Pr^{1/3}\, Re_x^{1/2}$$

$$\frac{h_x x}{k}\frac{D}{D} = 0.332\, Pr^{1/3}\, Re_D^{1/2}\left(\frac{x}{D}\right)^{1/2}$$

$$Nu_D = 0.332\, Pr^{1/3}\, Re_D^{1/2}\left(\frac{D}{x}\right)^{1/2}$$

$$= 0.332\,(8.13)^{1/3}(1757.5)^{1/2}\left(\frac{0.02\,m}{2.86\,m}\right)^{1/2}$$

$$= 2.34$$

4. Based on Churchill and Ozoe's correlation:

$$Gz = \frac{\pi}{4}\left(\frac{x/D}{Re_D\, Pr}\right)^{-1} = \frac{\pi/4}{0.01} = 78.54$$

$$\frac{Nu_D}{4.364\left[1 + \left(\frac{78.54}{29.6}\right)^2\right]^{1/6}} = 1.047 \qquad (6.52)$$

$$Nu_D = 6.47$$

By comparing these four estimates, we conclude that the graphic method is adequate, and that the extreme assumptions (methods 2 and 3) do not work very well. Method 2 works better downstream from $x = 2.86$m, while method 3 is most adequate for x's much smaller than 2.86m.

The heat flux and local wall temperature are

$$q''_w = \frac{r_o}{2}\rho\, c_P\, U \frac{d\,T_m}{dx} \qquad (6.39')$$

$$= \frac{1\text{ cm}}{2}\, 1\, \frac{\text{g}}{\text{cm}^3}\, 4.186\, \frac{\text{J}}{\text{g}\cdot\text{K}}\, 10\, \frac{\text{cm}}{\text{s}}\, \frac{(20 - 10)\text{ K}}{286\text{ cm}}$$

$$= 7396\, \frac{\text{W}}{\text{m}^2}$$

$$T_w(x) - T_m(x) = \frac{q''_w}{Nu_D}\frac{D}{k}$$

$$= \frac{1}{6.47} 7396 \frac{W}{m^2} \frac{0.02m}{0.59 \frac{W}{m \cdot K}} = 38.7°C$$

$$T_w(x) = T_m(x) + 38.7°C = 20°C + 38.7°C$$

$$= 58.7°C$$

The wall temperature increases from 10°C at the mouth of the pipe, to 58.7°C at $x = 2.86$m. In the entrance region, the wall temperature (T_w - 10°C) increases (roughly) as $x^{1/2}$. This means that the wall temperature averaged from 0 to x is

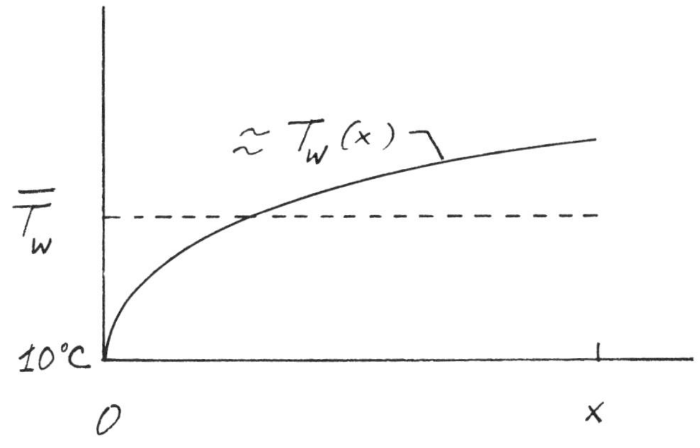

$$\overline{T}_w - 10°C = \frac{2}{3}[T_w(x) - 10°C]$$

$$= \frac{2}{3}(58.7 - 10)°C$$

in other words $\overline{T}_w \cong 42.5°C$.

Problem 6.9. It is convenient to introduce the dimensionless radial position

$$\gamma = \frac{r}{r_o}$$

as the independent variable on the right side of eq. (6.48). In view of the Hagen-Poiseuille profile (6.19), the energy equation becomes

$$\frac{4 q''_w r_o}{k \Delta T}(1 - \gamma^2) = -\frac{1}{\gamma}\frac{d}{d\gamma}(\gamma \phi')$$

where $\phi' = d\phi/d\gamma$. The first integration yields

$$\frac{4 q''_w r_o}{k \Delta T}\left(\frac{\gamma^2}{2} - \frac{\gamma^4}{4} + c_1\right) = -\gamma \phi'$$

which is the same as

$$\frac{4\,q_w'' r_o}{k\,\Delta T}\left(\frac{\gamma}{2} - \frac{\gamma^3}{4} + \frac{c_1}{\gamma}\right) = -\phi'$$

The boundary condition $\phi'(0) = 0$ requires that $c_1 = 0$. The second integration yields

$$\frac{4\,q_w'' r_o}{k\,\Delta T}\left(\frac{\gamma^2}{4} - \frac{\gamma^4}{16} + c_2\right) = -\phi$$

and, in view of the boundary condition $\phi(1) = 0$,

$$c_2 = -\frac{3}{16}$$

In conclusion, the radial dimensionless temperature distribution is

$$\phi = -\frac{q_w'' r_o}{k\,\Delta T}\left(\gamma^2 - \frac{\gamma^4}{4} - \frac{3}{4}\right)$$

The mean-temperature integral (6.50) is evaluated by writing, in order,

$$\frac{1}{U\pi r_o^2}\int_0^1 2U\,(1-\gamma^2)\,\frac{q_w'' r_o}{k\,\Delta T}\left(\frac{3}{4} - \gamma^2 + \frac{\gamma^4}{4}\right) 2\pi r_o^2\,\gamma\, d\gamma = 1$$

$$4\,\frac{q_w'' r_o}{k\,\Delta T}\int_0^1 (1-\gamma^2)\left(\frac{3}{4} - \gamma^2 + \frac{\gamma^4}{4}\right)\gamma\, d\gamma = 1$$

in which the integral turns out to be equal to 11/96. In conclusion, the Nusselt number is

$$Nu_D = \frac{q_w'' D}{k\,\Delta T} = 2\left(\frac{q_w'' r_o}{k\,\Delta T}\right) = 2\left(\frac{1}{4}\cdot\frac{96}{11}\right) = \frac{48}{11}$$

Problem 6.10. a) The properties of water at nearly atmospheric pressure and 25°C are

$$\nu = 0.00894 \frac{cm^2}{s} \qquad \rho = 0.997 \frac{g}{cm^3} \qquad Pr = 6.21$$

$$\alpha = 0.00144 \frac{cm^2}{s} \qquad c_P = 4.179 \frac{J}{g\,K}$$

The Reynolds number based on tube diameter is

$$Re_D = \frac{UD}{\nu} = 10 \frac{cm}{s}\, 0.5\, cm \frac{s}{0.00894\, cm^2}$$

$$= 559.3 \text{ (laminar flow, because Re < 2000)}$$

The hydrodynamic entrance length is approximately

$$X \cong 0.05\, D\, Re_D = 0.05 \times 0.5\, cm\, 559$$

$$\cong 14\, cm$$

This length is much shorter than the length of the tube (L = 4m), therefore the flow can be modelled as fully developed (Hagen-Poiseuille) over the entire length L.

The pressure drop can be calculated with the friction-factor formula (note: f = 16/Re)

$$\Delta P = f \frac{4L}{D} \frac{1}{2} \rho U^2$$

$$= \frac{16}{559.3} \frac{4 \times 4m}{0.5\, cm} \frac{1}{2} 0.997 \frac{g}{cm^3} (10 \frac{cm}{s})^2$$

$$= 4563 \frac{g}{cm\, s^2} = 456 \frac{N}{m^2} = 0.0045\, atm = 0.066 \frac{lbf}{in^2}$$

The pressure drop is negligible relative to the atmospheric pressure that prevails at the tube outlet. Consequently, the water pressure is nearly atmospheric all along the tube, as was noted during the evaluation of water properties at the start of this solution.

b) The thermal entrance length is approximately

$$X_T \cong 0.05\, D\, Re_D\, Pr \cong 0.05 \times 0.5\, cm\, 559.3 \times 6.21$$

$$\cong 86.8\, cm$$

This length is also considerably smaller than L (although not nearly as small as X), therefore we can model the temperature profile as fully developed all along the tube.

c) The water inlet temperature is $T_1 = 20°C$, and the constant wall temperature is $T_w = 30°C$. The water outlet temperature T_L is given by eq. (6.53)

$$\frac{T_w - T_L}{T_w - T_1} = \exp\left(-\frac{\alpha \, Nu_D \, L}{r_0^2 \, U}\right) \qquad (6.53)$$

$$= \exp\left[-0.00144 \frac{cm^2}{s} \frac{3.66 \times 4m}{(0.25 \text{ cm})^2 \, 10 \text{ cm/s}}\right]$$

$$= \exp(-3.373) = 0.0343$$

therefore

$$T_L = T_w - 0.0343 \, (T_w - T_1) = 29.7°C$$

The total heat transfer rate experienced by the water stream is equal to its enthalpy rise,

$$q = \dot{m} \, c_P \, (T_L - T_1)$$

in which $\dot{m}$ is the mass flowrate,

$$\dot{m} = \rho \frac{\pi D^2}{4} U = 0.997 \frac{g}{cm^3} \frac{\pi}{4} (0.5 \text{ cm})^2 \, 10 \frac{cm}{s}$$

$$= 1.96 \frac{g}{s}$$

In conclusion, the q value is

$$q = 1.96 \frac{g}{s} \, 4.179 \frac{J}{g°C} (29.7 - 20)°C$$

$$= 79.35 \text{ W}$$

Problem 6.11. a) We must solve

$$\frac{k}{r}\frac{d}{dr}\left(r\frac{dT}{dr}\right) + \mu\left(4U\frac{r}{r_0^2}\right)^2 = 0 \tag{1}$$

subject to two boundary conditions,

$$\frac{dT}{dr} = 0 \qquad \text{at} \qquad r = 0 \tag{2}$$

$$T = T_w \qquad \text{at} \qquad r = r_0 \tag{3}$$

The first integration of eq. (1) yields

$$\frac{dT}{dr} = -\frac{\mu}{k}\frac{4U^2}{r_0^4}r^3 + \frac{c_1}{r} \tag{4}$$

and, in view of eq. (2), $c_1 = 0$. The second integration produces

$$T = -\frac{\mu}{k}U^2\left(\frac{r}{r_0}\right)^4 + c_2 \tag{5}$$

and, after using eq. (3) to evaluate c_2,

$$T(r) - T_w = \frac{\mu U^2}{k}\left[1 - \left(\frac{r}{r_0}\right)^4\right] \tag{6}$$

The centerline-wall temperature difference is equal to $\mu U^2/k$. This conclusion makes sense physically, because it is reasonable to expect higher steady temperatures in more viscous fluids (μ), and in fluids that are extruded faster (U^2). The thermal conductivity plays a thermally stabilizing role: higher k values mean better thermal contact with the cooled wall, and, consequently, lower centerline temperatures.

b) The heat flux into the wall (in the positive r direction) is

$$q'' = k\left(-\frac{dT}{dr}\right)_{r=0} = 4\mu\frac{U^2}{r_0}, \qquad \text{(constant)} \tag{7}$$

The total cooling rate over a pipe length L is therefore

$$q = q''\, 2\pi r_0 L = 8\pi L\mu U^2$$

$$= 8\pi L\,\mu\, UU \tag{8}$$

The last U on the right side of eq. (8) can be replaced using eq. (6.18) from the text,

$$q = \underbrace{\pi r_0^2 U}_{\dot{m}/\rho} \underbrace{L\left(-\frac{dP}{dx}\right)}_{\Delta P} \tag{9}$$

therefore q is the same as the pumping power ($\dot{m}\, \Delta P/\rho$) needed in order to push the fluid through the pipe. We can say that the fluid flow converts ("dissipates") the mechanical power $\dot{m}\, \Delta P/\rho$ into the heat transfer rate q that is rejected to the pipe wall.

<u>Problem 6.12</u>. If the flow is undergoing transition in the fully developed region, it is also undergoing transition at the end of the entrance region of length X

$$\frac{X}{D} \cong 0.05\, Re_D = 0.05 \times 2000 = 100$$

The corresponding Reynolds number for the entrance boundary layer (Re_X) can be estimated by noting that

$$\frac{Re_X}{Re_D} = \frac{X}{D} = 100$$

therefore

$$Re_X \cong 100\, Re_D = 100 \times 2000$$

$$\cong 2 \times 10^5$$

This critical Re_X value is of the same order as the one listed in the criterion (5.85) for boundary layer flow.

Problem 6.13. The pumping power is proportional to the product $\dot{m}\,\Delta P$, namely

$$\mathbf{P} = \frac{1}{\rho}\,\dot{m}\,\Delta P \qquad (1)$$

where

$$\Delta P = f\,\frac{4L}{D}\,\frac{1}{2}\,\rho\,U^2 \qquad (2)$$

Since $\dot{m}$, L, D and the fluid do not change as the flow regime switches from laminar to turbulent, the Reynolds number and the mean velocity also do not change,

$$U = \frac{\dot{m}}{\rho\,\frac{\pi}{4}\,D^2} \qquad Re_D = \frac{UD}{\nu}$$

Equations (1) and (2) show that the pumping power changes in the same direction (and to the same degree) as the friction factor:

$$\frac{\mathbf{P}_{turb}}{\mathbf{P}_{lam}} = \frac{f_{turb}}{f_{lam}} \cong \frac{0.079\,Re_D^{-1/4}}{\frac{16}{Re_D}}$$

$$\cong 0.00494\,Re_D^{3/4}$$

At transition $Re_D \sim 2000$, therefore the pumping power experiences a jump of about 50 percent,

$$\frac{\mathbf{P}_{turb}}{\mathbf{P}_{lam}} \cong 1.48$$

Problem 6.14. In the following chain, we see that the temperature difference $(T_w - T_m)$ varies as the inverse of the Nusselt number, because q''_w and the fluid are fixed:

$$\frac{(T_w - T_m)_{turb}}{(T_w - T_m)_{lam}} = \frac{q''_w/h_{turb}}{q''_w/h_{lam}} = \frac{h_{lam}}{h_{turb}}$$

$$= \frac{Nu_{D,lam}}{Nu_{D,turb}} \cong \frac{4.364}{0.023\ Pr^{0.4}\ Re_D^{0.8}}$$

$$\cong 216.4\ Re_D^{-0.8} \qquad \text{(if } Pr = 0.72\text{)}$$

In the vicinity of $Re_D \sim 2500$, the temperature difference drops significantly if the laminar flow breaks down into turbulent flow:

$$\frac{(T_w - T_m)_{turb}}{(T_w - T_m)_{lam}} \cong 0.414 \qquad \text{(if } Pr = 0.72,\ Re_D = 2500\text{)}$$

Problem 6.15. The relevant properties of water at 20°C are

$$k = 0.59\ \frac{W}{m \cdot K} \qquad \rho = 0.998\ \frac{cm^2}{s}$$

$$Pr = 7.07 \qquad c_P = 4.182\ \frac{kJ}{kg \cdot K}$$

$$\nu = 0.01004\ \frac{cm^2}{s}$$

a) Pressure drop:

$$U = \frac{\dot{m}}{\rho \frac{\pi}{4} D^2} = \frac{0.5\ \frac{kg}{s}}{0.998\ \frac{g}{cm^3}\ \frac{\pi}{4}\ (0.02m)^2}$$

$$= 1.595\ \frac{m}{s}$$

$$Re_D = \frac{UD}{\nu} = 1.595\ \frac{m}{s}\ \frac{0.02m}{0.01004 \times 10^{-4}\ \frac{m^2}{s}}$$

$$= 31\ 773 \qquad \text{(turbulent)}$$

$$f \cong 0.046\, Re_D^{-1/5} = 0.00579 \qquad \text{[eq. (6.77): note } Re_D \text{ range]}$$

$$\Delta P = f \frac{4L}{D} \frac{1}{2} \rho U^2$$

$$= 0.00579 \frac{4 \times 10\text{m}}{0.02\text{m}} \frac{1}{2} 0.998 \frac{\text{g}}{\text{cm}^3} \left(1.595 \frac{\text{m}}{\text{s}}\right)^2$$

$$= 14\,690 \frac{\text{N}}{\text{m}^2} = 0.145 \text{ atm}$$

b) Heat transfer coefficient based on the Colburn analogy:

$$St = \frac{\frac{1}{2} f}{Pr^{2/3}} = \frac{\frac{1}{2} 0.00579}{(7.07)^{2/3}}$$

$$= 7.86 \times 10^{-4}$$

$$St = \frac{h}{\rho c_P U}$$

$$h = St\, \rho\, c_P\, U = 7.86 \times 10^{-4}\, 0.998 \frac{\text{g}}{\text{cm}^3} 4.182 \frac{\text{kJ}}{\text{kg·K}} 1.595 \frac{\text{m}}{\text{s}}$$

$$= 5230 \frac{\text{W}}{\text{m}^2\,\text{K}}$$

c) Heat transfer coefficient based on the Dittus-Boelter correlation (6.91):

$$Nu_D = 0.023\, Pr^{0.4}\, Re_D^{4/5} = 0.023\, (7.07)^{0.4}\, (31\,773)^{0.8}$$

$$= 201$$

$$h = Nu_D \frac{k}{D} = 201 \frac{0.59 \text{ W/m·K}}{0.02\text{m}}$$

$$= 5930 \frac{\text{W}}{\text{m}^2\,\text{K}}$$

The h estimate obtained in part (b) is 12 percent lower than Dittus-Boelter prediction, which is recommended.

d) Temperature difference across the stream

$$T_w - T_m = \frac{q''_w}{h} = \frac{5 \times 10^4 \text{ W/m}^2}{5930 \text{ W/m}^2 \text{ K}}$$

$$= 8.43 \text{ K}$$

e) Temperature increase in the longitudinal direction:

$$\dot{m}\, c_P \, d\, T_m = q''_w \, \pi D \, dx$$

$$\frac{d\, T_m}{dx} = \frac{q''_w \, \pi D}{\dot{m}\, c_P} = \frac{5 \times 10^4 \text{ W}}{\text{m}^2} \; \frac{\pi\; 0.02\text{m}}{0.5 \frac{\text{kg}}{\text{s}}} \; \frac{\text{kg}\cdot\text{K}}{4.182 \times 10^3 \text{ J}}$$

$$= 1.5 \frac{°\text{C}}{\text{m}}$$

$$T_{m,out} - T_{m,in} = \frac{d\, T_m}{dx} L = 1.5 \frac{°\text{C}}{\text{m}}\; 10 \text{ m}$$

$$= 15°\text{C}$$

Problem 6.16. We begin with the properties of air, which can be evaluated at the end-to-end averaged temperature (30°C + 90°C)/2 = 60°C:

$$\nu = 0.188 \frac{\text{cm}^2}{\text{s}} \qquad \rho = 1.06 \frac{\text{kg}}{\text{m}^3}$$

$$\text{Pr} = 0.7 \qquad c_P = 1.008 \frac{\text{kJ}}{\text{kg}\cdot\text{K}}$$

$$k = 0.028 \frac{\text{W}}{\text{m}\cdot\text{K}}$$

The required length of the pipe is given by eq. (6.103), for which we first calculate the heat transfer coefficient:

$$\text{Re}_D = \frac{UD}{\nu} = 5 \frac{\text{m}}{\text{s}}\; 0.04\text{m} \; \frac{\text{s}}{0.188 \times 10^{-4} \text{ m}^2}$$

$$= 1.064 \times 10^4 \quad \text{(turbulent)}$$

$$Nu_D = 0.023\ Pr^{0.4}\ Re_D^{4/5} \qquad \text{[eq. (6.91), with n = 0.4]}$$

$$= 0.023\ (0.7)^{0.4}\ (1.064 \times 10^4)^{0.8} = 33.21$$

$$h = Nu_D \frac{k}{D}$$

$$= 33.21\ \frac{0.028\ W}{m \cdot K}\ \frac{1}{0.04m} = 23.25\ \frac{W}{m^2\ K}$$

In eq. (6.103) we substitute

$$A = \frac{\pi}{4} D^2 \qquad \text{and} \qquad p = \pi D$$

and obtain

$$L = \frac{\rho\ DU\ c_P}{4h} \ln \frac{T_w - T_{in}}{T_w - T_{out}}$$

$$= \frac{1.06\ \frac{kg}{m^3}\ 0.04m\ 5\ \frac{m}{s}\ 1.008 \times 10^3\ \frac{J}{kg \cdot K}}{4 \times 23.25\ \frac{W}{m^2\ K}} \ln \frac{100 - 30}{100 - 90}$$

$$= 4.47m$$

Since the flow is turbulent, the entrance length

$$X\ (\text{or } X_T) \sim 10D = 0.4m$$

is less than one tenth of the pipe length L. In conclusion, the assumption of full development over the entire length L is justified.

Problem 6.17. The relevant properties of water at 20°C are

$$k = 0.59 \frac{W}{m \cdot K} \qquad \nu = 0.01004 \frac{cm^2}{s}$$

$$Pr = 7.07 \qquad \rho = 0.998 \frac{g}{cm^3}$$

We calculate, in order, the heat transfer coefficient, the Nusselt number, the Reynolds number, and finally the mass flowrate:

$$h = \frac{q''_w}{\Delta T} = \frac{10^4\,W}{m^2} \frac{1}{4\,K} = 2500 \frac{W}{m^2\,K}$$

$$Nu_D = \frac{hD}{k} = 2500 \frac{W}{m^2\,K} 0.025\,m \frac{m \cdot K}{0.59\,W}$$

$$= 105.93$$

$$Nu_D = 0.023\, Pr^{0.4}\, Re_D^{4/5} \quad \text{[eq. (6.91), with n = 0.4 for "heating"]}$$

$$Re_D^{4/5} = \frac{Nu_D}{0.023\,(7.07)^{0.4}} = 2106.4$$

$$Re_D = (2106.4)^{5/4} = 14\,270 \quad \text{[the flow is turbulent, and validates the use of eq. (6.91)]}$$

$$U = \frac{\nu}{D} Re_D = \frac{0.01004 \frac{cm^2}{s}}{0.025m} 14\,270$$

$$= 0.573 \frac{m}{s}$$

$$\dot{m} = \rho \frac{\pi}{4} D^2 U = 0.998 \frac{g}{cm^3} \frac{\pi}{4} (0.025m)^2\, 0.573 \frac{m}{s}$$

$$= 0.281 \frac{kg}{s}$$

Problem 6.18. Eliminating $d\overline{P}/dx$ between eqs. (6.66) and (6.68), and dividing by τ_w leads to the dimensionless equation

$$2\frac{r}{r_o} = \frac{d}{dr}\left(r\frac{\tau_{app}}{\tau_w}\right)$$

Integrating once, and dividing by r yields

$$\frac{r}{r_o} = \frac{\tau_{app}}{\tau_w} + \frac{C}{r}$$

The constant of integration C is determined finally by invoking the centerline condition $\tau_{app} = 0$ at $r = 0$. Its value must be $C = 0$, therefore the $\tau_{app}(r)$ distribution is linear:

$$\frac{\tau_{app}}{\tau_w} = \frac{r}{r_o} \qquad (6.69)$$

Problem 6.19. The information we know amounts to the following:

$U = 1$ m/s $\qquad \nu \cong 0.01\ cm^2/s$ (water at 20°C)

$D = 1$ cm $\qquad y^+ = 11.6$

The wall layer thickness that corresponds to $y^+ = 11.6$ is

$$y = y^+ \frac{\nu}{(\tau_w/\rho)^{1/2}}$$

or, after using eq. (6.75),

$$y = y^+ \frac{\nu}{U\,(f/2)^{1/2}} \qquad (1)$$

In order to determine f, we must first calculate the Reynolds number,

$$Re_D = \frac{UD}{\nu} = \frac{1m}{s}\,1\ cm\,\frac{s}{0.01\ cm^2} = 10^4 \qquad \text{(turbulent)}$$

This Re_D value recommends the use of eq. (6.76),

$$f = 0.079\ (10^4)^{-1/4} = 0.0079$$

and, in the end, eq. (1) yields

$$y = 11.6\,\frac{0.01\ cm^2}{s}\,\frac{s}{1\ m}\left(\frac{2}{0.0079}\right)^{1/2}$$

$$= 0.18\ mm$$

Problem 6.20. The analysis rests on eqs. (6.72)-(6.75). Substituting the velocity profile (6.72) in the integrand of the mean velocity definition (6.73) we obtain

$$U = \frac{1}{\pi r_0^2} \int_0^{2\pi} d\theta \int_0^{r_0} \bar{u}\, r\, dr$$

$$= \frac{1}{\pi r_0^2}\, 2\pi\, (8.7)\, \frac{(\tau_w/\rho)^{4/7}}{\nu^{1/7}}\, r_0^{15/7} \underbrace{\int_0^1 n^{1/7}(1-n)\, dn}_{49/120}$$

$$= \frac{98}{120}\, 8.7 \left(\frac{r_0}{\nu}\right)^{1/7} \left(\frac{\tau_w}{\rho}\right)^{4/7} \qquad (1)$$

The centerline condition (6.74) can be rearranged as

$$\bar{u}_c = 8.7 \left(\frac{r_0}{\nu}\right)^{1/7} \left(\frac{\tau_w}{\rho}\right)^{4/7} \qquad (2)$$

Eliminating τ_w/ρ between eqs. (1) and (2) we conclude that the centerline velocity is only 22 percent greater than the mean velocity (recall that in laminar flow $u_c/U = 2$):

$$\bar{u}_c = \frac{120}{98} U = 1.224\, U \qquad (3)$$

For the friction factor, we turn to eq. (6.75), in which we use eq. (2) to eliminate τ_w/ρ, and eq. (3) in order to eliminate $\bar{u}_c$:

$$f = \frac{2}{U^2} \frac{\tau_w}{\rho}$$

$$= \frac{2}{U^2} \left[\frac{\bar{u}_c}{8.7\, (r_0/\nu)^{1/7}} \right]^{7/4}$$

$$= (8.7)^{-7/4}\, 2^{5/4} \left(\frac{120}{98}\right)^{7/4} Re_D^{-1/4}$$

$$= 0.077\, Re_D^{-1/4} \qquad (6.76)$$

This result is almost identical to eq. (6.76) in the text. The 0.079 coefficient and the rest of eq. (6.76) are the original formula reported by H. Blasius in 1913.

Problem 6.21. The relation between the Stanton number and the Nusselt number is

$$\mathrm{St} = \frac{h}{\rho c_P U} = \frac{h}{\rho c_P U}\frac{D}{k}\frac{k}{D} = \mathrm{Nu}_D \frac{k/\rho c_P}{UD}$$

$$= \frac{\mathrm{Nu}_D}{\mathrm{Pe}_D} = \frac{\mathrm{Nu}_D}{\mathrm{Re}_D \mathrm{Pr}}$$

Substituting this and eq. (6.77) on both sides of eq. (6.89), we obtain, in order,

$$\frac{\mathrm{Nu}_D}{\mathrm{Re}_D \mathrm{Pr}} = \frac{0.046}{2} \mathrm{Re}_D^{-1/5} \mathrm{Pr}^{-2/3}$$

$$\mathrm{Nu}_D = 0.023\, \mathrm{Re}_D^{4/5} \mathrm{Pr}^{1/3} \tag{6.90}$$

Below $\mathrm{Re}_D \cong 2 \times 10^4$, more accurate is the friction factor formula (6.76): if we use it in Colburn's eq. (6.89) we obtain:

$$\frac{\mathrm{Nu}_D}{\mathrm{Re}_D \mathrm{Pr}} = \frac{0.079}{2} \mathrm{Re}_D^{-1/4} \mathrm{Pr}^{-2/3}$$

$$\mathrm{Nu}_D \cong 0.04\, \mathrm{Re}_D^{3/4} \mathrm{Pr}^{1/3}$$

Problem 6.22. a) There are two ways of estimating the total heat transfer rate through the isothermal wall of the pipe,

$$q = \int_0^L q''_w(x)\, \pi D\, dx \tag{1}$$

$$q = \dot{m}\, c_P (T_1 - T_L) \tag{2}$$

The second of these formulas yields

$$q = 100 \frac{g}{s}\, 4.182 \frac{J}{g\,K} (40 - 6)\, K$$

$$= 14\,219\ \mathrm{W}$$

b) Equation (6.53) shows that both sides of the equation do not change as the (D,L) design is replaced by the (D_1, L_1) design:

$$\frac{T_w - T_L}{T_w - T_1} = \exp\left(- \frac{2hL}{r_o \rho c_P U}\right) \tag{6.53}$$

Note that the left side of this equation is fixed and equal to (0 - 6)/(0 - 40) = 0.15. This means that the argument of the exponential is fixed also. The argument can be rewritten to highlight other quantities that remain fixed during the design change (namely, $\dot{m}$ and Nu_D):

$$\frac{2hL}{r_o \rho c_P U} = \frac{h\pi DL}{\dot{m}\, c_P} = \frac{\pi\, Nu_D\, k}{\dot{m}\, c_P} L$$

The last form shows that only L can influence the argument of the exponential. It follows that the pipe length in the new design must be equal to pipe length of the old design, if the argument of the exponential is to remain unchanged,

$$L_1 = L.$$

Problem 6.23. The limit $\Delta T_{in}/\Delta T_{out} \rightarrow 1$ is the same as writing

$$\Delta T_{in} = (1 + \varepsilon)\, \Delta T_{out} \qquad (1)$$

in which $\varepsilon \rightarrow 0$. Substituting eq. (1) in the general ΔT_{lm} expression (6.105), we obtain

$$\Delta T_{lm} = \frac{(1 + \varepsilon)\, \Delta T_{out} - \Delta T_{out}}{\ln \dfrac{(1 + \varepsilon)\, \Delta T_{out}}{\Delta T_{out}}} = \frac{1 + \varepsilon - 1}{\ln (1 + \varepsilon)} \Delta T_{out}$$

$$\cong \frac{\varepsilon}{\varepsilon} \Delta T_{out} = \Delta T_{out}, \qquad \text{as } \varepsilon \rightarrow 0 \qquad (2)$$

The combined message of eqs. (1) and (2) in the limit $\varepsilon \rightarrow 0$ is

$$\Delta T_{lm} = \Delta T_{out} = \Delta T_{in} \qquad (6.106)$$

Problem 6.24. a) The properties of water at 80°C and moderate (nearly atmospheric) pressures are

$$\nu = 0.00366 \frac{cm^2}{s} \qquad \rho = 0.9718 \frac{g}{cm^3}$$

The frictional pressure drop per unit length is given by

$$\frac{\Delta P}{L} = f \frac{p}{A} \frac{1}{2} \rho U^2$$
$$= f \frac{1}{D_h} 2\rho U^2 \qquad (1)$$

therefore, we must calculate in order

$$D_h = \frac{4A}{p} = 4 \frac{\pi D_o^2/4 - \pi D_i^2/4}{\pi D_o + \pi D_i} = D_o - D_i =$$
$$= (22 - 16)\text{ cm} = 6\text{ cm}$$

$$U = \frac{\dot{m}}{\rho A} = 100 \frac{10^3 kg}{3600 s} \frac{cm^3}{0.9718 g} \frac{1}{(\pi/4)\left(22^2 - 16^2\right) cm^2}$$
$$= 1.596 \frac{m}{s} = 159.6 \frac{cm}{s}$$

$$Re_{D_h} = \frac{U D_h}{\nu} = 159.6 \frac{cm}{s} 6\text{ cm} \frac{s}{0.00366\text{ cm}^2}$$
$$= 2.62 \times 10^5, \qquad \text{(turbulent)}$$

The friction factor is deduced from the Moody chart, where we know the value of Re_{D_h} on the abscissa, and the dimensionless roughness corresponding to "commercial steel",

$$\frac{k_s}{D_h} = \frac{0.05\text{ mm}}{60\text{ mm}} = 0.000833$$

therefore,

$$f \cong \frac{0.02}{4} = 0.005$$

We now have all the ingredients needed for evaluating numerically the right side of eq. (1):

$$\frac{\Delta P}{L} \cong \frac{0.005}{6\ \text{cm}} 2\,(0.9718)\ \frac{\text{g}}{\text{cm}^3}\ (159.6)^2\ \frac{\text{cm}^2}{\text{s}^2}$$

$$\cong 412.6\ \frac{\text{kg}}{\text{m}^2\text{s}^2} = 412.6\ \frac{\text{N/m}^2}{\text{m}} = 0.00407\ \frac{\text{atm}}{\text{m}}$$

If L = 200m, for example, the frictional pressure drop is 0.81 atm.

b) In the following calculations, we need two additional properties of water at 80°C and nearly atmospheric pressures:

$$c_P = 4.196\ \frac{\text{J}}{\text{gK}} \qquad\qquad Pr = 2.23$$

The mean temperature difference $\Delta T = T_w - T_m$ is given by the first law of thermodynamics, with reference to a duct element of length dx and annular flow area

$$\dot{m} c_P dT = h\,(\pi\, D_o\, dx)\, \Delta T$$

On the right side, πD_o is the portion of the wetted perimeter that is crossed by heat transfer (the inner wall of the annulus is insulated). In conclusion,

$$\Delta T = \left(\frac{dT}{dx}\right) \frac{\dot{m} c_P}{h \pi\ D_o}$$

which demands a value for the heat transfer coefficient h. Invoking the Colburn analogy,

$$St\, Pr^{2/3} = \frac{f}{2}$$

we learn first that

$$St = Pr^{-2/3}\,\frac{f}{2} = (2.23)^{-2/3}\ \frac{0.005}{2}$$

$$= 0.00146$$

therefore

$$h = St\, \rho\, c_P U$$

$$= 0.00146 \times\ 0.9718\ \frac{\text{g}}{\text{cm}^3}\ 4.196\ \frac{\text{J}}{\text{gK}}\ 159.6\ \frac{\text{cm}}{\text{s}}$$

$$= 0.95\ \frac{\text{W}}{\text{cm}^2\ \text{K}}$$

And, in view of the fact that dT/dx = 200°C/km, the ΔT formula yields

$$\Delta T = 200\ \frac{°\text{C}}{10^5\text{cm}}\ 100\ \frac{10^3\text{kg}}{3600\text{s}}\ 4.196\ \frac{\text{J}}{\text{gK}}\ \frac{\text{cm}^2\ \text{K}}{0.95\ \text{W}}\ \frac{1}{\pi\ 22\ \text{cm}}$$

$$= 3.55°\text{C}$$

Problem 6.25. The relevant properties of water (liquid) saturated at 200°C are:

$$\rho = 0.863 \frac{g}{cm^3} \qquad \nu = 0.0016 \frac{cm^2}{s}$$

$$c_P = 4.5 \frac{kJ}{kg \cdot K} \qquad k = 0.66 \frac{W}{m \cdot K}$$

$$\mu = 0.00139 \frac{g}{cm \cdot s} \qquad Pr = 0.94$$

The numerical data for the parallel-plate channel (crack in hot dry rock) are:

$L = 100$ m — $\dot{m} = 10$ kg/s

$W = 100$ m — $T_w = 200°C$

$D = 1$ cm, hence $D_h = 2$ cm — $T_{in} = 30°C$

a) The flow entrance length X is the same as the thermal entrance length X_T, because the Prandtl number is approximately equal to 1. We begin with eq. (6.4') written for hydraulic diameter,

$$\frac{X}{D_h} \cong 0.05 \, Re_{D_h}$$

and calculate in order

$$U = \frac{\dot{m}}{\rho DW} = \frac{10 \frac{kg}{s}}{0.863 \frac{g}{cm^3} 1 \text{ cm } 100 \text{ m}} = 1.16 \frac{cm}{s}$$

$$Re_{D_h} = \frac{U D_h}{\nu} = 1.16 \frac{cm}{s} 2 \text{ cm} \frac{s}{0.0016 \text{ cm}^2} = 1448 \qquad \text{(laminar flow)}$$

$$X \cong 0.05 \, D_h \, Re_{D_h} = 1.45 \text{ m}$$

The flow is hydrodynamically and thermally fully developed along most of the channel, because X is negligible relative to L.

b) The pressure drop ΔP could be calculated based on the friction factor listed in Table 6.1, however, it is more direct to use eq. (6.22):

$$-\frac{dP}{dx} = 12\frac{\mu U}{D^2} = 12 \times 0.00139 \frac{g}{cm{\cdot}s} 1.16 \frac{cm}{s} \frac{1}{cm^2}$$

$$= 0.0193 \frac{g}{cm^2 s^2}$$

$$\Delta P = \left(-\frac{dP}{dx}\right) L = 0.0193 \frac{g}{cm^2 s^2} 100\ m$$

$$= 19.3 \frac{N}{m^2} \cong 0.0002\ atm$$

c) For the total heat transfer rate q, we use the first law of thermodynamics for the water stream, namely eq. (6.100):

$$q = \dot{m}\, c_P \left(T_{out} - T_{in}\right)$$

The only unknown in this equation is T_{out}. We find it from eq. (6.103'):

$$\ln \frac{T_w - T_{in}}{T_w - T_{out}} = \frac{h A_w}{\dot{m}\, c_P}$$

for which we calculate, in order,

$$Nu_{D_h} = 7.54 \qquad \text{(Table 6.1)}$$

$$h = 7.54 \frac{k}{D_h} = 7.54 \times 0.66 \frac{W}{m{\cdot}K} \frac{1}{0.02\ m} = 248.8 \frac{W}{m^2K}$$

$$A_w = 2\,LW = 2 \times 10^4 m^2$$

$$\frac{h A_w}{\dot{m}\, c_P} = 248.8 \frac{W}{m^2K}\, 2 \times 10^4 m^2 \frac{s}{10\ kg} \frac{kg{\cdot}K}{4.5 \times 10^3 J} = 110.6$$

Equation (6.103') yields

$$T_{out} = T_w - \left(T_w - T_{in}\right) \underbrace{\exp(-110.6)}_{\cong\ zero}$$

$$\cong T_w = 200°C$$

and the total heat transfer rate q becomes

$$q = \dot{m}\, c_P \left(T_{out} - T_{in}\right) = 10\,\frac{kg}{s}\,4.5\,\frac{10^3 J}{kg\cdot K}\,(200 - 30)\,K$$

$$= 7.65 \times 10^6 W$$

Problem 6.26. We begin with estimating the total number of tubes n, and calculate in order

$$n = 11 \times 6 + 10 \times 5 = 116$$ (note: the first and last rows have 6 tubes each)

$$A_w = 116\,\pi\,DL$$

$$= 116\,\pi\,0.04m\;3m = 43.73\;m^2$$

$$6X_t = 42\;cm$$ (frontal width of tube bundle)

$$A = (6\,X_t)L = 42\;cm\;3m = 1.26\;m^2$$ (frontal area of tube bundle)

$$\dot{m} = \rho A\,U_\infty =$$

$$= 0.616\,\frac{kg}{m^3}\,1.26\;m^2\;2\,\frac{m}{s} = 1.555\,\frac{kg}{s}$$ (air at 300°C)

$$\frac{h\,A_w}{\dot{m}\,c_P} = \frac{62\;W}{m^2\,K}\,\frac{43.73\;m^2}{1.555\;kg/s}\,\frac{kg\cdot K}{1045\;J} = 1.668$$

$$\ln\frac{\Delta T_{in}}{\Delta T_{out}} = 1.668 \qquad (6.103')$$

$$\Delta T_{out} = 0.189\,\Delta T_{in}$$

$$= 0.189\,(300 - 30)°C \cong 51°C$$

$$T_{out} = 30°C + 51°C = 81°C$$

$$q = \dot{m}\,c_P\,(T_{in} - T_{out}) \qquad (6.100)$$

$$= 1.555\,\frac{kg}{s}\,1045\,\frac{J}{kg\,K}\,(300 - 81)\;K$$

$$\cong 356\;kW$$

Problem 6.27. If on either side of the hot blade the group $hA_w/\dot{m}\, c_P$ is much greater than 1 (say, greater than 3), eq. (6.103') shows that

$$\Delta T_{out} << \Delta T_{in}$$

and eq. (6.101) reduces to

$$q \cong \dot{m}\, c_P\, \Delta T_{in} \tag{1}$$

where ΔT_{in} and $T_w - T_0$, constant, We write eq. (1) for both sides of the blade,

$$q_1 = \dot{m}_1\, c_P\, \Delta T_{in} \tag{2}$$

$$q_2 = \dot{m}_2\, c_P\, \Delta T_{in} \tag{3}$$

In the case of fully developed laminar flow, we use eq. (6.22) twice

$$\dot{m}_1 = \frac{\rho W}{12\mu} \frac{\Delta P}{L} D_1^3 \tag{4}$$

$$\dot{m}_2 = \frac{\rho W}{12\mu} \frac{\Delta P}{L} D_2^3 \tag{5}$$

in which W is the width perpendicular to the plane $D \times L$. We write next

$$D_1 = yD \qquad \text{and} \qquad D_2 = (1 - y)D \tag{6}$$

and calculate the total heat transfer rate

$$q = q_1 + q_2$$

by using eqs. (2)-(6). The result is

$$\frac{q}{T_w - T_0} \frac{12\mu\, L}{\rho\, c_P\, W\, D^3\, \Delta P} = y^3 + (1 - y)^3$$

The following table shows that the thermal conductance $q/(T_w - T_0)$ is the smallest (i.e. the worst) when the hot blade is positioned in the middle of the channel. The thermal conductance is four times greater than this minimum value when the blade is attached to one of the walls of the channel:

y	$\frac{q}{T_w - T_0} \frac{12\mu L}{\rho c_P W D^3 \Delta P}$
0	1
0.25	0.44
0.5	0.25
0.75	0.44
1	1

Problem 6.28. The relation between the total heat transfer rate (q) and the largest temperature difference ($T_w - T_0$) is provided by a combination of eqs. (6.103')-(6.105),

$$q = \dot{m}\, c_P\, \Delta T_{in} \left[1 - \exp\left(-\frac{h A_w}{\dot{m}\, c_P}\right)\right] \quad (1)$$

where

$$\dot{m} = \frac{\rho W D^3}{12\mu} \frac{\Delta P}{L} \quad (2)$$

$$W = \text{width perpendicular to the plane } D \times L \quad (3)$$

$$\frac{h D_h}{k} = Nu = 4.86, \qquad \text{(Table 6.1)} \quad \text{(note: } D_h = 2D) \quad (4)$$

$$h = \frac{Nu}{2} \frac{k}{D}, \qquad A_w = L \times W \quad (5)$$

$$\frac{h A_w}{\dot{m}\, c_P} = \ldots = 6\, Nu \frac{\mu\alpha}{\Delta P \cdot L^2} \left(\frac{L}{D}\right)^4 \quad (6)$$

Next, we write the symbol Π for the dimensionless pressure drop

$$\Pi = \frac{\Delta P \cdot L^2}{\mu\alpha} \quad (7)$$

and substitute eqs. (2)-(7) in eq. (1):

$$q = \frac{W}{12} k (T_w - T_0)\, \Pi^{1/4}\, \delta^3 \left[1 - \exp\left(-29.16\, \delta^{-4}\right)\right] \quad (8)$$

where δ is the dimensionless spacing,

$$\delta = \frac{D}{L} \Pi^{1/4} \tag{9}$$

The maximization of q with respect to δ, namely $\partial q/\partial\delta = 0$, yields the equation

$$\exp(a) = 1 + \frac{4}{3} a, \quad \text{with} \quad a = 29.16\, \delta^{-4} \tag{10}$$

The solution of this equation is $a_{opt} = 0.5502$, which means that $\delta_{opt} = 2.70$, and

$$\frac{D_{opt}}{L} = 2.70\, \Pi^{-1/4} \tag{11}$$

By substituting eq. (11) in eq. (8), we obtain the corresponding maximum thermal conductance

$$\left(\frac{q}{T_w - T_0}\right)_{max} = 0.693\, Wk\, \Pi^{1/4} \tag{12}$$

Or, if we write $\overline{q''} = q/WL$ for the average heat flux, we obtain finally

$$\left(\frac{\overline{q''}}{T_w - T_0}\right)_{max} \frac{L}{k} = 0.693\, \Pi^{1/4} \tag{13}$$

In conclusion, when the pressure difference ΔP is specified, the optimal spacing has a certain value that varies as $\Delta P^{-1/4}$, and the maximum conductance increases as $\Delta P^{1/4}$.

Problem 6.29. a) The highest temperature of the constant-q" board occurs at the trailing edge,

$$T_h = T_w\,(x = L)$$

where the relationship between the wall temperature and the fluid outlet temperature is

$$\frac{q''}{T_h - T_{out}} \frac{D_h}{k} = Nu = 5.385 \qquad \text{(Table 6.1)}$$

We note that $D_h = 2D$, and conclude that

$$T_h - T_{out} = \frac{q''\, 2D}{k\, Nu} \tag{1}$$

$$\frac{d\,T_m}{dx} = \frac{p}{A}\,\frac{q''}{\rho\, c_P\, U} \tag{2}$$

where p = W, A = DW, W = width perpendicular to the plane D × L, and

$$U = \frac{\Delta P \cdot D^2}{12\,\mu L} \tag{3}$$

By integrating equation (2) from $T = T_0$ (at x = 0) to $T = T_{out}$ (at x = L) we obtain

$$T_{out} - T_0 = \frac{q'' L}{\rho\, c_P\, U D} \tag{4}$$

Next we eliminate T_{out} between eqs. (1) and (4), and then use eq. (3). The result can be arranged as follows

$$\frac{T_h - T_0}{q'' L/k} = 12\,\frac{\mu\alpha}{\Delta P \cdot L^2}\left(\frac{L}{D}\right)^3 + \frac{2}{Nu}\,\frac{D}{L} \tag{5}$$

This quantity (i.e. T_h) can be minimized with respect to the spacing D, but since L is fixed, it is easier to perform the minimization with respect to the dimensionless parameter $\delta = D/L$. The optimal value of this parameter is

$$\frac{D_{opt}}{L} = \left(18\,Nu\,\frac{\mu\alpha}{\Delta P \cdot L^2}\right)^{1/4} = 3.14\left(\frac{\mu\alpha}{\Delta P \cdot L^2}\right)^{1/4} \tag{6}$$

and, after substituting it in eq. (5), the corresponding minimum T_h is given by

$$\left(\frac{T_h - T_0}{q''}\right)_{min} \frac{k}{L} = 1.554\left(\frac{\mu\alpha}{\Delta P \cdot L^2}\right)^{1/4} \tag{7}$$

Turned upside down, this result spells out the maximum heat transfer rate that can be removed by the stream when the trailing temperature of the circuit board must not exceed the (specified) ceiling temperature T_h:

$$\left(\frac{q''}{T_h - T_0}\right)_{max} \frac{L}{k} = 0.644\left(\frac{\Delta P \cdot L^2}{\mu\alpha}\right)^{1/4} \tag{8}$$

b) The corresponding result for the board with uniform temperature T_w is (see the preceding problem)

$$\left(\frac{\overline{q}''}{T_w - T_0}\right)_{min} \frac{L}{k} = 0.693 \left(\frac{\Delta P \cdot L^2}{\mu \alpha}\right)^{1/4} \qquad (9)$$

When the board ceiling temperature is fixed in both designs, $T_w = T_h$, the maximum heat transfer of the isothermal board exceeds by 8 percent the maximum heat transfer made possible by the board with uniform heat flux. The reason for the 8 percent difference is the leading section of the isothermal board, which is considerably warmer (i.e. higher above T_0) than the leading section of the constant-flux board.

Project 6.1. a) The pressure drop per unit length contributed by the annular space is

$$\left(\frac{\Delta P}{L}\right)_a = \frac{f_a}{D_h} 2\rho U_a^2 \tag{1}$$

where U_a is the mean velocity

$$U_a = \frac{\dot{m}}{\rho (\pi/4) (D_o^2 - D_i^2)} \tag{2}$$

and where D_h is the hydraulic diameter of the annular cross-section,

$$D_h = \frac{4A}{p} = 4 \frac{(\pi/4) (D_o^2 - D_i^2)}{\pi (D_o + D_i)} = D_o - D_i \tag{3}$$

Combined, eqs. (1-3) yield

$$\left(\frac{\Delta P}{L}\right)_a = \frac{2\rho}{D_o^5} \left(\frac{\dot{m}}{\rho}\right)^2 \left(\frac{4}{\pi}\right)^2 \frac{f_a}{(1 - r)^3 (1 + r)^2} \tag{4}$$

in which r is the dimensionless diameter ratio

$$r = \frac{D_i}{D_o} \tag{5}$$

Similarly, the pressure drop per unit length contributed by the adjacent flow through the inner pipe is

$$\left(\frac{\Delta P}{L}\right)_i = \frac{f_i}{D_i} 2\rho U_i^2 \tag{6}$$

with

$$U_i = \frac{\dot{m}}{\rho (\pi/4) D_i^2} \tag{7}$$

Equation (6) can be written as

$$\left(\frac{\Delta P}{L}\right)_i = \frac{2\rho}{D_o^5} \left(\frac{\dot{m}}{\rho}\right)^2 \left(\frac{4}{\pi}\right)^2 \frac{f_i}{r^5} \tag{8}$$

therefore, the total pressure drop contributed by each unit length of coaxial heat exchanger is

$$\frac{\Delta P}{L} = \left(\frac{\Delta P}{L}\right)_a + \left(\frac{\Delta P}{L}\right)_i$$

$$= \frac{2\rho}{D_o^5}\left(\frac{\dot{m}}{\rho}\right)^2 \left(\frac{4}{\pi}\right)^2 \underbrace{\left[\frac{f_a}{(1 - r)^3 (1 + r)^2} + \frac{f_i}{r^5}\right]}_{F(r)} \tag{9}$$

The position of the wall of diameter D_i is governed by the parameter r. It is easy to see that F(0) = F(1) = ∞, which means that F and ΔP/L have a minimum at a special, intermediate r value. Minimizing F(r) numerically in the case where $f_a = f_i$ = constant, we learn that the special r value is

$$r = 0.653 \tag{10}$$

b) If the flow is laminar, the friction factors are affected strongly by the respective Reynolds numbers. For the inner pipe, that relationship is

$$f_i = \frac{16}{Re_{D_i}} \tag{11}$$

The f_a relationship for the annular cross-section is considerably more complicated*; when D_i is not much smaller than D_o, however, the annulus is similar to the space between two parallel plates positioned $(D_o - D_i)/2$ apart, therefore

$$f_a \cong \frac{24}{Re_{D_h}} \tag{12}$$

where $D_h = D_o - D_i$. Substituting eqs. (11,12) in the total pressure drop formula (9), we obtain

$$\frac{\Delta P}{L} = \frac{8}{\pi}\frac{\dot{m}}{\rho}\frac{\mu}{D_o^4}\left[\frac{24}{(1 - r)^3 (1 + r)} + \frac{16}{r^4}\right] \tag{13}$$

Numerically, we find that the quantity listed inside the square brackets reaches its minimum at

$$r = 0.621 \tag{14}$$

This result is nearly the same as the one obtained in the fully-rough limit of the turbulent regime, eq. (10).

---

*A. Bejan, Convection Heat Transfer, Wiley, New York, 1988, Problem 2, p. 104.

Project 6.2. a) The energy equation for the liquid in the relative-motion gap reduces to

$$\frac{\partial^2 T}{\partial y^2} = 0 \tag{1}$$

with the general solution

$$T = f_1 y + f_2$$

where f_1 and f_2 are at most functions of x. They are determined from the boundary conditions

$$T = T_m + \Delta T \quad \text{at} \quad y = 0 \tag{2}$$

$$T = T_m \quad \text{at} \quad y = \delta \tag{3}$$

so that the temperature distribution emerges as a function of y only:

$$T = T_m + \left(1 - \frac{y}{\delta}\right)\Delta T \tag{4}$$

The energy balance for a control volume of infinitesimal thickness drawn around the melting front requires

$$\underbrace{L\rho\, V h_s}_{\substack{\text{enthalpy}\\ \text{inflow from}\\ \text{above}}} + \underbrace{Lk\frac{\Delta T}{\delta}}_{\substack{\text{conduction}\\ \text{heat transfer}\\ \text{from below}}} = \underbrace{L\rho V h_f}_{\substack{\text{enthalpy}\\ \text{outflow,}\\ \text{downward}}} \tag{5}$$

The second term on the left side comes from using eq. (4) in the definition of the heat transfer rate oriented in the y direction,

$$q' = Lk\left(-\frac{\partial T}{\partial y}\right)_{y=\delta} \tag{6}$$

Equation (5) constitutes one relationship between the chief unknowns of this problem, the melting speed V and the liquid gap thickness δ:

$$V\delta = \frac{k\,\Delta T}{\rho\, h_{sf}} \tag{7}$$

The second relationship between V and δ is provided by the fluid mechanics part of the liquid flow. The momentum equation reduces to

$$\frac{dP}{dx} = \mu \frac{\partial^2 u}{\partial y^2} \tag{8}$$

and this yields

$$u = \frac{1}{2\mu}\left(\frac{dP}{dx}\right) y^2 + f_3 y + f_4 \tag{9}$$

The functions $f_3(x)$ and $f_4(x)$ are determined by invoking the boundary conditions

$$u = U \qquad \text{at} \qquad y = 0 \tag{10}$$

$$u = 0 \qquad \text{at} \qquad y = \delta \tag{11}$$

and the resulting expression for u(x,y) is

$$u = \frac{1}{2\mu}\left(\frac{dP}{dx}\right)\left(y^2 - y\delta\right) + U\left(1 - \frac{y}{\delta}\right) \tag{12}$$

This solution cannot be used yet because it involves the unknown pressure gradient dP/dx. The pressure must be related to the normal force with which the melting block is pushed downward, and to the fact that P = 0 at both ends of the liquid channel, x = 0 and x = L.

We determine the pressure distribution by first calculating the liquid "flowrate"

$$Q = \int_0^\delta u \, dy \tag{13}$$

which after using eq. (12) becomes

$$Q = \frac{1}{12\mu}\left(-\frac{dP}{dx}\right)\delta^3 + \frac{1}{2} U\delta \tag{14}$$

Next, we recognize the mass conservation statement

$$\frac{\partial u}{\partial x} + \frac{\partial v}{\partial y} = 0 \tag{15}$$

which can be integrated across the liquid gap to yield, in order,

$$\underbrace{\frac{d}{dx}\int_0^{\delta} u\,dy}_{Q} + \underbrace{(v)_{y=\delta}}_{-V} - \underbrace{(v)_{y=0}}_{\text{zero}} = 0 \tag{16}$$

$$\frac{dQ}{dx} - V = 0 \tag{17}$$

By eliminating Q between eqs. (14) and (17) we obtain an equation for P(x),

$$\frac{1}{12\mu}\left(-\frac{d^2P}{dx^2}\right)\delta^3 = V \tag{18}$$

which yields

$$P = -\frac{6\mu V}{\delta^3}x^2 + C_1 x + C_2 \tag{19}$$

The constants C_1 and C_2 follow from the end conditions (P = 0 at x = 0 and x = L), therefore the pressure distribution is parabolic:

$$P(x) = \frac{6\mu V}{\delta^3}\left(Lx - x^2\right) \tag{20}$$

Integrated along the liquid gap, this yields the second relationship between V and δ,

$$F_n = \int_0^L P\,dx = \;....\; = \mu V\left(\frac{L}{\delta}\right)^3 \tag{21}$$

namely

$$\delta = L\left(\frac{\mu V}{F_n}\right)^{1/3} \tag{22}$$

The answer for the melting speed is obtained by eliminating δ between eqs. (7) and (22):

$$V = \left(\frac{F_n}{\mu}\right)^{1/4}\left(\frac{k\,\Delta T}{\rho\, h_{sf} L}\right)^{3/4} \tag{23}$$

The tangential force experienced by the melting block is equal to the negative of the shear force experienced by the hot slider:

$$F_t = -\int_0^L \mu \left(\frac{\partial u}{\partial y}\right)_{y=0} dx = \; .... \; =$$

$$= L\mu \frac{U}{\delta} \tag{24}$$

The coefficient of friction is therefore

$$\mu_f = \frac{F_t}{F_n} = \; ... \; = \frac{U}{V^{1/3}} \left(\frac{\mu}{F_n}\right)^{2/3} \tag{25}$$

or, after using eq. (23),

$$\mu_f = U \left(\frac{\mu}{F_n}\right)^{3/4} \left(\frac{\rho\, h_{sf} L}{k\, \Delta T}\right)^{1/4} \tag{26}$$

This shows that the coefficient of friction decreases as the normal force increases.

Project 6.3. a) In the limit $D \to 0$ each channel becomes slender enough for the flow and heat transfer to be in the fully developed regime. Equation (6.22) yields the average longitudinal velocity when the pressure drop ΔP is specified,

$$U = \frac{D^2}{12\mu} \frac{\Delta P}{L} \tag{6.22}$$

The corresponding mass flowrate (per unit length normal to the $H \times L$ plane) is

$$\dot{m}' = \rho U H = \rho H \frac{D^2}{12\mu} \frac{\Delta P}{L}$$

In the same limit, the right hand side of eq. (6.103') blows up, because in the laminar regime $h \sim k/D$,

$$\ln \frac{\Delta T_{in}}{\Delta T_{out}} = \frac{h\, A_w}{c_P\, \dot{m}} \sim \frac{k}{D} \frac{L}{c_P} \frac{1}{\rho H \dfrac{D^2}{\mu} \dfrac{\Delta P}{L}}$$

(blows up as $1/D^3$)

This means that $\Delta T_{out}/\Delta T_{in} \rightarrow 0$, or that the mean outlet temperature is practically the same as the board temperature T_0:

$$T_{out} \cong T_0$$

The total heat transfer rate removed by forced convection from all the boards is

$$q'_a = \dot{m}' c_P (T_{out} - T_{in}) = \dot{m}' c_P (T_0 - T_\infty)$$

$$= \rho H \frac{D^2}{12\mu} \frac{\Delta P}{L} c_P (T_0 - T_\infty) \qquad \text{(a)}$$

In conclusion, in the $D \rightarrow 0$ limit the total heat transfer rate decreases as D^2. This trend is illustrated in the attached sketch.

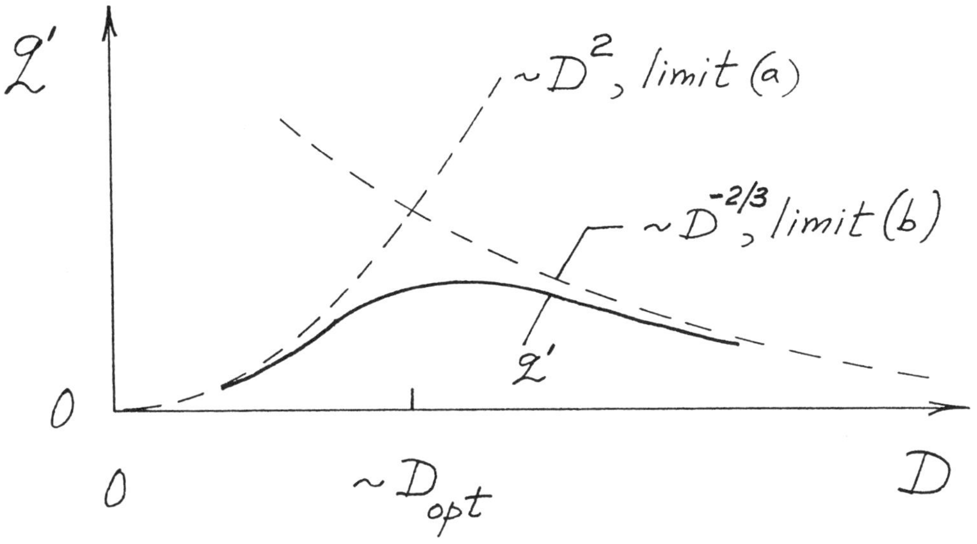

b) In the opposite limit ($D \rightarrow \infty$) the boundary layer that lines each surface becomes "distinct", i.e. each channel has the features of flow and heat transfer associated with the beginning of the entrance region. Since the overall pressure drop is fixed, the important question here is what is the "free stream" velocity U_∞ that sweeps these boundary layers?

The overall force balance on the control volume H × L requires

$$\Delta P \cdot H = n \cdot 2 \cdot \overline{\tau}_w L$$

in which n = H/D is the number of channels, and $\overline{\tau}_w$ is the L-averaged wall shear stress,

$$\overline{\tau}_w = 1.328\, Re_L^{-1/2} \frac{1}{2} \rho U_\infty^2 \qquad (5.55)$$

Combined, those last two equations yield:

$$U_\infty = \left(\frac{1}{1.328} \frac{\Delta P \cdot H}{n L^{1/2} \rho \nu^{1/2}} \right)^{2/3}$$

The total heat transfer rate from one of the L-long surfaces is given by eq. (5.82b)

$$\frac{\overline{h} L}{k} = \frac{\overline{q''}}{T_0 - T_\infty} \frac{L}{k} = 0.664\, Pr^{1/3} \left(\frac{U_\infty L}{\nu} \right)^{1/2}$$

in other words

$$q_1' = \overline{q''} L = k \left(T_0 - T_\infty \right) 0.664\, Pr^{1/3} \left(\frac{U_\infty L}{\nu} \right)^{1/2}$$

The total heat transfer rate is 2n larger than this,

$$q_b' = 2n q_1' = 2n\, k \left(T_0 - T_\infty \right) 0.664\, Pr^{1/3} \left(\frac{U_\infty L}{\nu} \right)^{1/2}$$

or, after using n = H/D and the U_∞ expression derived above,

$$q_b' = 1.208\, k \left(T_0 - T_\infty \right) H \frac{Pr^{1/3} L^{1/3} \Delta P^{1/3}}{\rho^{1/3} \nu^{2/3} D^{2/3}} \qquad (b)$$

The second conclusion is that in the large-D limit the total heat transfer rate decreases as $D^{2/3}$. This second trend has been added to the same graph, to suggest that the maximum of the actual (unknown) curve q'(D) occurs at an optimal spacing D_{opt} that is of the same order as the D value obtained by intersecting the asymptotes (a) and (b):

$$q'_a \sim q'_b$$

$$\vdots$$

$$\frac{D_{opt}}{L} \sim 2.73 \left(\frac{\mu\alpha}{\Delta P \cdot L^2} \right)^{1/4}$$

c) The optimal spacing increases as $L^{1/2}$, and decreases as $\Delta P^{-1/4}$. The fluid that is used as coolant affects the design through the group $(\mu\alpha)^{1/4}$. For example, when ΔP and L are fixed, the switch from air (1 atm, 100°C) to Freon 12 (near-saturated liquid at 367 K, or 94°C) recommends a factor of 4 decrease in the optimal spacing:

$$\frac{D_{opt,\ Freon\ 12}}{D_{opt,\ air}} = \frac{(\mu\alpha)^{1/4}_{Freon\ 12}}{(\mu\alpha)^{1/4}_{air}}$$

$$= \left(\frac{1.16 \times 10^{-4}\ \mathrm{kg/s \cdot m}}{2.18 \times 10^{-5}\ \mathrm{kg/s \cdot m}} \right)^{1/4} \left(\frac{2.89 \times 10^{-4}\ \mathrm{cm^2/s}}{0.328\ \mathrm{cm^2/s}} \right)^{1/4}$$

$$\cong 0.26$$

d) Other issues to consider:

- under what conditions (constraint) is the <u>laminar</u> boundary layer assumption valid, and how this affects the D_{opt} formula,
- what is the corresponding D_{opt} formula for turbulent flow, and
- how the maximum heat transfer rate $q'(D_{opt})$ scales with L, H and the fluid properties.

For example, the order of magnitude of the maximum total heat transfer rate can be estimated by substituting D_{opt} into eq. (a):

$$q'_{max} \sim \left(\frac{\rho}{Pr} \right)^{1/2} c_P \left(T_0 - T_\infty \right) H \left(\Delta P \right)^{1/2}$$

The corresponding maximum "density" of heat-generating electronics is

$$q'''_{max} = \frac{q'_{max}}{HL \left(T_0 - T_\infty \right)} \sim \frac{(\Delta P)^{1/2}}{L} c_P \left(\frac{\rho}{Pr} \right)^{1/2}$$

This result shows that the density can be increased by dividing each board into shorter lengths, while each channel is supplied with $(\Delta P, T_\infty)$ fluid. The relative goodness of each coolant is described by the size of the group $c_P\,(\rho/Pr)^{1/2}$. Continuing with the numerical example of the preceding paragraph, if $(T_0 - T_\infty)$, ΔP and L are fixed, the switch from air to Freon 12 results in a 20-fold increase in the maximum heat generating density,

$$\frac{q'''_{max,Freon\ 12}}{q'''_{max,air}} = \frac{\left[c_P\,(\rho/Pr)^{1/2}\right]_{Freon\ 12}}{\left[c_P\,(\rho/Pr)^{1/2}\right]_{air}} \cong 20$$

Project 6.4. Examine the cross-section through the flow passages and fins, which is shown in the attached figure. Assume that the heat transfer from the wall (T_w) to the fluid flow (bulk temperature T_f) is due mainly to the fins. In other words, assume that the heat transfer through the unfinned wall patches of width D is negligible.

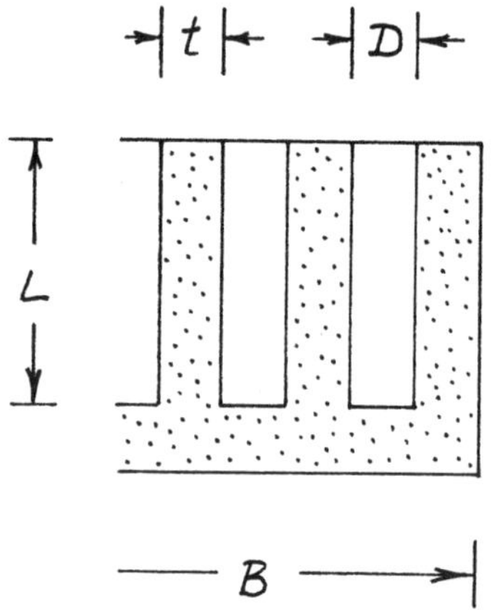

For each fin, the heat transfer rate (per unit length normal to the figure) can be calculated with the formula (2.115)

$$q'_{fin} = h \times 2 \times L\,(T_w - T_f)\,\eta \qquad (1)$$

where the factor 2 means "two sides", i.e. the wetted perimeter per unit length normal to the figure. The fin efficiency is

$$\eta = \frac{\tanh\,(mL)}{mL} \qquad (2)$$

where

$$mL = \left(\frac{hp}{k_w A_c}\right)^{1/2} L \tag{3}$$

In equation (3) we substitute $p = 2W$ and $A_c = tW$, where W is the dimension normal to the figure. We also assume that the flow is laminar and fully developed, and that $L >> D$. According to Table 6.1, the Nusselt number

$$Nu_{D_h} = \frac{h\, D_h}{k_f} \tag{4}$$

varies between 7.54 and 8.235 depending on how we model (approximate) the thermal boundary condition over the lateral surfaces of each fin. The Nu_h value can be even smaller (of order 5) if L/D is not very large (the case L/D = 4 is listed in Table 6.1). The key here is that Nu_{D_h} is a constant (called Nu for short), whose order of magnitude is known. Remembering that $D_h = 2D$, we conclude that

$$h = Nu\, \frac{k_f}{2D} \tag{5}$$

and

$$mL = \left(Nu\, \frac{k_f}{k_w}\right)^{1/2} \frac{L}{(tD)^{1/2}} \tag{6}$$

The total number of fins spread over the wall of breadth B is B/(t + D). Therefore the total heat transfer rate released from the wall to all the channel streams is

$$\begin{aligned} q' &= \frac{B}{t+D}\, q'_{fin} \\ &= \frac{B}{t+D}\, Nu\, k_f \frac{L}{D} (T_w - T_f)\, \eta \end{aligned} \tag{7}$$

It is most convenient if we nondimensionalize the total heat transfer rate by dividing eq. (7) by $k_w(T_w - T_f)B/L$, and call it Q,

$$\frac{q'}{k_w\,(T_w - T_f)\, B/L} = Q \tag{8}$$

Equation (7) can be written sequentially as

$$Q = Nu \frac{k_f}{k_w} \frac{(L/D)^2}{1 + \frac{t}{D}} \eta$$

$$= \frac{L}{D} \left(Nu \frac{k_f}{k_w} \right)^{1/2} \frac{(t/D)^{1/2}}{1 + \frac{t}{D}} \tanh \left[\frac{L}{D} \left(Nu \frac{k_f}{k_w} \right)^{1/2} \left(\frac{D}{t} \right)^{1/2} \right] \tag{9}$$

or, more succinctly,

$$Q = b \frac{x^{1/2}}{1 + x} \tanh \left(\frac{b}{x^{1/2}} \right) \tag{10}$$

where b is a number in the 0.1-10 range (because $L/D > 1$ and $k_f/k_w < 1$)

$$b = \frac{L}{D} \left(Nu \frac{k_f}{k_w} \right)^{1/2} \tag{11}$$

and x is the dimensionless fin thickness,

$$x = \frac{t}{D} \tag{12}$$

The fin thickness t (or x) is the dimension that must be chosen optimally, while holding the other parameters fixed. That Q has an optimum with respect to x can be seen by considering the following extremes. When $x \ll 1$, Q increases as $x^{1/2}$, and when $x \gg 1$, Q decreases $1/x$. The function Q of eq. (10) can be maximized numerically for a given b, and the results are

b	x_{opt}	Q_{max}	η
0.1	0.057	0.0089	0.945
0.2	0.113	0.0322	0.896
0.5	0.270	0.152	0.775
1	0.498	0.419	0.627
2	0.809	0.971	0.439
4	0.989	1.999	0.248
10	0.999	5	0.1
20	0.999	10	0.05

In conclusion, when $b \gtrsim 2$ the optimal fin thickness is approximately the same as the channel spacing D. When b is smaller, say $b \lesssim 1$, the optimal fin thickness is approximately proportional to b,

$$\frac{x_{opt}}{b} \cong 0.057 \qquad (13)$$

which means that the optimal fin thickness is proportional to the L,

$$\frac{t_{opt}}{L} \cong 0.057 \left(Nu \frac{k_f}{k_w} \right)^{1/2} \qquad (14)$$

Said another way, when $b \lesssim 1$ the optimal fin thickness is such that the slenderness of the fin profile equals the quantity listed on the right side of eq. (14).

Project 6.5. When the board is positioned off center, say, $D_1 > D_2$, the $\dot{m}_1$ stream will be larger (and a better coolant) than the $\dot{m}_2$ stream. The board surface cooled by the $\dot{m}_1$ stream (surface no. 1) will have a temperature $T_1(x)$ that is lower than the corresponding (aligned) temperature of the surface cooled by the $\dot{m}_2$ stream, $T_2(x)$. The local temperature difference $T_2 - T_1$ drives a conduction heat flux through the board, from surface no. 2 to surface no. 1,

$$q_c'' = \frac{k_w}{t}(T_2 - T_1) \qquad (1)$$

The heat flux generated by the electronics mounted on each surface is q'' = constant. The heat flux removed by the $\dot{m}_1$ stream is larger than this, because of the q_c'' contribution,

$$q_1'' = q'' + q_c'' \qquad (2)$$

For the same reason, only a portion of the electrically generated q'' is removed from surface no. 2 by the $\dot{m}_2$ stream,

$$q_2'' = q'' - q_c'' \qquad (3)$$

The heat fluxes (q_1'', q_2'') and surface temperatures (T_1, T_2) are functions of longitudinal position, x. The temperatures increase with x, reaching their highest values at the trailing end (x = L),

$$T_1(L) = T_{h,1}, \qquad T_2(L) = T_{h,2} \qquad (4)$$

The larger of these two values is the most critical item in the design: our objective is to minimize it.

We obtain the temperature distributions $T_1(x)$ and $T_2(x)$ by making the simplifying assumption that the temperature increase along each surface [for example, $T_1(L) - T_1(0)$] is considerably greater than the local temperature difference between the surface and the corresponding stream. This assumption becomes better as the $D \times L$ channel becomes more slender. It means that we approximate the local stream temperature as being almost equal to the temperature of the neighboring spot on the surface. Consequently, the first law of thermodynamics for a dx slice of the D_1 subchannel is written as

$$\dot{m}_1 c_P \, dT_1 = q_1'' \, W \, dx \tag{5}$$

in which W is the width perpendicular to the $D \times L$ plane. Note that only one surface of subchannel no. 1 transfers heat to the stream. The corresponding first-law statement for the second subchannel is

$$\dot{m}_2 c_P \, dT_2 = q_2'' \, W \, dx \tag{6}$$

The mass flowrates $\dot{m}_1$ and $\dot{m}_2$ are driven by the same pressure drop ΔP, which is maintained across the entire assembly. Assuming that the flow is laminar and fully developed along most of the length L in both subchannels, we recognize eq. (6.22) and $\dot{m}_1 = \rho \, U_1 D_1 W$, and write

$$\dot{m}_1 = \frac{\rho W}{12\mu} \frac{\Delta P}{L} D_1^3 \tag{7}$$

$$\dot{m}_2 = \frac{\rho W}{12\mu} \frac{\Delta P}{L} D_2^3 \tag{8}$$

At this stage, we have all the ingredients that are necessary for integrating eqs. (5) and (6) away from the entrance, where

$$T_1(0) = T_2(0) \cong T_0 \tag{9}$$

This operation is considerably simpler (and safer) if we restate the problem (1)-(8) in terms of the following dimensionless variables:

$\xi = \frac{x}{L}$	longitudinal coordinate
$\theta_1 = \frac{T_1 - T_0}{\Delta T_{scale}}$	temperature of surface and stream no. 1
$\theta_2 = \frac{T_2 - T_0}{\Delta T_{scale}}$	temperature of surface and stream no. 2
$\Delta T_{scale} = \frac{12\mu L^2 q''}{\rho \, c_P \, \Delta P \, D^3}$	scale of longitudinal temperature rise [picked like this in order to "clean up" eqs. (5,6) to appear as in eqs. (10,11)]
$y = \frac{D_1}{D}$	spacing of subchannel no. 1

$$1 - y = \frac{D_2}{D} \qquad \text{spacing of subchannel no. 2}$$

The problem reduces to integrating for $\theta_1(\xi)$ and $\theta_2(\xi)$ the two equations

$$y^3 \frac{d\theta_1}{d\xi} = 1 + B\left(\theta_2 - \theta_1\right) \tag{10}$$

$$(1 - y)^3 \frac{d\theta_2}{d\xi} = 1 - B\left(\theta_2 - \theta_1\right) \tag{11}$$

by starting from the inlet, where $\theta_1(0) = \theta_2(0) = 0$. The dimensionless group B accounts for the transversal conductance of the board (i.e. its substrate),

$$B = \frac{k_w \, \Delta T_{scale}}{t \, q''} = 12 \, \frac{k_w}{k} \, \frac{\mu \, \alpha \, L^2}{\Delta P \cdot D^3 t} \tag{12}$$

First, we eliminate θ_1 by adding eqs. (10) and (11),

$$\theta_1 = \frac{2}{y^3} \xi - \left(\frac{1 - y}{y}\right)^3 \theta_2 \tag{13}$$

We substitute this into eq. (11) and obtain a single equation for θ_2,

$$\frac{d\theta_2}{d\xi} + B\left[\frac{1}{(1 - y)^3} + \frac{1}{y^3}\right] \theta_2 - \frac{1}{(1 - y)^3} - \frac{2 B \xi}{(1 - y)^3 \, y^3} = 0 \tag{14}$$

The solution to this first-order linear ordinary differential equation is (recall that $\theta_2 = 0$ at $\xi = 0$)

$$\theta_2(\xi) = \left(\frac{a}{p} - \frac{b}{p^2}\right)\left[1 - \exp\left(- p\xi\right)\right] + \frac{b}{p} \xi \tag{15}$$

with the shorthand notation

$$p = B\left[\frac{1}{(1 - y)^3} + \frac{1}{y^3}\right] \tag{16}$$

$$a = \frac{1}{(1 - y)^3} \tag{17}$$

$$b = \frac{2B}{(1-y)^3 y^3} \tag{18}$$

The highest temperature of surface no. 2 is at the trailing edge, $\theta_{h,2} = \theta_2(1)$, namely

$$\theta_{h,2} = \left(\frac{a}{p} - \frac{b}{p^2}\right)[1 - \exp(-p)] + \frac{b}{p} \tag{19}$$

This temperature is a function the position of the board (y) and the board conductance number (B), as shown in the attached figure.

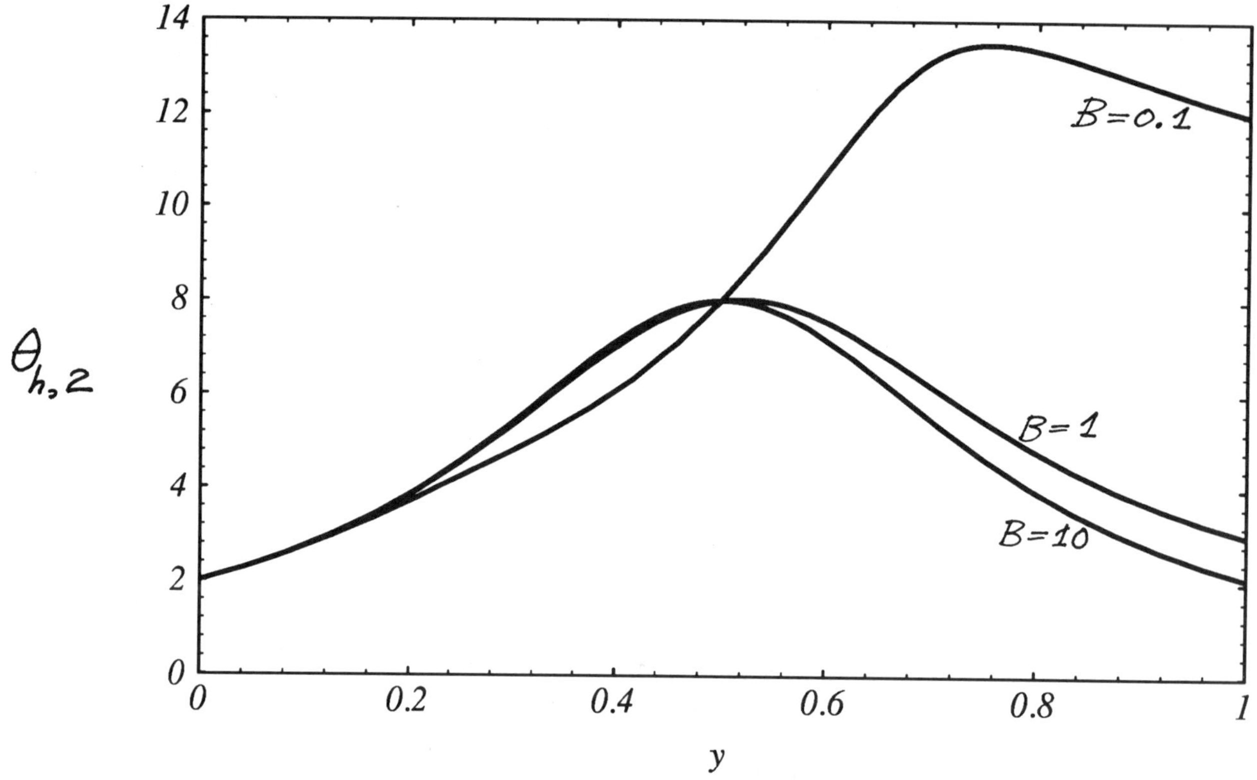

The highest temperature for surface no. 1, $\theta_{h,1}$, is obtained by switching y and (1 – y) in the $\theta_{h,2}$ solution (19). Graphically, this is the same as superimposing on the attached figure another set of curves (for $\theta_{h,1}$) that are the mirror image of the $\theta_{h,2}$ curves (the mirror is the y = 1/2 vertical line). On the composite graph that results, we seek the design (board position y) that results in the lowest $\theta_{h,1}$ and $\theta_{h,2}$, when B is specified. The answer depends on whether the board substrate is a good conductor:

a) When B is of the order of 1 or larger, the $\theta_{h,1}$ and $\theta_{h,2}$ curves are bell shaped and fall on top of each other. The lowest temperatures are registered at y = 0 and y = 1, i.e. when the board is positioned close to one of the insulated walls of the channel. The worst position is in the middle of the channel, y = 1/2, where the highest temperature rise ($\theta_{h,1}$, or $\theta_{h,2}$) is about four times greater than when the board is mounted close to one of the insulated walls.

b) When the board is a poor thermal conductor, such that B is smaller than the order of 1, then $\theta_{h,1}$ and $\theta_{h,2}$ curves intersect forming a cusp (a V-shaped valley) at $y = 1/2$. That intersection corresponds to the lowest ($\theta_{h,1} = \theta_{h,2}$) values, indicating that the best position for the board is along the midplane of the $D \times L$ channel.

The student may wish to examine eq. (19) more closely, to determine the exact B that marks the transition from the optimal design for

a) highly conducting boards, $y_{opt} = 0,1$

to

b) poorly conducting boards, $y_{opt} = 0.5$

That "critical" B value is obtained by setting $\theta_{h,2}(1/2) = \theta_{h,2}(1)$, and solving for B.

It is absolutely fascinating that the optimal design for boards of type (b) [namely $y = 1/2$] is the same as the worst possible design for boards of type (a)! This observation stresses the importance of the dimensionless number B: this must be calculated early, to determine the problem character, (a) versus (b).

Chapter 7

NATURAL CONVECTION

Problem 7.1. a) The $\rho = \rho(T,P)$ equation of state of air as an ideal gas is

$$\rho = \frac{P}{RT}$$

therefore

$$\frac{1}{\rho}\left(\frac{\partial \rho}{\partial T}\right)_P = -\frac{1}{T}, \qquad \text{and} \qquad \frac{1}{\rho}\left(\frac{\partial \rho}{\partial P}\right)_T = \frac{1}{P}$$

Near the reference state ρ_∞ (T_∞, P_0) these quantities are respectively equal to

$$-\frac{1}{T_\infty}, \qquad \text{and} \qquad \frac{1}{P_0}$$

These results can be substituted in eq. (7.13), which after division by ρ_∞ reads

$$\frac{\rho_\infty}{\rho} \cong 1 - \frac{T - T_\infty}{T_\infty} + \frac{P - P_0}{P_0} \tag{a}$$

b) The hydrostatic temperature difference between the floor and ceiling of a room of height H = 4m filled with 20°C air is

$$P - P_0 \sim \rho g H = 1.205 \times 10^{-3} \frac{g}{cm^3} (9.81) \frac{m}{s^2} 4m$$

$$= 473 \frac{g}{cm\ s^2} = 47.3 \frac{N}{m^2}$$

$$= 0.0047 \text{ atm}$$

When $T_\infty = (273.15 + 20)$K, $T - T_\infty \sim 20$ K and $P_0 = 1$ atm, eq. (a) becomes

$$\frac{\rho}{\rho_\infty} \cong 1 - \frac{20}{293.15} + \frac{0.0047}{1}$$

$$\cong 1 - \underbrace{0.068}_{\Delta T \text{ effect}} + \underbrace{0.0047}_{\Delta P \text{ effect}} \tag{b}$$

Equation (b) shows that the ΔP effect on density is between one and two orders of magnitude smaller than the ΔT effect. Worth keeping in mind is that ΔP is proportional to the height H, therefore in configurations where the heated (or cooled) walls are shorter than 4m, the ΔP effect is even smaller than in eq. (b).

Problem 7.2. To the left side of eq. (7.17) we add the left side of eq. (7.2) multiplied by v,

$$v\frac{\partial u}{\partial x} + v\frac{\partial v}{\partial y} +$$

$$u\frac{\partial v}{\partial x} + v\frac{\partial v}{\partial y} = \nu\frac{\partial^2 v}{\partial x^2} + g\beta(T - T_\infty)$$

and the resulting boundary layer-simplified momentum equation is

$$\frac{\partial(uv)}{\partial x} + \frac{\partial(v^2)}{\partial y} = \nu\frac{\partial^2 v}{\partial x^2} + g\beta(T - T_\infty)$$

Integrating this equation from x = 0 to x = δ,

$$(uv)_\delta - (uv)_0 + \int_0^\delta \frac{\partial(v^2)}{\partial y}\,dx = \nu\left(\frac{\partial v}{\partial x}\right)_\delta - \nu\left(\frac{\partial v}{\partial x}\right)_0$$

$$+ \int_0^\delta g\beta(T - T_\infty)\,dx$$

we note that $v_\delta = 0$, $u_0 = 0$ and $(\partial v/\partial x)_\delta = 0$, therefore

$$\int_0^\delta \frac{\partial(v^2)}{\partial y}\,dx = -\nu\left(\frac{\partial v}{\partial x}\right)_0 + \int_0^\delta g\beta(T - T_\infty)\,dx$$

According to Leibniz' formula for differentiating an integral, the left side of the above equation is equal to

$$\int_0^\delta \frac{\partial(v^2)}{\partial y}\,dx = \frac{d}{dy}\int_0^\delta v^2dx - \frac{d\delta}{\delta x}(v^2)_\delta$$

$$= \frac{d}{dy}\int_0^\delta v^2dx\,,$$

because $(v^2)_\delta = 0$. In conclusion, the integral form of the momentum equation for boundary layer natural convection along a vertical wall is

$$\frac{d}{dy}\int_0^\delta v^2 dx = -\nu\left(\frac{\partial v}{\partial x}\right)_0 + \int_0^\delta g\beta(T - T_\infty)\, dx$$

Problem 7.3. To the left side of eq. (7.11) we add the left side of the mass conservation eq. (7.2) multiplied by T,

$$T\frac{\partial u}{\partial x} + T\frac{\partial v}{\partial y}$$

$$+ u\frac{\partial T}{\partial x} + v\frac{\partial T}{\partial y} = \alpha\frac{\partial^2 T}{\partial x^2}$$

and obtain

$$\frac{\partial(uT)}{\partial x} + \frac{\partial(vT)}{\partial y} = \alpha\frac{\partial^2 T}{\partial x^2}$$

Integrating this equation from $x = 0$ to $x = \delta_T$, and noting that $u_0 = 0$, $(uT)_{\delta_T} = u_{\delta_T} T_\infty$ and $(\partial T/\partial x)_{\delta_T} = 0$, we obtain, in order,

$$(uT)_{\delta_T} - (uT)_0 + \int_0^{\delta_T} \frac{\partial}{\partial y}(vT)\, dx = \alpha\left(\frac{\partial T}{\partial x}\right)_{\delta_T} - \alpha\left(\frac{\partial T}{\partial x}\right)_0$$

$$u_{\delta_T} T_\infty + \int_0^{\delta_T} \frac{\partial}{\partial y}(vT)\, dx = -\alpha\left(\frac{\partial T}{\partial x}\right)_0 \qquad (1)$$

The integral appearing on the left side is, cf. Leibniz' formula,

$$\int_0^{\delta_T} \frac{\partial}{\partial y}(vT)\, dx = \frac{d}{dy}\int_0^{\delta_T} vT\, dx - \frac{d\delta_T}{dy}(vT)_{\delta_T}$$

$$= \frac{d}{dy}\int_0^{\delta_T} vT\, dx - \frac{d\delta_T}{dy} v_{\delta_T} T_\infty \qquad (2)$$

The horizontal entrainment velocity u_{δ_T} of the first term of eq. (1) is obtained by integrating eq. (7.2) across the thermal boundary layer,

$$u_{\delta_T} - u_0 + \int_0^{\delta_T} \frac{\partial v}{\partial y}\, dx = 0 \qquad (3)$$

where $u_0 = 0$ and

$$\int_0^{\delta_T} \frac{\partial v}{\partial y}\, dx = \frac{d}{dy}\int_0^{\delta_T} v\, dx - \frac{d\delta_T}{dy} v_{\delta_T} \qquad (4)$$

In conclusion,

$$u_{\delta_T} = -\frac{d}{dy}\int_0^{\delta_T} v\, dx + \frac{d\delta_T}{dy} v_{\delta_T} \qquad (5)$$

and, in view of eqs. (2) and (5), the integral energy equation (1) becomes,

$$-\frac{d}{dy}\int_0^{\delta_T} v\, T_\infty\, dx + T_\infty \frac{d\delta_T}{dy} v_{\delta_T} +$$

$$+\frac{d}{dy}\int_0^{\delta_T} v\, T\, dx - T_\infty \frac{d\delta_T}{dy} v_{\delta_T} = -\alpha \left(\frac{\partial T}{\partial x}\right)_0$$

in other words,

$$\frac{d}{dy}\int_0^{\delta_T} v\,(T - T_\infty)\, dx = -\alpha \left(\frac{\partial T}{\partial x}\right)_0$$

Problem 7.4. The two equations to solve are the integral momentum and energy equations listed in the statements of Problems 7.2 and 7.3. Substituting the assumed velocity profiles in these equations we obtain

$$\frac{d}{dy}\left(\frac{V^2\delta}{105}\right) = -\nu\frac{V}{\delta} + \frac{g\beta\Delta T}{3}\delta \qquad \text{(M)}$$

$$\frac{d}{dy}(V\delta) = \frac{60}{\delta} \qquad \text{(E)}$$

The unknowns are V(y) and δ(y). Their y dependence is already known from the scale analysis (subsection 7.2.2.),

$$V = C_1\, y^{1/2}, \qquad \delta = C_2\, y^{1/4}$$

where C_1 and C_2 are two constants that follow from eqs. (M) and (E):

$$C_1 = 5.17\,\nu\left(Pr + \frac{20}{21}\right)^{-1/2}\left(\frac{g\beta\Delta T}{\nu^2}\right)^{1/2}$$

$$C_2 = 3.93\,Pr^{-1/2}\left(Pr + \frac{20}{21}\right)^{1/4}\left(\frac{g\beta\Delta T}{\nu^2}\right)^{-1/4}$$

In particular, the complete δ(y) solution reads:

$$\frac{\delta}{y} = 3.93\left(\frac{20/21}{Pr} + 1\right)^{1/4} Ra_y^{-1/4}$$

The local Nusselt number is inversely proportional to the slenderness ratio δ/y:

$$Nu_y = \frac{q''_{w,y}}{\Delta T}\frac{y}{k} = \frac{-k(\partial T/\partial x)_0}{\Delta T}\frac{y}{k} = 2\frac{y}{\delta}$$

$$= 0.508\left(\frac{20/21}{Pr} + 1\right)^{-1/4} Ra_y^{1/4}$$

This integral analysis is known as Squire's solution*. Even though the $\delta_T = \delta$ assumption is valid only when Pr ~ 1, the "heat transfer part" of the solution (Nu_y) behaves properly over the entire Pr domain:

---

* H. B. Squire, solution published in Modern Developments in Fluid Dynamics, Vol. II, S. Goldstein, ed., Dover, New York, 1965, pp. 641-643.

$$Nu_y = 0.508\ Ra_y^{1/4}, \qquad Pr >> 1$$

$$Nu_y = 0.514\ (Ra_y Pr)^{1/4}, \qquad Pr << 1$$

The flow part of the solution (e.g. the vertical velocity profile) fails in the two extremes because its assumed shape is based on only one length scale (δ). Recall that, in general, the v profile has two length scales that in subsection 7.2.2. were labeled as:

	inner thickness	outer thickness
$Pr >> 1$	δ_T	δ
$Pr << 1$	δ_s	δ_T

Problem 7.5. We evaluate the needed properties of air at the film temperature (20°C + 30°C)/2 = 25°C,

$$Pr = 0.72 \qquad k = 0.0255\ \frac{W}{m \cdot K} \qquad \frac{g\beta}{\alpha\nu} = \frac{98.9}{cm^3 K}$$

and calculate in order

$$Ra_H = \frac{g\beta}{\alpha\nu} H^3 \left(T_\infty - T_w\right)$$

$$= \frac{98.9}{cm^3 K} (100\ cm)^3\ (30 - 20)\ K = 9.89 \times 10^8 \qquad \text{(laminar, barely)}$$

$$\overline{Nu}_H = 0.517\ Ra_H^{1/4} = 91.68$$

$$\overline{Nu}_H = \frac{\bar{q}''_w}{\Delta T} \frac{H}{k}$$

$$\bar{q}''_w = 91.68 \times 10K\ 0.0255\ \frac{W}{m \cdot K} \frac{1}{1m} = 23.4\ \frac{W}{m^2}$$

$$A = 1m \times 0.6m = 0.6m^2$$

$$q = \bar{q}''_w A = 23.4\ \frac{W}{m^2}\ 0.6\ m^2 \cong 14\ W$$

Problem 7.6. The relevant properties of water at the film temperature $(55°C + 15°C)/2 = 35°C$ are

$$Pr = 4.87 \qquad k = 0.62 \frac{W}{m \cdot K} \qquad \frac{g\beta}{\alpha\nu} = \frac{30.88 \times 10^3}{cm^3 K}$$

We begin with the calculation of the Rayleigh number, and the identification of the flow regime:

$$Ra_H = \frac{g\beta}{\alpha\nu} H^3 \left(T_w - T_\infty\right)$$

$$= \frac{30.88 \times 10^3}{cm^3\ K} (8\ cm)^3\ (55 - 15)\ K = 6.32 \times 10^8$$

$$Gr_H = \frac{Ra_H}{Pr} = \frac{6.32 \times 10^8}{4.87} = 1.3 \times 10^8 \quad \text{(laminar)}$$

For the overall Nusselt number we can use the correlation (7.62) with $Pr = 4.87$:

$$\overline{Nu_H} = 0.68 + 0.601\ Ra_H^{1/4} = 95.97$$

$$\overline{Nu_H} = \frac{\overline{q''_w}}{\Delta T} \frac{H}{k}$$

$$\overline{q''_w} = 95.97\ (55 - 15)\ K\ 0.62 \frac{W}{m \cdot K} \frac{1}{0.08m} = 29\ 751 \frac{W}{m^2}$$

$$A = (2\ \text{sides})\ 0.08m\ \ 0.15m = 0.024\ m^2$$

$$q = \overline{q''_w}\ A = 29\ 751 \frac{W}{m^2}\ 0.024\ m^2 = 714\ W$$

Problem 7.7. The relevant equations are the integral momentum and energy equations listed in the statements of Problems 7.2 and 7.3, and the additional constraint

$$q''_w = -k\left(\frac{\partial T}{\partial x}\right)_{x=0} = \frac{2k}{\delta(y)}\left[T_w(y) - T_\infty\right] = \text{constant}$$

Substituting the assumed T and v profiles into the integral equations we obtain

$$\frac{d}{dy}\left(\frac{V^2\delta}{105}\right) = -\nu\frac{V}{\delta} + \frac{g\beta}{3}\delta(T_w - T_\infty) \qquad \text{(M)}$$

$$\frac{d}{dy}\left[V\delta(T_w - T_\infty)\right] = 60\,\frac{T_w - T_\infty}{\delta} \qquad \text{(E)}$$

The functions $\delta(y)$, $T_w(y)$ and $V(y)$ are determined based on the three equations listed above. The dependence of δ and $(T_w - T_\infty)$ on y has been determined in subsection 7.2.5.,

$$\delta = C_1\, y^{1/5}, \qquad T_w - T_\infty = C_2\, y^{1/5} \qquad (1,2)$$

The relationship between V and y can be determined by recalling that in $Pr \geq 1$ fluids v is of order $(\alpha/y)\, Ra_y^{1/2}$, where Ra_y was based on y and temperature difference. It follows that V varies as $(1/y)(T_w - T_\infty)^{1/2}(y^3)^{1/2}$, and, since $(T_w - T_\infty)$ is proportional to $y^{1/5}$, the conclusion is

$$V = C_3\, y^{3/5} \qquad (3)$$

The constants C_1, C_2 and C_3 follow from substituting the assumptions (1-3) into q_w = constant, (M) and (E). Of particular interest are the C_1 and C_2 values, which substituted in eqs. (1,2) yield

$$\frac{\delta}{y} = 360^{1/5}\left(\frac{4/5}{Pr} + 1\right)^{1/5} Ra_y^{*\,-1/5}$$

$$T_w - T_\infty = \frac{360^{1/5}}{2}\,\frac{q''_w\, y}{k}\left(\frac{4/5}{Pr} + 1\right)^{1/5} Ra_y^{*\,-1/5}$$

where

$$Ra_y^* = \frac{g\beta y^4 q''_w}{\alpha\nu k}$$

The corresponding local Nusselt number is

$$Nu_y = \frac{q''_w}{T_w - T_\infty} \frac{y}{k} = \frac{2}{360^{1/5}} \left(\frac{4/5}{Pr} + 1 \right)^{-1/5} Ra_y^{*\,1/5}$$

where $2/360^{1/5} = 0.616$. This integral solution was reported by E. M. Sparrow (NACA TN 3508, July 1955).

Problem 7.8. Let q''_y represent the local heat flux from the natural convection boundary layer to the solid-liquid interface T_m,

$$q''_y = \frac{k_f}{y} \left(T_\infty - T_m \right) Nu_y \qquad \text{(a)}$$

where y is measured downward, from the upper edge of the wall, and Nu_y is the local Nusselt number. The same heat flux must penetrate by conduction the solidified layer of local thickness L(y),

$$q''_y = k_s \frac{T_m - T_w}{L} \qquad \text{(b)}$$

The solid-layer thickness is therefore equal to

$$L = \frac{k_s}{k_f} \frac{T_m - T_w}{T_\infty - T_m} \frac{y}{Nu_y} \qquad \text{(c)}$$

for which eq. (7.51) recommends (recall that Pr = 55.9):

$$Nu_y = 0.487\, Ra_y^{1/4} \qquad \text{(d)}$$

Equations (c) and (d) demonstrate that L is proportional to the local boundary layer thickness $y\, Ra_y^{1/4}$.

Before substituting numerical values in eq. (c) it is convenient to first nondimensionalize this equation by dividing both sides by H,

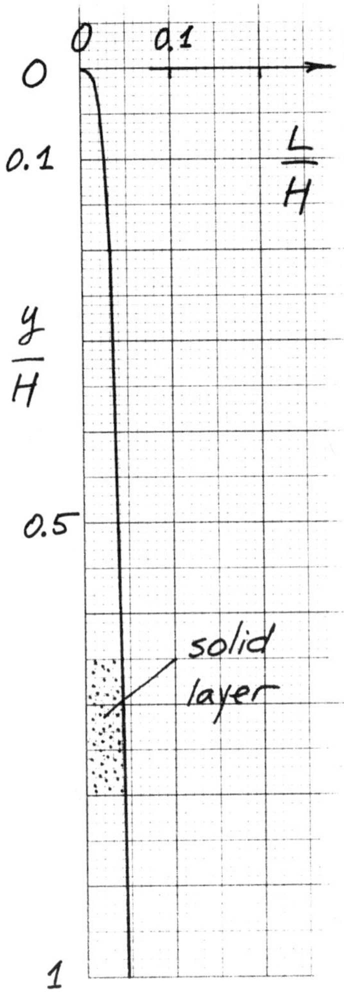

$$\frac{L}{H} = \frac{k_s}{k_f}\,\frac{T_m - T_w}{T_\infty - T_m}\,\frac{1}{0.487\,Ra_H^{1/4}}\left(\frac{y}{H}\right)^{1/4} \qquad \text{(e)}$$

The Rayleigh number based on the overall height is

$$Ra_H = \frac{g\beta\left(T_\infty - T_m\right)H^3}{\alpha\nu} = \frac{9.81\,m}{s^2}\;\frac{8.5\times10^{-4}\,K^{-1}\;(35 - 27.5)\,K\;(0.1m)^3}{9\times10^{-4}\,cm^2/s \qquad 0.05\,cm^2/s}$$

$$= 1.39\times10^8, \qquad \text{(laminar flow)}$$

which means that eq. (e) becomes

$$\frac{L}{H} = \frac{0.36}{0.15}\;\frac{27.5 - 20}{35 - 27.5}\;\frac{1}{0.487\left[1.39\times10^8\right]^{1/4}}\left(\frac{y}{H}\right)^{1/4}$$

$$= 0.045\left(\frac{y}{H}\right)^{1/4}$$

The shape of the L-thin solidified paraffin layer has been drawn to scale in the attached figure.

Problem 7.9. Let us assume that the average wall temperature is sufficiently higher than $T_\infty = 25°C$ so that the film temperature is 30°C. We will validate this assumption later. The relevant properties of water at 30°C are

$$Pr = 5.49 \qquad k = 0.61\,\frac{W}{m\cdot K} \qquad \frac{g\beta}{\alpha\nu} = \frac{25\,130}{cm^3\,K}$$

The flux Rayleigh number suggests that the flow is turbulent, although we will verify the flow regime later:

$$q = 1000\,W$$

$$q''_w = \frac{q}{A} = \frac{1000\,W}{0.5m\;0.5m} = 4000\,W$$

$$Ra_H^* = \frac{g\beta}{\alpha\nu} H^4 \frac{q_w''}{k}$$

$$= \frac{25\ 130}{cm^3\ K} (50\ cm)^4\ 4000 \frac{W}{m^2} \frac{m \cdot K}{0.61\ W} = 1.03 \times 10^{13}$$

For the overall Nusselt number we use eq. (7.70) in which we substitute Pr = 5.49 and neglect the 0.825 term on the right side:

$$\overline{Nu}_H = \left(... + 0.363\ Ra_H^{1/6} \right)^2 = 0.132\ Ra_H^{1/3} \qquad (1)$$

Next, we make the observation that

$$Ra_H = \frac{Ra_H^*}{\overline{Nu}_H}$$

and continue with eq. (1), which yields

$$\overline{Nu}_H^{4/3} = 0.132\ Ra_H^{*\ 1/3}$$

$$\overline{Nu}_H = 0.219\ Ra_H^{*\ 1/4} = 392.3$$

$$\overline{Nu}_H = \frac{q_w''}{\overline{\Delta T}} \frac{H}{k}$$

$$\overline{\Delta T} = \frac{q_w''}{\overline{Nu}_H} \frac{H}{k}$$

$$= \frac{1}{392.3}\ 4000 \frac{W}{m^2} \frac{0.5 m}{0.61 \frac{W}{m \cdot K}} = 8.4°C$$

$$\overline{T}_w = 25°C + 8.4°C = 33.4°C$$

The correct film temperature (25°C + 33.4°C)/2 = 29.2°C is nearly the same as the value assumed at the start.

In order to verify that the boundary layer is indeed turbulent, we evaluate the $\overline{\Delta T}$-based Rayleigh number Ra_H:

$$Ra_H = \frac{Ra_H^*}{\overline{Nu}_H} = \frac{1.03 \times 10^{13}}{392.3} = 2.6 \times 10^{10}$$

and the corresponding Grashof number,

$$Gr_H = \frac{Ra_H}{Pr} = \frac{2.6 \times 10^{10}}{5.49} = 4.8 \times 10^9 \quad \text{(turbulent, barely)}$$

Problem 7.10. The film temperature of the air flow is (25°C + 15°C)/2 = 20°C. The air property needed for calculating the air Rayleigh number is $g\beta/(\alpha\nu)$, which has been calculated and tabulated in Appendix D,

$$\frac{g\beta}{\alpha\nu} \cong 107 \text{ cm}^{-3}\text{ K}^{-1}$$

The air Rayleigh number,

$$Ra_{air} = \frac{g\beta}{\alpha\nu}\Delta T\, H_{air}^3 = \frac{107}{\text{cm}^3\text{ K}}\, 10\text{K}\,(300\text{ cm})^3$$

$$= 2.89 \times 10^{10}$$

is close to the transition regime (Fig. 7.7),

$$Gr_{air} = \left(\frac{Ra}{Pr}\right)_{air} = \frac{2.89 \times 10^{10}}{0.72} = 4 \times 10^{10}$$

The Rayleigh number in the water experiment must be the same as Ra_{air},

$$Ra_{water} = \frac{g\beta}{\alpha\nu}\Delta T\, H_{water}^3 = \frac{14.45 \times 10^3}{\text{cm}^3\text{ K}}\, 10\text{K}\, H_{water}^3$$

$$= 1.445 \times 10^5 \left(\frac{H_{water}}{\text{cm}}\right)^3$$

therefore, setting $Ra_{water} = Ra_{air}$ we obtain

$$H = 58.5 \text{ cm}$$

Problem 7.11. a) The Rayleigh number based on the average temperature difference $(\overline{T}_w - T_\infty)$ can be rearranged as follows:

$$Ra_y = \frac{g\beta y^3 (\overline{T}_w - T_\infty)}{\alpha\nu}$$

$$= \frac{g\beta y^3 (\overline{T}_w - T_\infty)}{\alpha\nu} \frac{q''_w y}{k} \frac{k}{q''_w y}$$

$$= Ra^*_y \frac{k(\overline{T}_w - T_\infty)}{q''_w y}$$

$$= \frac{Ra^*_y}{\overline{Nu}_y}$$

b) Using the $\overline{Nu}_y$ formula (7.67), in which $Ra^*_y = 10^{13}$, we obtain

$$\overline{Nu}_y = 0.75 \, (10^{13})^{1/5} = 298.6$$

The Ra_y value that corresponds to $Ra^*_y = 10^{13}$ is therefore

$$Ra_y = \frac{10^{13}}{298.6} = 3.35 \times 10^{10}$$

This value is comparable with the 7×10^9 value, which is obtained after substituting $Pr \cong 7$ (water) in eq. (7.57).

Problem 7.12. In the $Pr \rightarrow \infty$ limit, eq. (7.55) yields

$$\overline{Nu_y} = 0.671\ Ra_y^{1/4} \qquad (1)$$

This result holds for an isothermal wall (T_w) exposed to an isothermal reservoir (T_∞). It should hold approximately also in the case of the linearly stratified reservoir of Fig. 7.9, provided the $\overline{Nu_y}$ and Ra_y of eq. (1) are based on the average wall-reservoir temperature difference:

$$\frac{\overline{q''_{w,y}}}{\Delta T_{avg}} \frac{y}{k} \cong 0.671 \left(\frac{g\beta y^3\, \Delta T_{avg}}{\alpha \nu}\right)^{1/4} \qquad (2)$$

In order to form the groups $\overline{Nu_H}$ and Ra_H defined by eqs. (7.72a,b) in the text, in the above equation we set y = H, and multiply and divide by ΔT_{max}:

$$\frac{\Delta T_{max}}{\Delta T_{avg}} \frac{\overline{q''_{w,H}}}{\Delta T_{max}} \frac{H}{k} \cong 0.671 \left(\frac{g\beta H^3\, \Delta T_{max}}{\alpha \nu}\right)^{1/4} \left(\frac{\Delta T_{avg}}{\Delta T_{max}}\right)^{1/4} \qquad (3)$$

In short,

$$\frac{\Delta T_{max}}{\Delta T_{avg}} \overline{Nu_H} \cong 0.671\ Ra_H^{1/4} \left(\frac{\Delta T_{avg}}{\Delta T_{max}}\right)^{1/4} \qquad (4)$$

which means that

$$\overline{Nu_H} \cong 0.671 \left(\frac{\Delta T_{avg}}{\Delta T_{max}}\right)^{5/4} Ra_H^{1/4} \qquad (5)$$

Using eq. (7.71), we reach the conclusion that the ratio $\overline{Nu_H}/Ra_H^{1/4}$ must indeed decrease as the stratification parameter b increases:

$$\overline{Nu_H} \cong 0.671 \left(1 - \frac{b}{2}\right)^{5/4} Ra_H^{1/4} \qquad (6)$$

The curve $\overline{Nu_H}/Ra_H^{1/4}$ recommended by eq. (6) falls below the $Pr \rightarrow \infty$ curve of Fig. 7.9., the largest discrepancy (12.8 percent) occuring at the right extremity of the graph (b = 1).

Problem 7.13. The film temperature is (10°C + 20°C)/2 = 15°C. The air properties that will be needed are

$$\mathrm{Pr} = 0.72 \qquad k = 0.025\,\frac{\mathrm{W}}{\mathrm{m \cdot K}} \qquad \frac{g\beta}{\alpha\nu} = \frac{116}{\mathrm{cm^3\,K}}$$

The 5°C/m stratification of the room air means that the top and bottom edges of the window see the following air temperatures:

$$T_\infty\,(\mathrm{top}) = 20°\mathrm{C} + 0.5\mathrm{m}\ 5\,\frac{°\mathrm{C}}{\mathrm{m}} = 22.5°\mathrm{C}$$

$$T_\infty\,(\mathrm{bottom}) = 20°\mathrm{C} - 0.5\mathrm{m}\ 5\,\frac{°\mathrm{C}}{\mathrm{m}} = 17.5°\mathrm{C}$$

The corresponding temperature differences are

$$\Delta T\,(\mathrm{top}) = 22.5°\mathrm{C} - 10°\mathrm{C} = 12.5°\mathrm{C} = \Delta T_{max}$$

$$\Delta T\,(\mathrm{bottom}) = 17.5°\mathrm{C} - 10°\mathrm{C} = 7.5°\mathrm{C} = \Delta T_{min}$$

$$b = \frac{\Delta T_{max} - \Delta T_{min}}{\Delta T_{max}} = \frac{12.5 - 7.5}{12.5} = 0.4$$

The Pr = 0.7 curve of Fig. 7.9 shows that

$$\frac{\overline{Nu}_H}{Ra_H^{1/4}} \cong 0.44 \qquad (1)$$

in which

$$Ra_H = \frac{g\beta}{\alpha\nu}\,H^3\,\Delta T_{max}$$

$$= \frac{116}{\mathrm{cm^3 K}}\,(100\ \mathrm{cm})^3\ 12.5\ \mathrm{K} = 1.45 \times 10^9$$

The flow is transitional (barely laminar), therefore we continue to rely on eq. (1) and Fig. 7.9 while keeping in mind that the actual heat transfer rate may be slightly larger than the purely laminar estimate produced below:

$$\overline{Nu}_H = 0.44\left(1.45 \times 10^9\right)^{1/4} = 85.86$$

$$\overline{Nu}_H = \frac{\overline{q}''_w}{\Delta T_{max}}\,\frac{H}{k}$$

$$\overline{q}''_w = 85.86 \frac{12.5°C}{1m} 0.025 \frac{W}{m \cdot K} = 26.83 \frac{W}{m^2}$$

$$A = 0.8m \; 1m = 0.8 \; m^2$$

$$q = \overline{q}''_w \, A = 26.83 \frac{W}{m^2} 0.8 \; m^2 = 21.5 \; W$$

Problem 7.14. The film temperature is (43.1°C + 23.6°C)/2 = 33.4°C; after interpolating linearly in Appendix C we find the needed properties of water:

$$k = 0.62 \frac{W}{m \cdot K} \qquad \frac{g\beta}{\alpha\nu} = \frac{28\;982}{cm^3 \; K}$$

The characteristic length of the upward facing disc is

$$L = \frac{A}{p} = \frac{(\pi/4)\; D^2}{\pi D} = \frac{D}{4} = \frac{8.7 \; cm}{4} = 2.18 \; cm$$

with the corresponding Rayleigh number

$$Ra_L = \frac{g\beta}{\alpha\nu} L^3 \left(T_w - T_\infty\right)$$

$$= \frac{28\;982}{cm^3 \; K} (2.18 \; cm)^3 \; (43.1 - 23.6) \; K = 5.86 \times 10^6$$

The appropriate heat transfer correlation is eq. (7.77a):

$$\overline{Nu_L} = 0.54 \; Ra_L^{1/4}$$

$$= 0.54 \left(5.86 \times 10^6\right)^{1/4} = 26.56$$

$$\overline{h} = \overline{Nu_L} \frac{k}{L}$$

$$= 26.56 \times 0.62 \frac{W}{m \cdot K} \frac{1}{0.0218m} = 755.5 \frac{W}{m^2 K}$$

$$q = \overline{h} \left(\frac{\pi}{4} D^2\right) \left(T_w - T_\infty\right)$$

$$= 755.5 \frac{W}{m^2 \; K} \frac{\pi}{4} (0.087 \; m)^2 \; (43.1 - 23.6) \; K = 87.6 \; W$$

Problem 7.15. The air properties evaluated at the film temperature $(20°C + 40°C)/2 = 30°C$ are

$$Pr = 0.72 \qquad k = 0.026 \frac{W}{m \cdot K} \qquad \frac{g\beta}{\alpha\nu} = \frac{90.7}{cm^3\,K}$$

a) Vertical plate of height H = 4 cm:

$$Ra_H = \frac{g\beta}{\alpha\nu} H^3 \left(T_w - T_\infty\right)$$

$$= \frac{90.7}{cm^3\,K}(4\ cm)^3\ (40 - 20)\ K = 1.16 \times 10^5$$

The Nusselt number correlation is eq. (7.62'):

$$\overline{Nu}_H = 0.68 + 0.515\ Ra_H^{1/4}$$

$$= 0.68 + 0.515\left(1.16 \times 10^5\right)^{1/4} = 10.19$$

$$\bar{h} = \overline{Nu}_H \frac{k}{H}$$

$$= 10.19 \times 0.026 \frac{W}{m \cdot K}\ \frac{1}{0.04\ m} = 6.62 \frac{W}{m^2\,K}$$

This $\bar{h}$ value applies to both sides of the plate. The total heat transfer rate is

$$q' = \bar{h}\ (2\ \text{sides})\ H\left(T_w - T_\infty\right)$$

$$= 6.62 \frac{W}{m^2\,K}\ 2 \times 0.04m\ (40 - 20)\ K = 10.6 \frac{W}{m}$$

b) Plate inclined at 45°C: the calculation of q' consists of repeating part (a) in which g is replaced by g cos (45°) = 0.707 g. The results are, in order,

$$Ra_H = 8.2 \times 10^4$$

$$\overline{Nu}_H = 0.68 + 0.515\ Ra_H^{1/4} = 9.4$$

$$\bar{h} = 6.1 \frac{W}{m^2\,K} \qquad \text{(this applies to both sides of the plate)}$$

$$q' = \bar{h}\ (2\ \text{sides})\ H\left(T_w - T_\infty\right) = 9.77 \frac{W}{m}$$

c) Horizontal plate: in this case we distinguish between the upper surface (hot, facing upward), and the lower surface of the plate (cold, facing downward).

For the upper surface, we begin with the characteristic length of the rectangular surface:

$$L = \frac{A}{p} = \frac{H \times \text{length}}{2 \times \text{length}} = 2 \text{ cm}$$

$$Ra_L = \frac{g\beta}{\alpha\nu} L^3 \left(T_w - T_\infty\right) = 1.45 \times 10^4$$

Equation (7.77a) delivers the Nusselt number:

$$\overline{Nu_L} = 0.54\, Ra_L^{1/4} = 5.93$$

$$\bar{h}_{upper} = \overline{Nu_L}\, \frac{k}{L}$$

$$= 5.93 \times 0.026 \frac{W}{m \cdot K} \frac{1}{0.02 \text{ m}} = 7.7 \frac{W}{m^2 K}$$

The lower surface has the same L and Ra_L. As Nusselt number correlation we use eq. (7.78): this choice is an approximate one because eq. (7.78) is more accurate when Ra_L exceeds 10^5. We obtain:

$$\overline{Nu_L} = 0.27\, Ra_L^{1/4} = 2.96$$

$$\bar{h}_{lower} = \overline{Nu_L}\, \frac{k}{L} = 3.85 \frac{W}{m^2 K}$$

$$q' = q'_{upper} + q'_{lower}$$

$$= \bar{h}_{upper} H \left(T_w - T_\infty\right) + \bar{h}_{lower} H \left(T_w - T_\infty\right)$$

$$= \left(\bar{h}_{upper} + \bar{h}_{lower}\right) H \left(T_w - T_\infty\right)$$

$$= (7.7 + 3.85) \frac{W}{m^2 K}\, 0.04 \text{ m } (40 - 20) \text{ K}$$

$$= 9.24 \frac{W}{m}$$

In conclusion, the best position for heat transfer is the vertical one, and the worst the horizontal. The difference between these extremes is relatively small, only 13 percent.

Problem 7.16. The properties of air at the film temperature $(10°C + 30°C)/2 = 20°C$ are

$$Pr = 0.72 \qquad k = 0.025 \frac{W}{m \cdot K} \qquad \frac{g\beta}{\alpha\nu} = \frac{107}{cm^3 K}$$

a) Sphere with the diameter D = 3m

$$Ra_D = \frac{g\beta}{\alpha\nu} = D^3 \left(T_\infty - T_w\right)$$

$$= \frac{107}{cm^3 K} (300 \text{ cm})^3 (30 - 10) \text{ K} = 5.78 \times 10^{10}$$

Next, we substitute this Ra_D and Pr = 0.72 in eq. (7.81):

$$\overline{Nu}_D = 2 + 0.455 \, Ra_D^{1/4} = 225.2$$

$$\bar{h} = \overline{Nu}_D \frac{k}{D} =$$

$$= 225.2 \times 0.025 \frac{W}{m \cdot K} \frac{1}{3m} = 1.88 \frac{W}{m^2 K}$$

$$q = \bar{h} \, \pi D^2 \left(T_\infty - T_w\right)$$

$$= 1.88 \frac{W}{m^2 K} \pi (3m)^2 (30 - 10) \text{ K} = 1.06 \text{ kW}$$

b) Horizontal cylinder with the diameter d = 1.5m. To have the same volume as the D-sphere, the cylinder must have the length:

$$L = \frac{\frac{4}{3}\pi \left(\frac{D}{2}\right)^3}{\pi \left(\frac{d}{2}\right)^2} = \frac{2}{3} \frac{D^3}{d^2} = 8m$$

$$Ra_d = \frac{g\beta}{\alpha\nu} d^3 \left(T_\infty - T_w\right) = 7.22 \times 10^9$$

The overall Nusselt number is supplied by eq. (7.79), in which Pr = 0.72:

$$\overline{Nu}_d = \left(0.6 + 0.322\, Ra_d^{1/6}\right)^2 = 217.3$$

$$\bar{h} = \overline{Nu}_d \frac{k}{d}$$

$$= 217.3 \times 0.025 \frac{W}{m{\cdot}K} \frac{1}{1.5m} = 3.62 \frac{W}{m^2\,K}$$

The total area of the horizontal cylinder is

$$A = \pi dL + 2\pi \left(\frac{d}{2}\right)^2$$

$$= \pi d \left(L + \frac{d}{2}\right)$$

$$= \pi\ 1.5m\ (8m + 0.75m) = 41.23\ m^2$$

Since the area of the two end discs is only 8.6 percent of the total area, it is safe (i.e. a good approximation) to assume that the same $\bar{h}$ holds over the entire area:

$$q = \bar{h} A \left(T_\infty - T_w\right)$$

$$= 3.62 \frac{W}{m{\cdot}K}\ 41.23\ m^2\ (30 - 10)\ K \cong 3\ kW$$

c) The cylinder will be heated three times faster than the sphere, therefore the preferred shape is the sphere.

Problem 7.17. The film temperature and the water properties that will be needed are

$$T_{film} = (80°C + 20°C)/2 = 50°C$$

$$Pr = 3.57 \qquad k = 0.64 \frac{W}{m{\cdot}K} \qquad \frac{g\beta}{\alpha\nu} = \frac{51\ 410}{cm^3\ K}$$

a) Focusing on eq. (7.84), we calculate in order

$$l = 1 \text{ cm} + 2 \text{ cm} + 1 \text{ cm} = 4 \text{ cm}$$

$$Ra_l = \frac{g\beta}{\alpha\nu} l^3 \left(T_w - T_\infty\right)$$

$$= \frac{51\ 410}{\text{cm}^3\text{ K}} (4 \text{ cm})^3 (80 - 20) \text{ K} = 1.974 \times 10^8 \quad \text{(laminar)}$$

$$\overline{Nu_l} = 0.52\ Ra_l^{1/4} = 61.64$$

$$\overline{Nu_l} = \frac{\bar{h}\, l}{k}$$

$$\bar{h} = 61.64 \frac{0.64 \frac{\text{W}}{\text{m·K}}}{0.04\text{m}} = 986.2 \frac{\text{W}}{\text{m}^2\text{ K}}$$

b) If we rely on eq. (7.86) and Table 7.1, we proceed as follows:

$$\underline{L} = A^{1/2} = \left[6\,(2 \text{ cm})^2\right]^{1/2} = 4.9 \text{ cm}$$

$$Ra_{\underline{L}} = \frac{g\beta}{\alpha\nu} \underline{L}^3 \left(T_w - T_\infty\right) = 3.63 \times 10^8 \quad \text{(laminar)}$$

$$\overline{Nu_{\underline{L}}} = 3.388 + \frac{0.67 \times 0.951\ Ra_{\underline{L}}^{1/4}}{\left[1 + \left(\frac{0.492}{3.57}\right)^{9/16}\right]^{4/9}}$$

$$= 3.388 + 0.562\ Ra_{\underline{L}}^{1/4} = 80.95$$

$$\bar{h} = 80.95 \frac{0.64 \frac{\text{W}}{\text{m·K}}}{0.049\text{m}} = 1057.4 \frac{\text{W}}{\text{m}^2\text{ K}}$$

c) On the other hand, if we use the simpler formula (7.87), we obtain

$$\overline{Nu_{\underline{L}}} = 3.47 + 0.51 \left(3.63 \times 10^8\right)^{1/4}$$

$$= 73.87$$

$$\bar{h} = 73.87 \frac{0.64 \frac{\text{W}}{\text{m·K}}}{0.049\text{m}} = 964.8 \frac{\text{W}}{\text{m}^2\text{ K}}$$

If we regard estimate (b) as reference, we see that estimate (a) is off (smaller) by only 6.7 percent , while estimate (c) is off by 8.8 percent. These deviations are insignificant in heat transfer calculations (think of the uncertainty in the values used for water properties), therefore all three methods are satisfactory.

Problem 7.18. The six surfaces of the parallelepiped are of four types (A_1, A_2, A_3, A_4), therefore

$$q = \overline{q_1''} A_1 + \overline{q_2''}\, 2A_2 + \overline{q_3''}\, 2A_3 + \overline{q_4''} A_4$$

$$= \left(\overline{h_1} A_1 + 2\overline{h_2} A_2 + 2\overline{h_3} A_3 + \overline{h_4} A_4\right) \Delta T$$

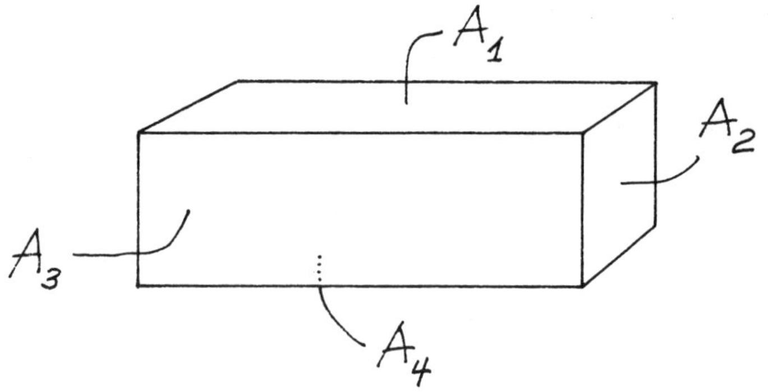

All the air properties are evaluated at the film temperature (0°C + 20°C)/2 = 10°C; namely:

$$k \cong 2.5 \times 10^{-4} \frac{W}{cm\,K}, \qquad \frac{g\beta}{\alpha\nu} \cong \frac{125}{cm^3\,K}$$

The A_1 surface is cold and faces upward. The heat transfer coefficient follows from eq. (7.78)

$$\overline{Nu_L} = 0.27\, Ra_L^{1/4} \qquad (7.78)$$

in which the length scale L is

$$L = \frac{A_1}{p_1} = \frac{30\text{ cm } 1\text{m}}{2\,(1\text{m} + 30\text{ cm})} = \frac{30}{2\,(130)}\text{ m}$$

$$= 11.54\text{ cm}$$

Therefore, the Rayleigh number based on L is

$$Ra_L = \frac{g\beta L^3 \Delta T}{\alpha\nu} = \frac{125}{cm^3\,K}\,(11.54)^3\, cm^3\; 20\text{ K}$$

$$= 3.84 \times 10^6,$$

and, from eq. (7.78), $\overline{Nu_L} = 11.95$. The $\overline{h_1}$ coefficient is

$$\overline{h_1} = \overline{Nu_L}\,\frac{k}{L} = 11.95\,\frac{(2.5)\,10^{-4}W}{cm\,K}\,\frac{1}{11.54\,cm}$$

$$= 2.59 \times 10^{-4}\,\frac{W}{cm^2\,K} = 2.59\,\frac{W}{m^2\,K}$$

The surfaces A_2 and A_3 are both vertical of height H = 0.3m, therefore

$$\overline{h_2} = \overline{h_3} = \overline{Nu_H}\,\frac{k}{H} \tag{1}$$

The average Nusselt number is given by eq. (7.62),

$$\overline{Nu_H} = 0.68 + \frac{0.67\,Ra_H^{1/4}}{\left[1 + (0.492/Pr)^{9/16}\right]^{4/9}} \tag{7.62}$$

which in the case of air (Pr = 0.72) reduces to

$$\overline{Nu_H} = 0.68 + 0.515\,Ra_H^{1/4}$$

The Rayleigh number based on H is

$$Ra_H = \frac{125}{cm^3\,K}\,30^3\,cm^3\,20\,K = 6.75 \times 10^7$$

therefore $\overline{Nu_H} = 47.36$, and

$$\overline{h_2} = \overline{h_3} = 47.36\,\frac{2.5 \times 10^{-4}\,W}{cm\,K}\,\frac{1}{30\,cm}$$

$$= 3.95 \times 10^{-4}\,\frac{W}{cm^2\,K} = 3.95\,\frac{W}{m^2\,K}$$

Finally, the A_4 surface is cold and faces downward. It has the same L = 11.54 cm and $Ra_L = (3.84)\,10^6$ as the A_1 surface, therefore eq. (7.77a) yields

$$\overline{Nu_L} = 0.54\,Ra_L^{1/4} \tag{7.77a}$$

$$= 23.9$$

which translates into the average heat transfer coefficient

$$\overline{h_4} = \overline{Nu_L}\,\frac{k}{L} = 23.9\,\frac{2.5 \times 10^{-4}\,W}{cm\,K}\,\frac{1}{11.54\,cm}$$

$$= 5.18 \times 10^{-4} \frac{W}{cm^2\,K} = 5.18 \frac{W}{m^2\,K}$$

In summary, the q expression listed at the start of this solution becomes

$$q = [2.59 \times 1 \times 0.3 + 2 \times 3.95 \times (0.3)^2 +$$

$$+ 2 \times 3.95 \times 1 \times 0.3 + 5.18 \times 1 \times 0.3] \frac{W}{m^2\,K} m^2\, 20°C$$

$$= 108.24\ W$$

The instantaneous melting rate is proportional to q,

$$q = \dot{m}\, h_{sf}$$

where the latent heat of melting is h_{sf} = 333.4 kJ/kg, therefore

$$\dot{m} = \frac{q}{h_{sf}} = \frac{108.24\ W}{333.4\ kJ/kg} = 0.325 \frac{W}{J/g}$$

$$= 0.325 \frac{g}{s}$$

Problem 7.19. In view of the elongated shape of the ice block, the length l used in Lienhard's formula

$$\overline{Nu_l} \cong 0.52\ Ra_l^{1/4} \qquad (7.84)$$

can be approximated by measuring half of the perimeter of the smaller (square) cross-section:

$$l = \frac{0.3m}{2} + 0.3m + \frac{0.3m}{2} = 0.6m$$

Evaluating all the air properties at 10°C, we calculate in order

$$Ra_l = \frac{125}{cm^3\,K}\, 20°C\, (60)^3\, cm^3 = 5.4 \times 10^8$$

$$\overline{Nu_l} \cong 0.52 \left[(5.4)\, 10^8\right]^{1/4} = 79.27$$

$$\overline{h} = \overline{Nu}_l \frac{k}{l} = 79.27 \frac{(2.5)\,10^{-4}W}{cm\,K} \frac{1}{60\,cm}$$

$$= 3.3 \frac{W}{m^2K}$$

The total area of the ice block is

$$A = 2(0.3m)^2 + 4(0.3)(1)m^2 = 1.38m^2$$

and the total heat transfer rate can now be evaluated by simply writing

$$q = \overline{h}A\Delta T = 3.3 \frac{W}{m^2K} 1.38m^2\ 20°C$$

$$= 91.1\ W$$

This "quick" estimate is only 16 percent lower than the more rigorous (and tedious) result of the preceding problem.

Problem 7.20. Let ΔT be the unknown of the problem, i.e. the temperature difference between wire and ambient air. The conservation of energy in each cross-section requires

$$q' = \overline{h}\,\pi D\,\Delta T = \overline{Nu}_D\,\pi k\,\Delta T$$

therefore ΔT can be calculated with the formula

$$\Delta T = \frac{q'}{\pi\, k\, \overline{Nu}_D} \qquad (1)$$

The average Nusselt number is given by eq. (7.79),

$$\overline{Nu}_D = \left\{0.6 + \frac{0.387\ Ra_D^{1/6}}{\left[1 + (0.559/Pr)^{9/16}\right]^{8/27}}\right\}^2 \qquad (7.79)$$

which in the case of air ($Pr = 0.72$) reduces to

$$\overline{Nu}_D = \left(0.6 + 0.332\ Ra_D^{1/6}\right)^2 \qquad (2)$$

The Rayleigh number depends on the unknown ΔT,

$$Ra_D = \frac{g\beta}{\alpha\nu} D^3 \Delta T \qquad (3)$$

If we do not have access to a programmable calculator, the temperature difference ΔT can be calculated by trial and error, executing the following steps:

i) assume a ΔT value (labeled ΔT_a),

ii) calculate sequentially Ra_D and $\overline{Nu_D}$, using eqs. (3) and (2),

iii) calculate the ΔT value (labeled ΔT_c), and compare it with the ΔT_a guess.

iv) if ΔT_c differs greatly from ΔT_a, go back and repeat the (i)-(iv) sequence.

In the first iteration of this calculation, we assume

$$\Delta T_a = 10°C$$

and calculate the air properties at the reservoir temperature of 20°C,

$$k \cong 2.5 \times 10^{-4} \frac{W}{cm\ K}, \qquad \frac{g\beta}{\alpha\nu} \cong \frac{107}{cm^3\ K}$$

Normally, we would be evaluating these properties at the film temperature (20°C + 30°C)/2 = 25°C, however, this first iteration is "rough", and the above properties are accurate enough. From eqs. (3) and (2) we obtain:

$$Ra_D = 1.07, \qquad \text{and} \qquad \overline{Nu_D} = 0.857$$

Finally, eq. (1) yields

$$\Delta T_c = \frac{0.01\ W}{cm} \frac{cm\ K}{\pi\ 2.5 \times 10^{-4}\ W\ 0.857}$$

$$= 14.86\ K = 14.86°C$$

In conclusion, $\Delta T_c > \Delta T_a$; the second iteration begins with assuming a larger ΔT_a value:

$$\Delta T_a = 20°C$$

This time we evaluate the properties at the film temperature, (20° + 40°)/2 = 30°C:

$$k \cong 2.6 \times 10^{-4} \frac{W}{cm\ K}, \qquad \frac{g\beta}{\alpha\nu} \cong \frac{88}{cm^3\ K}$$

Equations (3), (2) and (1) yield in order

$Ra_D = 1.76, \qquad \overline{Nu}_D = 0.91, \qquad \Delta T_c = 13.46°C$

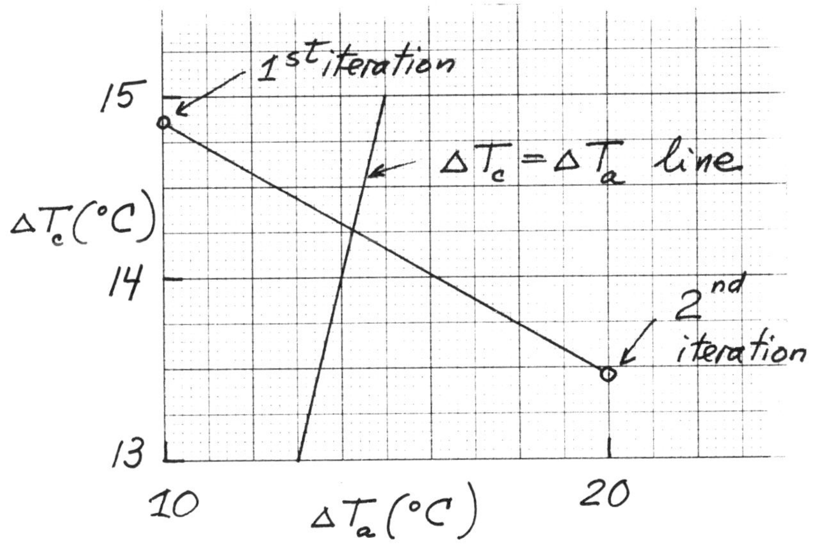

This time, the calculated ΔT is smaller than the assumed value, ΔT_a. In summary, the two iterations carried out so far allow us to draw a line in the plane (ΔT_a, ΔT_c). Intersecting this line with the theoretical line $\Delta T_a = \Delta T_c$ (see the attached sketch), we obtain the answer to the problem:

$$\Delta T \cong 14.3°C$$

If we have access to a programmable calculator, we can solve eq. (2) directly for ΔT. We still have to assume the film temperature, in order to evaluate the physical properties. To begin with, we evaluate all the properties at 20°C, and eq. (2) becomes

$$\frac{12.73°C}{\Delta T} = \left[0.6 + 0.332\left(\frac{\Delta T}{9.35°C}\right)^{1/6}\right]^2$$

with the solution ΔT = 13.96°C. This answer suggests that the film temperature is higher than the assumed 20°C. Therefore, as a second try we evaluate the air properties at 30°C; eq. (2) changes slightly,

$$\frac{12.24°C}{\Delta T} = \left[0.6 + 0.332\left(\frac{\Delta T}{11.36°C}\right)^{1/6}\right]^2$$

and the new solution is ΔT = 13.77°C. In summary, the table below shows that the true film temperature is approximately 20°C + ΔT/2 ≅ 26.9°C.

T_{film}	20°C	30°C
ΔT	13.96°C	13.77°C

Interpolating between the values listed in the table, we obtain the final answer, ΔT = 13.83°C, which is not far off the trial-and-error value determined in the first part of this solution, ΔT ≅ 14.3°C.

Problem 7.21. Since each cylinder is surrounded by its own boundary layer, the heat transfer is impeded first by the boundary layer resistance between T_1 and the water reservoir [$T_\infty = (T_1 + T_2)/2 = 25°C$], and later by the corresponding boundary layer resistance between T_∞ and T_2. The two resistances are equal and in series. They are equal because of symmetry and the assumption that the water properties are the same (well, nearly the same) in the two boundary layers.

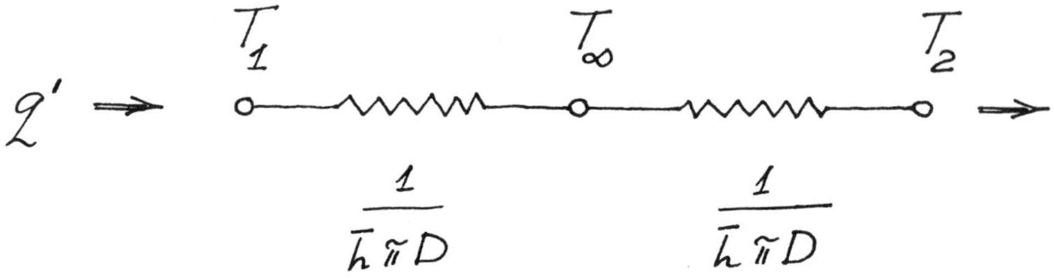

The heat transfer rate from the cylinder T_1 to the cylinder T_2, expressed per unit length in the direction perpendicular to the cross-section, is

$$q' = \frac{T_1 - T_2}{\dfrac{1}{\overline{h}\,\pi D} + \dfrac{1}{\overline{h}\,\pi D}} = \frac{\pi}{2} k\, \overline{Nu}_D\, (T_1 - T_2)$$

where $\overline{Nu}_D$ can be estimated based on eq. (7.79). The Rayleigh number in that correlation, Ra_D, is based on the temperature difference across one boundary layer, e.g. $\Delta T = T_1 - T_\infty = 5°C$, and on properties evaluated at the film temperature [$(30°C + 25°C)/2 = 27.5°C$, or $(25°C + 20°C)/2 = 22.5°C$]. We assume that the two sets of film properties are approximated well by the properties evaluated at the in-between temperature of $(27.5°C + 22.5°C)/2 = 25°C$,

$$k = 0.6\,\frac{W}{m \cdot K} \qquad Pr = 6.21 \qquad \frac{g\beta}{\alpha\nu} = \frac{19.81 \times 10^3}{cm^3\,K}$$

and calculate, in order,

$$Ra_D = \frac{g\beta}{\alpha\nu} D^3\, \Delta T$$

$$= \frac{19.81 \times 10^3}{cm^3\,K} (4\ cm)^3\, 5K = 6.34 \times 10^6$$

$$\overline{Nu}_D = \cdots = \left\{0.6 + \frac{0.387 \times 13.6}{1.07}\right\}^2 = 30.45 \qquad (7.79)$$

$$q' = \frac{\pi}{2}\, 0.6\,\frac{W}{m \cdot K}\, 30.45\, (30 - 20)\ K = 287\,\frac{W}{m}$$

Problem 7.22. The following are the needed properties of water at 95°C, 1 atm and sea level:

$$\frac{g\beta}{\alpha\nu} \cong 132\ \text{cm}^{-3}\ \text{K}^{-1}, \qquad Pr = 1.88, \qquad k = 0.68\ \frac{W}{m\cdot K}$$

a) We calculate, in order, the Rayleigh number based on hot dog diameter,

$$Ra_D = \frac{g\beta}{\alpha\nu}(T_\infty - T_w)\,D^3$$
$$= \frac{132}{cm^3\ K}(95 - 20)\ K\ (1.7\ cm)^3 = 4.86 \times 10^4$$

the overall Nusselt number cf. eq. (7.79),

$$\overline{Nu_D} = \left\{0.6 + \frac{0.387\ Ra_D^{1/6}}{\left[1 + \left(\frac{0.559}{Pr}\right)^{9/16}\right]^{8/27}}\right\}^2$$
$$= \left(0.6 + 0.343\ Ra_D^{1/6}\right)^2 = 7.134$$

the average heat transfer coefficient,

$$\overline{h} = \frac{k}{D}\overline{Nu_D} =$$
$$= 0.68\ \frac{W}{m\cdot K}\ \frac{1}{0.017m}\ 7.134 = 285\ \frac{W}{m^2\ K}$$

and the instantaneous heat transfer rate,

$$q' = \pi D\ \overline{q''_{w,D}} = \pi D\ \overline{h}\left(T_\infty - T_w\right) =$$
$$= \pi\ 0.017m\ 285\ \frac{W}{m^2\ K}\ (95 - 20)\ K$$
$$= 1143\ \frac{W}{m}$$

the lateral heat transfer rate,

$$q_{lat} = q'L = 1143\ \frac{W}{m}\ 0.1m = 114.3\ W$$

the heat transfer through one end (assuming the same $\overline{h}$ value over the end surface)

$$q_{end} = \frac{\pi}{4} D^2 \, \overline{h} \, (T_\infty - T_w)$$

$$= \frac{\pi}{4} (0.017 \text{m})^2 \, 285 \, \frac{\text{W}}{\text{m}^2 \text{K}} (95 - 20) \text{ K} = 4.85 \text{W}$$

and, finally, the total heat transfer rate,

$$q = q_{lat} + 2 q_{end}$$

$$= 114.3 \text{ W} + 2 \times 4.85 \text{ W} = 124 \text{ W}$$

b) As the time increases and the hot dog internal temperature becomes nonuniform (colder in the center), the radial meat conduction contributes a thermal resistance that is greater than (and must be "added" to) the resistance due to external natural convection. Both resistances are time dependent, and must be accounted for according to the method of section 4.4. To prove that this is necessary, we calculate the Biot number (based on the hot dog radius), and find that Bi is greater than 1,

$$Bi = \frac{\overline{h} \, r_o}{k_{meat}}$$

$$= 285 \, \frac{\text{W}}{\text{m}^2 \text{K}} \, \frac{0.0085 \text{m}}{(\sim 0.4) \frac{\text{W}}{\text{m} \cdot \text{K}}} \cong 6$$

Problem 7.23. The cake mix surface is cold (T_w = 22°C) and faces upward in warmer air (T_∞ = 28°C). The calculation is guided by eq. (7.78), and consists of the following steps:

$$A = (46 \text{ cm})^2 = 2116 \text{ cm}^2$$

$$L = \frac{A}{p} = \frac{2116 \text{ cm}^2}{4 \times 46 \text{ cm}} = 11.5 \text{ cm} \qquad (7.76)$$

$$T_{film} = \frac{1}{2} (T_w + T_\infty) = 25°\text{C}$$

$$\left(\frac{g\beta}{\alpha\nu} \right)_{25°\text{C air}} = 98.85 \text{ cm}^{-3} \text{ K}^{-1} \qquad (\text{App. D})$$

$$Ra_L = \frac{g\beta}{\alpha\nu} L^3 (T_\infty - T_w) = 9.02 \times 10^5$$

$$\overline{Nu_L} = 0.27 \, Ra_L^{1/4}$$

$$= 8.32 = \frac{\overline{h} L}{k} \qquad (7.78)$$

$$k_{25^\circ C\ air} = 0.0255 \frac{W}{m \cdot K} \qquad \text{(App. D)}$$

$$h = 8.32 \times 0.0255 \frac{W}{m \cdot K} \frac{1}{0.115 m}$$

$$= 1.85 \frac{W}{m^2\, K}$$

$$q = hA\left(T_\infty - T_w\right)$$

$$= 1.85 \frac{W}{m^2\, K} 2116 \times 10^{-4}\, m^2\, (28 - 22)\, K$$

$$= 2.35\ W$$

The heat transfer rate q is proportional to $h \cdot [T_\infty - T_w(t)]$, where h is proportional to $[T_\infty - T_w(t)]^{1/4}$ cf. eq. (7.78). Therefore q is proportional to the instantaneous temperature difference raised to the power 5/4, and decreases as t increases, because $T_w(t)$ approaches T_∞.

Problem 7.24. Consider a narrow vertical channel of height H, cross-sectional area A, wetted perimeter p, and hydraulic diameter $D_h = 4A/p$. Let V be the average vertical velocity through the channel. Recognizing T_∞ and T_w as the inlet and outlet mean temperatures of the stream, we can easily calculate the total heat transfer rate from the channel walls (T_w) to the stream:

$$q = \rho V A\, c_P\, (T_w - T_\infty) \qquad (1)$$

Defining the channel-averaged heat flux

$$\overline{q''} = \frac{q}{pH} \qquad (2)$$

we rearrange eq. (1) as an average Nusselt number (note $\Delta T = T_w - T_\infty$):

$$\overline{Nu_H} = \frac{\overline{q''}}{\Delta T} \frac{H}{k} = \frac{VA}{\alpha p} = \frac{V D_h}{4\alpha} \qquad (3)$$

The problem reduces to determining the average velocity V. Comparing eqs. (6.20) and (7.91) for fully developed duct flow,

$$\nabla^2 v = \frac{1}{\mu}\frac{dP}{dy}, \qquad \text{(forced convection)} \qquad (6.20)$$

$$\nabla^2 v = -\frac{g\beta\Delta T}{\nu}, \qquad \text{(natural convection)} \qquad (7.91)$$

we learn that the two flows are equivalent if

$$-\frac{dP}{dy} \equiv \rho g \beta \Delta T \tag{4}$$

The left side of this equation can be evaluated based on the fully developed flow solutions summarized in Table 6.1,

$$-\frac{dP}{dy} = f \frac{p}{A} \frac{1}{2} \rho V^2 \tag{5}$$

in other words, using eq. (4),

$$V^2 = \frac{g\beta\Delta T D_h}{2f} = \frac{g\beta\Delta T D_h}{2\left(f\,Re_{D_h}\right)} \frac{V D_h}{\nu} \tag{6}$$

or

$$V = \frac{g\beta\Delta T D_h^2}{2\nu} \frac{1}{\left(f\,Re_{D_h}\right)} \tag{7}$$

Substituting this conclusion into the $\overline{Nu_H}$ expression (3) leads to

$$\overline{Nu_H} = \frac{1}{8\left(f\,Re_{D_h}\right)} Ra_{D_h} \tag{8}$$

in which the group $(f\,Re_{D_h})$ is a constant listed in Table 6.1. Those constants can be used to construct a new table:

Cross-section shape	$\overline{Nu_H}/Ra_{D_h}$
parallel plates	$\frac{1}{8 \times 24} = \frac{1}{192}$
circular	$\frac{1}{8 \times 16} = \frac{1}{128}$
square	$\frac{1}{8 \times 14.2} = \frac{1}{113.6}$
equilateral triangle	$\frac{1}{8 \times 13.3} = \frac{1}{106.4}$

Problem 7.25. For the warm side of the single-pane window, eq. (7.55) recommends the following formula for the height-averaged heat flux (note: Pr = 0.72 for air)

$$\overline{Nu}_H = 0.517\, Ra_H^{1/4}, \qquad (Pr = 0.72) \qquad (7.55)$$

Recall that both $\overline{Nu}_H$ and Ra_H are based on the reservoir-wall temperature difference, which in this case is $T_h - T_w$:

$$\frac{\overline{q''}}{T_h - T_w}\frac{H}{k} = 0.517\left[\frac{g\beta(T_h - T_w)\, H^3}{\alpha\nu}\right]^{1/4}$$

Symmetry requires that the glass temperature T_w be situated half-way between T_h and T_c, therefore

$$T_h - T_w = \frac{1}{2}(T_h - T_c)$$

and the average heat flux formula based on $T_h - T_c$ becomes

$$\frac{\overline{q''}}{T_h - T_c}\frac{H}{k} = 0.217\left[\frac{g\beta(T_h - T_c)\, H^3}{\alpha\nu}\right]^{1/4}$$

Problem 7.26. For the warm side of the glass layer, eq. (7.66) recommends the local Nusselt number:

$$\frac{q''}{T_h - T_w(y)}\frac{y}{k} = 0.530\left(\frac{g\beta q''\, y4}{\alpha\nu k}\right)^{1/5}, \qquad (Pr = 0.72) \qquad (7.66)$$

The corresponding overall Nusselt number is obtained by replacing y with H, $T_w(y)$ with $\overline{T}_w$, and the coefficient 0.53 with

$$\frac{0.53}{1 + (-1/5)} = 0.663$$

where (- 1/5) is the exponent n in the y-dependence of the local heat transfer coefficient,

$$\frac{q''}{T_h - T_w(y)} \sim y^n$$

In conclusion, eq. (7.66) becomes

$$\frac{q''}{T_h - \overline{T}_w}\frac{H}{k} = 0.663\left(\frac{g\beta q'' H^4}{\alpha\nu k}\right)^{1/5}$$

or, after some algebra,

$$\frac{q''}{T_h - \overline{T}_w}\frac{H}{k} = 0.598\left[\frac{g\beta(T_h - \overline{T}_w) H^3}{\alpha\nu}\right]^{1/4}$$

The average temperature difference $T_h - \overline{T}_w$ can only be equal to half of the side-to-side temperature difference,

$$T_h - \overline{T}_w = \frac{1}{2}(T_h - T_c)$$

therefore, the wanted relationship between q" and $(T_h - T_c)$ is

$$\frac{q''}{T_h - T_c}\frac{H}{k} = 0.252\left[\frac{g\beta(T_h - T_c) H^3}{\alpha\nu}\right]^{1/4}$$

Note that the 0.252 coefficient is only 16 percent greater than the 0.217 coefficient derived based on the constant-T_w model (see the preceding problem).

Problem 7.27. An approximate way to determine the relationship between the heat flux from T_h to T_c (namely q", uniform) and the difference $T_h - T_c$ is to invoke the conclusion of the preceding problem (single-pane window, uniform q"):

$$\frac{q''}{T_h - \overline{T}_{core}}\frac{H}{k} \cong 0.252\left[\frac{g\beta(T_h - \overline{T}_{core}) H^3}{\alpha\nu}\right]^{1/4} \quad (1)$$

Here $\overline{T}_{core}$ is the average temperature of the air core contained between the two boundary layers inside the cavity. In real life, T_{core} increases from T_c at the bottom of the cavity, to T_h at the very top. Its value at midheight is clearly

$$T_{core}\left(y = \frac{H}{2}\right) = \frac{1}{2}(T_h + T_c) \quad (2)$$

The average core temperature $\overline{T}_{core}$ is also equal to $(T_h + T_c)/2$, therefore

$$T_h - \overline{T}_{core} = \frac{1}{2}(T_h - T_c) \quad (3)$$

and, in terms of $T_h - T_c$, eq. (1) becomes

$$\frac{q''}{T_h - T_c}\frac{H}{k} \cong 0.106 \left[\frac{g\beta(T_h - T_c)\, H^3}{\alpha\nu}\right]^{1/4} \qquad (4)$$

This formula applies when the cavity is wide enough so that the innermost boundary layers are distinct. When estimating the thickness of one of the inner boundary layers, in order to see whether they are distinct,

$$\delta_T \sim H \left[\frac{g\beta(\overline{T}_w - \overline{T}_{core})\, H^3}{\alpha\nu}\right]^{-1/4} \qquad (5)$$

keep in mind that the glass-core temperature difference is approximately one fourth of the overall temperature difference,

$$\overline{T}_w - \overline{T}_{core} \cong \frac{1}{4}(T_h - T_c) \qquad (6)$$

Problem 7.28. The temperature difference between the warm milk and the cold tap water is 60°C - 10°C = 50°C, therefore the bottle wall has an average temperature of order 10°C + 50°C/2 = 35°C, and the temperature difference between the wall and each of the fluids is approximately 25°C. The film temperature on the cold side of the wall is (10°C + 35°C)/2 = 22.5°C, and on the warm side is (35°C + 60°C)/2 = 47.5°C. The hierarchy of temperatures across the vertical wall of the bottle is:

cold fluid	cold film	wall	warm film	warm fluid
10°C	22.5°C	≅ 35°C	47.5°C	60°C

Next, we assume that the milk properties are similar to those tabulated for water. The relevant properties of water on both sides of the wall are:

	at the cold film temperature 22.5°C	at the warm film temperature 47.5°C
$\frac{g\beta}{\alpha\nu}$	$17.1 \times 10^3\ cm^{-3}\ K^{-1}$	$47.9 \times 10^3\ cm^{-3}\ K^{-1}$
k	0.6 W/m·K	0.64 W/m·K
Pr	6.64	3.76

The Rayleigh numbers based on H = 6 cm and ΔT = 25°C on both sides of the wall are

$$Ra_{cold} = \frac{g\beta}{\alpha\nu} H^3 \Delta T = 17.1 \times 10^3 \frac{6^3 \ cm^3 \ 25K}{cm^3 \ K} \cong 9.2 \times 10^7$$

$$Ra_{warm} = \frac{g\beta}{\alpha\nu} H^3 \Delta T = 47.9 \times 10^3 \frac{6^3 \ cm^3 \ 25K}{cm^3 \ K} = 2.6 \times 10^8$$

Invoking eq. (7.62), we calculate the H-averaged Nusselt numbers for both sides

$$\overline{Nu}_{cold} = 0.68 + 0.611 \ Ra_{cold}^{1/4} = 60.5$$

$$\overline{Nu}_{warm} = 0.68 + 0.593 \ Ra_{warm}^{1/4} = 75.9$$

From the definition $\overline{Nu} = hH/k$, we deduce the respective H-averaged heat transfer coefficients for both sides,

$$h_{cold} = \frac{(k \ \overline{Nu})_{cold}}{H} = \frac{0.6 \ W}{m \cdot K} \frac{60.5}{0.06m} \cong 605 \frac{W}{m^2K}$$

$$h_{warm} = \frac{(k \ \overline{Nu})_{warm}}{H} = \frac{0.64 \ W}{m \cdot K} \frac{75.9}{0.06m} \cong 810 \frac{W}{m^2K}$$

Writing U for the overall heat transfer coefficient between the warm milk and the cold tap water, and assuming that the bottle wall is thin enough so that the heat transfer area has the same size on both sides of the wall, we have

$$\frac{1}{U} = \frac{1}{h_{cold}} + \frac{1}{h_{warm}}$$

in other words:

$$U = \frac{h_{cold} \ h_{warm}}{h_{cold} + h_{warm}} = \frac{605 \times 810}{605 + 810} \frac{W}{m^2K}$$

$$= 346 \frac{W}{m^2K}$$

We evaluated h_{cold} and h_{warm} by claiming that the two sides of the vertical wall are lined by distinct (thin) boundary layers. For example, the thermal boundary layer thickness on the milk side is

$$\delta_{warm} \sim H \ Ra_{warm}^{-1/4} \sim 6 \ cm \ 0.0079 \cong 0.5 \ mm$$

The milk boundary layer is "thin" because 0.5 mm is much smaller than the wall height (6 cm) and the bottle diameter (4 cm).

Problem 7.29. When the temperature difference is applied so that in each cell the hot wall is positioned below the cold wall, the circulation is similar to the roll inside a square enclosure of height H = 10 cm. We can approximate the average heat flux using the $\overline{Nu}_H$ formula (7.100), because the Rayleigh number is high enough to satisfy the inequality listed under eq. (7.100):

$$Ra_H = \left(\frac{g\beta}{\alpha\nu}\right)_{20°C} H^3 \Delta T = \frac{107}{cm^3 K} (10\ cm)^3\ 20\ K$$

$$= 2.14 \times 10^6$$

The average Nusselt number is

$$\overline{Nu}_H = 0.18 \left(\frac{Pr}{0.2 + Pr} Ra_H\right)^{0.29} \left(\frac{L}{H}\right)^{-0.13}$$

$$= 0.18 \left(\frac{0.72}{0.2 + 0.72} 2.14 \times 10^6\right)^{0.29} \left(\frac{10}{10}\right)^{-0.13}$$

$$\cong 11.5$$

Invoking the $\overline{Nu}_H$ definition,

$$\overline{Nu}_H = \frac{\overline{q''}\ H}{\Delta T\ k}$$

we calculate finally the average heat flux

$$\overline{q''} = \frac{\Delta T\ k}{H} \overline{Nu}_H = \frac{20\ K}{0.1\ m} 0.025 \frac{W}{m \cdot K} 57.5$$

$$\cong 57.5 \frac{W}{m^2}$$

On the other hand, when the hot segment is above the cold segment across each cell, the air tends to be stably stratified and the flow circulation is considerably less intense. If we model the air as perfectly motionless, the heat transfer between the vertical surfaces of the double wall is by pure conduction, and the average heat flux is roughly

$$\overline{q''} = k \frac{\Delta T}{L} = 0.025 \frac{W}{m \cdot K} \frac{20\ K}{0.1\ m}$$

$$= 5 \frac{W}{m^2}$$

In conclusion, these extreme models of the operating regimes of the wall show that in the "on" mode the heat flux is approximately 10 times greater than in the "off" mode.

Project 7.1. The procedure for executing this project is presented in detail in the text.

Project 7.2. a) Consider first the limit of vanishingly small fin-to-fin spacing, $D \to 0$. In this limit we can use with confidence eq. (7.94) for the overall heat transfer rate extracted from the two surfaces of one channel,

$$q_1 = q'W = \frac{\rho g \beta c_P (\Delta T)^2 D^3}{12 \nu} W$$

where $\Delta T = T_0 - T_\infty$. Note that eq. (7.94) delivers only the per-unit-length heat transfer rate q'. The total number of channels is L/D, therefore, the total heat transfer rate removed from the assembly is

$$q_a = q_1 \frac{L}{D} = \frac{\rho g \beta c_P (\Delta T)^2 D^3}{12 \nu} W \frac{L}{D} \qquad (1)$$

This shows that in the $D \to 0$ limit the total heat transfer rate decreases as D^2.

b) Consider now the opposite limit, in which the spacing D is "large", for example, larger than the thickness of the air boundary layer formed on one of the vertical surfaces,

$$D > H \, Ra_H^{-1/4}$$

where $Ra_H = g\beta H^3 \Delta T/(\alpha\nu)$, and $\Delta T = T_0 - T_\infty$. In this limit the boundary layers are distinct (thin compared with D), and the center of the fin-to-fin spacing is occupied by T_∞-air. The number of air boundary layers is 2(L/D), because there are two for each D spacing. The corresponding formula for the total heat transfer rate is

$$q_b = 2 \frac{L}{D} q_{BL}$$

where q_{BL} is the heat transfer rate through only one boundary layer (one fin surface $W \times H$):

$$q_{BL} = \bar{h}_H WH \Delta T$$

The average heat transfer coefficient is given by eq. (7.55'):

$$\bar{h}_H = \frac{k}{H} 0.517 \, Ra_H^{1/4}$$

In conclusion, in the large-D limit the total heat transfer rate is proportional to 1/D:

$$q_b = 2 \frac{L}{D} WH \Delta T \frac{k}{H} 0.517 \, Ra_H^{1/4} \qquad (2)$$

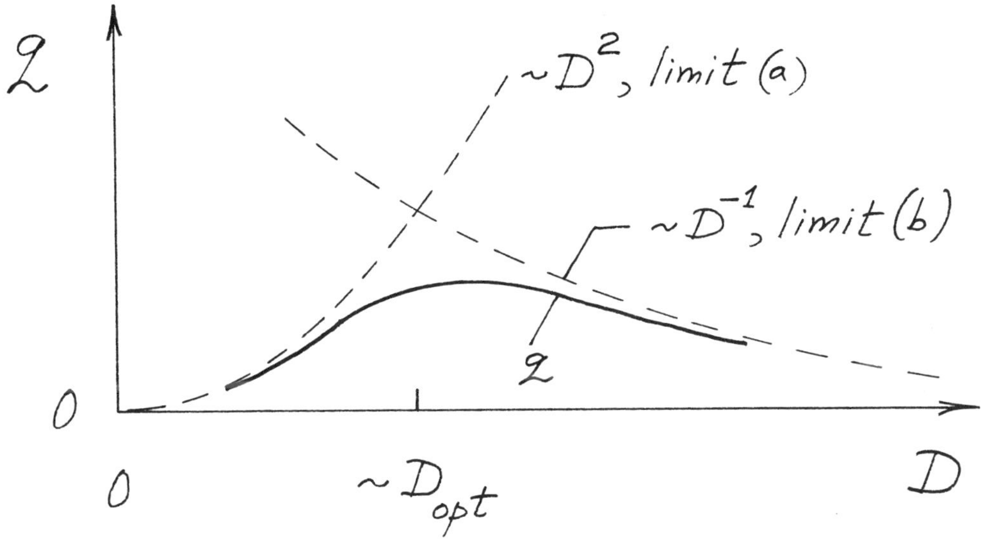

What we have determined so far are the two asymptotes of the real (unknown) curve of q versus D. The attached drawing shows that the asymptotes intersect above what would be the q maximum of the actual curve. The optimum spacing (D_{opt}) can be estimated (approximately) as the D value where the two asymptotes intersect,

$$q_a \cong q_b$$

By using eqs. (1) and (2), and setting $D = D_{opt}$, we obtain in order

$$Ra_{D_{opt}} \sim 12.4\ Ra_H^{1/4}$$

$$\frac{D_{opt}^3}{H^3} Ra_H \sim 12.4\ Ra_H^{1/4}$$

$$\frac{D_{opt}^3}{H^3} \sim 12.4\ Ra_H^{-3/4}$$

$$\frac{D_{opt}}{H} \sim (12.4)^{1/3}\ Ra_H^{-1/4}$$

$$D_{opt} \sim 2.3\ H\ Ra_H^{-1/4}$$

Project 7.3. The thickness of the window glass varies linearly with altitude,

$$\delta = \bar{\delta} + b\left(\frac{1}{2} - \xi\right) \tag{1}$$

where y is measured downward,

$$\xi = \frac{y}{H} \qquad b = -\frac{d\delta}{d\xi} \tag{2}$$

The H-averaged thickness $\bar{\delta}$ is fixed, in other words, the total volume (or weight) of the window is fixed. The taper parameter b is one variable in the design of the window.

The local heat flux must overcome two resistances in series, the air boundary layer on the room side (1/h), and the glass pane itself (δ/k_w):

$$q'' = \frac{\Delta T}{\frac{1}{h} + \frac{\delta}{k_w}} \tag{3}$$

The total heat leak through the window (per unit length in the lateral direction) is

$$q' = \int_0^H \frac{\Delta T\, dy}{\frac{1}{h} + \frac{\delta}{k_w}} \tag{4}$$

Assume now that the heat transfer coefficient decreases in the downward direction, as in laminar boundary layer natural convection, eq. (7.51),

$$h = h_{min}\left(\frac{y}{H}\right)^{-n} \tag{5}$$

In this equation, n = 1/4, and h_{min} is the value of h at the bottom of the window, y = H (i.e. the smallest value). By combining eqs. (1)-(5), we can nondimensionalize the total heat leak as

$$\tilde{q}' = \int_0^1 \frac{d\xi}{\xi^n + Bi\left[1 + S\left(\frac{1}{2} - \xi\right)\right]} \tag{6}$$

in which

$$\tilde{q}' = \frac{q'}{h_{min} H \, \Delta T} \tag{7}$$

$$Bi = \frac{h_{min} \, \bar{\delta}}{k_w} \tag{8}$$

$$S = - \frac{H}{\bar{\delta}} \frac{d\delta}{dy} \tag{9}$$

The objective is to minimize $\tilde{q}'$ with respect to the "shape" (dimensionless-taper) parameter S, while Bi is fixed by glass volume and air natural convection constraints. The integral (6) can be evaluated numerically (recall that n = 1/4):

S	$\tilde{q}'$		
	Bi = 0.01	Bi = 0.1	Bi = 1
0	1.31005	1.16184	0.5605
0.5	1.30860	1.15306	0.55605
0.535			***0.55603**
0.6			0.55610
1.0	1.30716	1.14540	
1.5	1.30576	1.13876	
2.0	***1.30437**	***1.13307**	

The asterisks indicate the smallest value reached by $\tilde{q}'$ as the taper parameter S changes. At low Biot numbers, $\tilde{q}'$ decreases monotonically as the taper increases, so that

$$\tilde{q}'_{min} = \tilde{q}' \, (S = 2)$$

The smallest heat leak occurs when the taper is the most accentuated (S = 2), i.e. when the glass thickness at the top of the window is $2\bar{\delta}$, and zero at the bottom.

At higher Biot numbers (e.g. Bi = 1) the $\tilde{q}'$ behavior changes. The minimum heat leak occurs at an intermediate (optimal) taper, in this case at S = 0.535.

The top line in the table (S = 0) shows the reference case, i.e. the heat leak through the glass pane of constant thickness. The relative merit of the tapered glass design is indicated by the ratio:

	Bi = 0.01	Bi = 0.1	Bi = 1
$\dfrac{\tilde{q}'_{min}}{\tilde{q}'_{ref}\,(S = 0)}$	0.9957	0.9752	0.9920
% reduction	0.43	2.48	0.8

The bottom line in this table shows that the q' reduction associated with the tapering of the glass is minimal. This is especially true at small Biot numbers, which is the normal range of operation of regular-size windows. In order to see this, consider the following order of magnitude calculation:

glass pane data:

$$\bar{\delta} = 0.5 \text{ cm}$$

$$k_w = 0.81 \frac{W}{m \cdot K}$$

air side data:

$$Ra_H \sim 10^8$$

$$H = 1 \text{ m}$$

$$k_{air} = 0.025 \frac{W}{m \cdot K}$$

equation (7.52a):

$$\frac{h_{min} H}{k_{air}} \sim 0.5\, Ra_H^{1/4} \sim 50$$

$$h_{min} \sim 1.25 \frac{W}{m^2\, K}$$

Biot number:

$$Bi = \frac{h_{min}\, \bar{\delta}}{k_w} \sim 0.008$$

In conclusion, if we consider the added difficulty of manufacturing tapered glass, and the unrealistic assumption that the bottom edge of the window can have a knife edge (zero thickness, or S = 2), the marginal heat leak reduction calculated above does not recommend the tapered-glass window as a viable energy conservation feature for building design.

Project 7.4. i) The total thermal resistance between the bare horizontal cylinder (T_i, r_i) and the surrounding fluid reservoir (T_∞) is

$$R_t = \frac{\ln \frac{r_o}{r_i}}{2\pi k L} + \frac{1}{2\pi\, r_o\, L h} \tag{1}$$

In laminar boundary layer natural convection we can expect $\overline{Nu_D} \sim Ra_D^{1/4}$, in other words

$$\frac{hD}{k_{fluid}} \sim \left[\frac{g\beta\,(T_o - T_\infty)\, D^3}{\alpha\nu}\right]^{1/4} \tag{2}$$

where T_o is the outer wall temperature (at $r = r_o$), and D is equal to $2\, r_o$. We conclude that h varies with D and ($T_o - T_\infty$), according to

$$h = C\,(T_o - T_\infty)^{1/4}\, D^{-1/4} \tag{3}$$

in which C is a constant. We use eq. (3) to define the "reference" heat transfer coefficient by imagining the case where the wall is bare ($T_o = T_i$, $r_o = r_i$):

$$h_* = C\,(T_i - T_\infty)^{1/4}\,(2\, r_i)^{-1/4} \tag{4}$$

Dividing eqs. (3) and (4) side by side we obtain

$$\frac{h}{h_*} = \left(\frac{T_o - T_\infty}{T_i - T_\infty}\right)^{1/4} \left(\frac{r_o}{r_i}\right)^{-1/4} \tag{5}$$

which invites us to define the dimensionless outer temperature difference

$$\theta = \frac{T_o - T_\infty}{T_i - T_\infty} \qquad \text{(note: } 0 < \theta < 1) \tag{6}$$

and the radii ratio:

$$\rho = \frac{r_o}{r_i} \qquad \text{(note: } \rho > 1) \tag{7}$$

In terms of θ and ρ, the total heat transfer rate from T_i to T_∞ can be written in two ways,

$$q = \frac{2\pi\, kL}{\ln \frac{r_o}{r_i}} (T_i - T_o)$$

$$= 2\pi\, kL \left(T_i - T_\infty\right) \frac{1-\theta}{\ln \rho} \qquad (8)$$

and

$$q = 2\pi\, r_o\, Lh \left(T_o - T_\infty\right)$$

$$= 2\pi\, r_i\, Lh_* \left(T_i - T_\infty\right) \theta^{5/4}\, \rho^{3/4} \qquad (9)$$

Equating (8) and (9) we obtain the relationship between the outer temperature (T_o, or θ) and the outer radius (r_o, or ρ),

$$\frac{1-\theta}{\theta^{5/4}} = Bi\, \rho^{3/4} \ln \rho \qquad (10)$$

where Bi is the Biot number based on r_i and h_*,

$$Bi = \frac{h_*\, r_i}{k} \qquad (11)$$

ii) Equation (10) can be solved by trial and error to find θ as a function of ρ and Bi, in other words, $\theta = \theta\,(\rho, Bi)$. This result is then substituted in one of eqs. (8,9), say eq. (8),

$$\frac{q}{2\pi\, kL \left(T_i - T_\infty\right)} = \frac{1 - \theta\,(\rho, Bi)}{\ln \rho} \qquad (12)$$

to obtain the total heat transfer rate (left-hand side, dimensionless) as a function of ρ and Bi,

$$Q = Q\,(\rho, Bi) \qquad (13)$$

where

$$Q = \frac{q}{2\pi\, kL \left(T_i - T_\infty\right)} \qquad (14)$$

Note that Q is the inverse of a nondimensionalized version of the thermal resistance shown in eq. (1). The behavior of Q vs. ρ can be studied numerically. One finds that when Bi is small enough, Q reaches a relatively flat maximum at a critical outer radius $r_{o,c}$ represented by $\rho_c = r_{o,c}/r_i$. The following table shows a sample of my calculations:

Bi	ρ_c, [calculated based on eq. (13)]	$\rho_c = \frac{1}{Bi}$, [eq. (15)]
0.5	1.34	2
0.25	3.92	4

iii) It is useful to compare the calculated ρ_c values with the classical result (2.44) based on the constant-h assumption (h = h*),

$$r_{o,c} = \frac{k}{h_*} \tag{2.44}$$

In the present notation, eq. (2.44) is the same as writing

$$\rho_c = \frac{1}{Bi} \tag{15}$$

The table shows that the discrepancy between the calculated rc value and the classical prediction based on eq. (2.44) decreases as Bi becomes small.

Project 7.5. The original fluid layer has the height H, and the bottom-to-top temperature difference ΔT. The horizontal partition divides this layer into two sublayers, with the following thicknesses and bottom-to-top temperature differences (see the figure)

$$H_1 = x\,H \qquad \Delta T_1 = x\,\Delta T \tag{1}$$

$$H_2 = (1 - x)\,H \qquad \Delta T_2 = (1 - x)\,\Delta T \tag{2}$$

The number x is between 0 and 1, and describes the relative position of the partition. For example, x = 1/2 means that the partition is inserted at midheight. It is important to note also that the ΔT_1 and ΔT_2 expressions written above are based on the assumption that there is no convection in either sublayer.

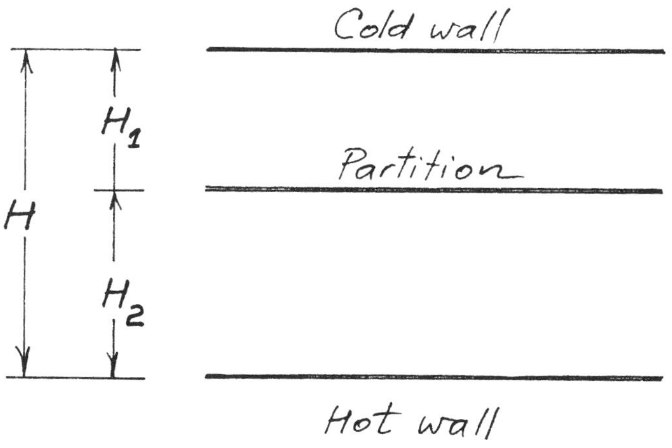

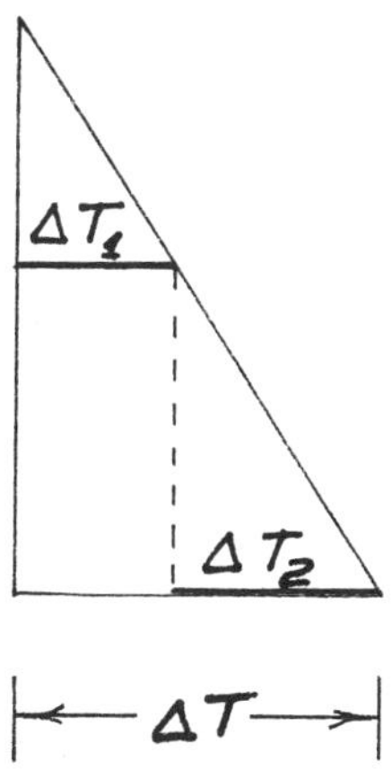

Our objective is to determine the optimal partition location x so that the regime of pure conduction (no convection) is extended to the largest possible Rayleigh number

$$Ra = \frac{g\,\beta}{\alpha\,\nu} H^3\,\Delta T \tag{3}$$

This is the "external" Rayleigh number, which is based on the overall temperature difference ΔT. The objective then is to maximize the overall ΔT while preserving the state of pure conduction, i.e. least heat flux in the vertical direction. Recall that the formation of convection cells augments the heat flux relative to the heat flux present in the no-convection state.

If we assume (for simplicity) that the partition is isothermal at some intermediate temperature between the top wall and bottom wall temperatures, the top sublayer is without convection currents as long as its own Rayleigh number is smaller than the critical value,

$$\frac{g\,\beta}{\alpha\,\nu} H_1^3\,\Delta T_1 < 1708 \tag{4}$$

The same can be said about the avoidance of convection in the lower sublayer,

$$\frac{g\,\beta}{\alpha\,\nu} H_2^3\,\Delta T_2 < 1708 \tag{5}$$

Convection is suppressed in the entire layer (H) when conditions (4) and (5) are satisfied simultaneously. Now it is a simple matter to use the x-based formulas (1,2) and the external-Ra definition (3) to rewrite conditions (4) and (5) as

$$Ra < \frac{1708}{x^4} \tag{6}$$

$$Ra < \frac{1708}{(1-x)^4} \tag{7}$$

Conditions (6) and (7) are satisfied simultaneously by all the (Ra, x) points situated <u>under</u> the roof-shaped line described by $Ra = 1708/x^4$ and $Ra = 1708/(1-x)^4$ in the attached figure. It is clear that the design with the maximum external Rayleigh number for "no convection" corresponds to

$$x_{opt} = \frac{1}{2} \tag{8}$$

i.e. to a partition installed at midheight. That maximum critical external Rayleigh number is (approximately, of course)

$$Ra_{c,max} = \frac{1708}{(1/2)^4} \sim 2.7 \times 10^3 \qquad (9)$$

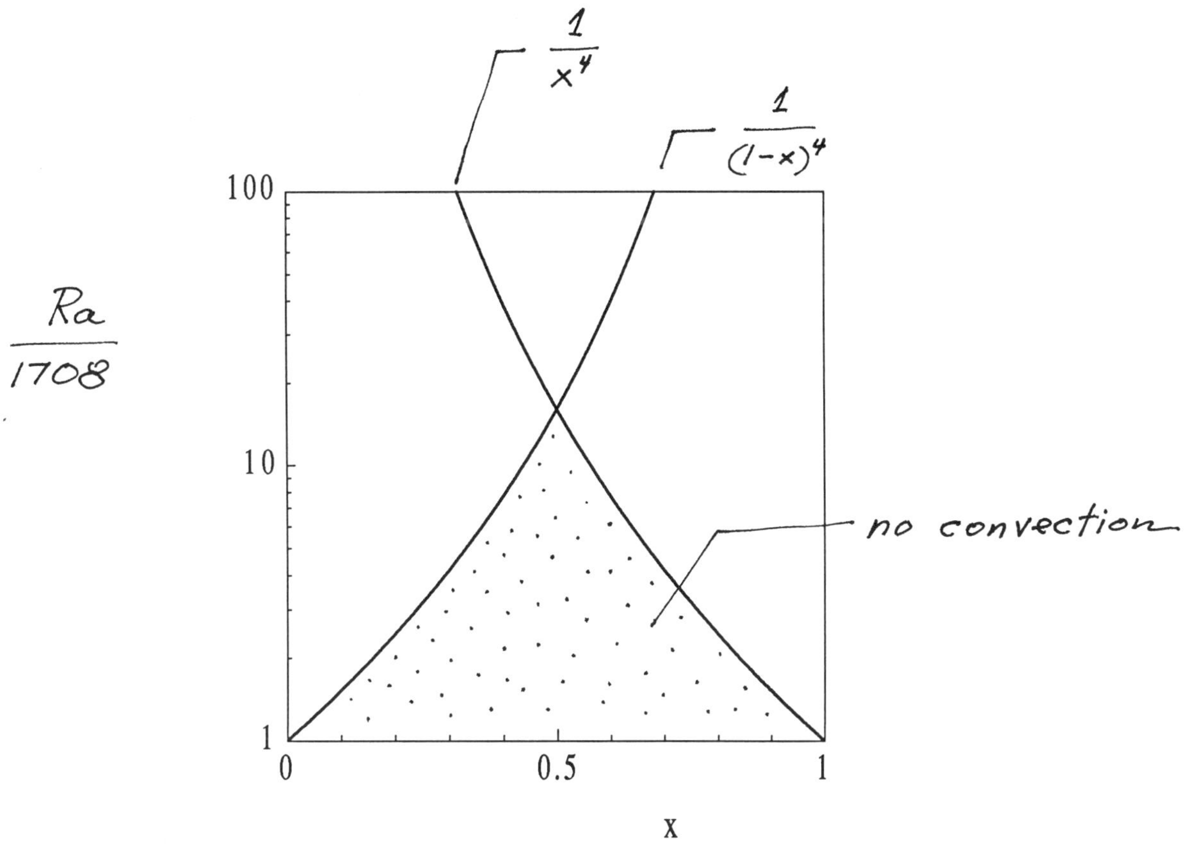

The modelling of the thermal boundary condition on the partition does not affect the optimal location determined in eq. (8). For example, if the partition is modelled as a wall with uniform flux (instead of uniform temperature), the only item that changes in the analysis is the numerical value (of the order of 10^3) that must appear on the right-hand side of the inequalities (4) and (5). This value, however, drops out in the steps that follow en route to eq. (8).

The design can be refined by accounting for the temperature dependence of the property group $g\beta/\alpha\nu$. If this dependence is significant, the optimal location of the partition will differ from $x_{opt} = 1/2$.

Chapter 8

CONVECTION WITH CHANGE OF PHASE

Problem 8.1. If we use the δ(y) formula (8.12) on the right side of eq. (8.4), we obtain

$$\Gamma(y) = 0.943 \left(\frac{g\Delta\rho}{\nu_l}\right)^{1/4} \left(\frac{k_l \,\Delta T\, y}{h'_{fg}}\right)^{3/4} \tag{1}$$

in which $\Delta\rho = \rho_l - \rho_v$ and $\Delta T = T_{sat} - T_w$. According to eqs. (8.20) and (8.15), the total cooling rate is

$$q' = 0.943\, k_l\, \Delta T \left(\frac{L^3 h'_{fg}\, g\, \Delta\rho}{k_l\, \nu_l\, \Delta T}\right)^{1/4}$$

$$= 0.943\, (k_l\, \Delta T)^{3/4} L^{3/4} \left(h'_{fg}\right)^{1/4} \left(\frac{g\Delta\rho}{\nu_l}\right)^{1/4} \tag{2}$$

By substituting y = L in eq. (1), and dividing eqs. (1) and (2), we obtain finally

$$\frac{\Gamma(L)}{q'} = \frac{0.943 \left(\frac{g\Delta\rho}{\nu_l}\right)^{1/4} L^{3/4} (k_l\, \Delta T)^{3/4} \left(h'_{fg}\right)^{-3/4}}{0.943 \left(\frac{g\Delta\rho}{\nu_l}\right)^{1/4} L^{3/4} (k_l\, \Delta T)^{3/4} \left(h'_{fg}\right)^{1/4}}$$

$$= \frac{1}{h'_{fg}} \tag{8.22}$$

Problem 8.2. With reference to the drawing that accompanies the problem statement, we write the first law for the control volume,

$$0 = h_g\, \Gamma(L) - H(L) - q' \tag{1}$$

and then use eq. (8.7) in order to evaluate the enthalpy flowrate through the bottom end of the film,

$$H(L) = \left[h_f - \frac{3}{8} c_{P,l} (T_{sat} - T_w)\right] \Gamma(L) \tag{2}$$

Combining eqs. (1) and (2) we obtain

$$q' = \left[h_{fg} + \frac{3}{8} c_{P,l} (T_{sat} - T_w)\right] \Gamma (L)$$

$$= h'_{fg} \Gamma (L) \qquad (3)$$

The actual numerical coefficient that will figure in front of $c_{P,l}(T_{sat} - T_w)$ will depend on the shapes of the u and T profiles across the bottom end of the film. Aside from this, the $q' \sim \Gamma(L)$ proportionality written as eq. (3) has general validity.

Problem 8.3. We first note that

$$q' = \overline{h}_L L (T_{sat} - T_w) \qquad (8.20)$$

$$\Gamma (L) = \frac{q'}{h'_{fg}} \qquad (8.21)$$

and that the elimination of q' yields

$$L = \frac{h'_{fg} \Gamma (L)}{\overline{h}_L (T_{sat} - T_w)} \qquad (i)$$

Next, we evaluate $\overline{h}_L$ based on eq. (8.15), and eventually eliminate L using eq. (i). The sequence of these operations is:

$$\overline{h}_L = \frac{k_l}{L} 0.943 \left[\frac{L^3 h'_{fg} g \overbrace{(\rho_l - \rho_v)}^{\approx \rho_l}}{k_l \nu_l (T_{sat} - T_w)}\right]^{1/4} \qquad (8.15)$$

$$\overline{h}_L^4 = k_l^3 (0.943)^4 \frac{1}{L} \frac{h'_{fg} g \rho_l}{\nu_l (T_{sat} - T_w)}$$

$$= k_l^3 (0.943)^4 \frac{\overline{h}_L (T_{sat} - T_w)}{h'_{fg} \Gamma} \frac{h'_{fg} g \rho_l}{\nu_l (T_{sat} - T_w)}$$

$$\overline{h}_L^3 = k_l^3 (0.943)^4 \frac{g \rho_l}{\Gamma \nu_l}$$

$$\frac{\overline{h}_L^3}{k_l^3}\frac{\nu_l^2}{g} = (0.943)^4\, 4\, \frac{\mu_l}{4\Gamma}$$

$$\frac{\overline{h}_L}{k_l}\left(\frac{\nu_l^2}{g}\right)^{1/3} = \underbrace{(0.943)^{4/3}\, 4^{1/3}}_{1.468}\, Re_L^{-1/3} \tag{8.24}$$

Problem 8.4. The following relations hold for the laminar vertical film analyzed in subsection 8.1.1,

$$v(y,\delta) = v_{max}(y) = \frac{g}{\mu_l}(\rho_l - \rho_v)\,\delta^2\left(1 - \frac{1}{2}\right) \tag{8.3}$$

$$= \frac{g}{\mu_l}(\rho_l - \rho_v)\frac{\delta^2}{2} \tag{i}$$

$$\Gamma(y) = \frac{g\,\rho_l}{3\,\mu_l}(\rho_l - \rho_v)\,\delta^3 \tag{8.4}$$

Therefore the Re_y definition (8.22) can be rewritten as

$$Re_y = \frac{4}{\mu_l}\Gamma(y) = \frac{4}{\mu_l}\frac{g\,\rho_l}{3\,\mu_l}(\rho_l - \rho_v)\,\delta\,\delta^2 \tag{ii}$$

for which the last δ^2 factor on the right side is delivered by the v_{max} expression (i),

$$\delta^2 = \frac{2\,\mu_l\, v_{max}(y)}{g\,(\rho_l - \rho_v)}$$

Equation (ii) becomes

$$Re_y = \frac{4}{\mu_l}\frac{g\,\rho_l}{3\,\mu_l}(\rho_l - \rho_v)\,\delta\,\frac{2\,\mu_l\, v_{max}(y)}{g\,(\rho_l - \rho_v)}$$

$$= \frac{8}{3}\,\frac{\rho_l\, v_{max}\,\delta}{\mu_l}$$

in which both v_{max} and δ are functions of y.

Problem 8.5. We begin with the laminar-film solutions

$$\frac{\bar{h}_L L}{k_l} = 0.943 \left(L^3 C\right)^{1/4} \qquad (8.15)$$

$$\frac{\bar{h}_D D}{k_l} = 0.729 \left(D^3 C\right)^{1/4} \qquad (8.30)$$

in which C represents the physical constant

$$C = \frac{h'_{fg}\, g\, (\rho_l - \rho_v)}{k_l\, \nu_l\, (T_{sat} - T_w)}$$

Next, we recall that q' and Γ are proportional. Therefore we equate the cooling rate of the two-sided slab with the cooling rate provided by the single horizontal cylinder:

$$q'_{slab} = q'_{cylinder}$$

$$\bar{h}_L\, 2L\, (T_{sat} - T_w) = \bar{h}_D\, \pi D\, (T_{sat} - T_w)$$

$$0.943\, k_l\, L^{-1/4} C^{1/4}\, 2L\, (T_{sat} - T_w) = 0.729\, k_l D^{-1/4} C^{1/4}\, \pi D\, (T_{sat} - T_w)$$

$$(0.943)\, 2\, L^{3/4} = (0.729)\, \pi\, D^{3/4}$$

$$\frac{L}{D} = 1.296$$

In conclusion, the special diameter is D = 0.772L.

Problem 8.6. a) The film temperature of the condensate is (80°C + 100°C)/2 = 90°C, at which the pertinent physical properties have the following values:

$$c_{P,l} = 4.205 \frac{kJ}{kg} \qquad \mu_l = 3.16 \times 10^{-4} \frac{kg}{m \cdot s}$$

$$k_l = 0.67 \frac{W}{m \cdot K} \qquad \nu_l = 3.27 \times 10^{-7} \frac{m^2}{s}$$

$$\rho_l = 965.3 \frac{kg}{m^3} \qquad Pr_l = 1.98$$

Note further that h_{fg} = 2257 kJ/kg, therefore

$$Ja = \frac{c_{P,l}(T_{sat} - T_w)}{h_{fg}} = 0.0373$$

$$h'_{fg} = h_{fg}(1 + 0.68 \times 0.0373) \cong 2314 \frac{kJ}{kg}$$

The dimensionless group that appears in the Nusselt number formula for laminar film condensation is (note $\rho_l >> \rho_v$),

$$\frac{L^3 h'_{fg} g \rho_l}{k_l \nu_l (T_{sat} - T_w)} = \frac{1\ m^3\ 2314 \frac{kJ}{kg}\ 9.81 \frac{m}{s^2}\ 965.3 \frac{kg}{m^3}}{0.67 \frac{W}{m \cdot K}\ (3.27)\ 10^{-7} \frac{m^2}{s}\ 20\ K}$$

$$= 5 \times 10^{15}$$

therefore, using eq. (8.15) we obtain in order

$$\overline{h}_L = \frac{k_l}{L} 0.943 \left[5 \times 10^{15}\right]^{1/4} = 5313 \frac{W}{m^2K}$$

$$q' = \overline{h}_L L (T_{sat} - T_w) = 1.06 \times 10^5 \frac{W}{m}$$

$$\Gamma(L) = \frac{q'}{h'_{fg}} = 0.046 \frac{kg/s}{m}$$

$$Re_L = \frac{4}{\mu_l} \Gamma(L) = 581$$

b) The Reynolds number shows that the film is dominated by the wavy flow regime (Fig. 8.5). This means that a more accurate condensation rate estimate can be obtained using the chart of Fig. 8.6, in which

$$B = L (T_{sat} - T_w) \frac{4 k_l}{\mu_l h'_{fg}} \left(\frac{g}{\nu_l^2}\right)^{1/3} = 3306$$

By entering B = 3306 and $Pr_l \cong 2$ in Fig. 8.6, we obtain in order

$$Re_L \cong 800$$

$$\Gamma(L) = 800 \frac{\mu_l}{4} = 0.063 \frac{kg/s}{m}$$

This condensation rate is 37 percent larger than the estimate based on the assumption that the film is laminar everywhere.

Problem 8.7. We begin with the laminar-film solutions

$$\frac{\bar{h}_L L}{k_l} = 0.943 \left(L^3 C\right)^{1/4} \qquad (8.15)$$

$$\frac{\bar{h}_D D}{k_l} = 0.729 \left(D^3 C\right)^{1/4} \qquad (8.30)$$

in which C represents the constant

$$C = \frac{h'_{fg}\, g\, (\rho_l - \rho_v)}{k_l\, \nu_l\, (T_{sat} - T_w)}$$

In the present case the perimeter of the flattened cross-section equals the original perimeter, therefore

$$2L = \pi D, \qquad \text{or} \qquad \frac{L}{D} = \frac{\pi}{2}$$

Next, we calculate the relative change in the total condensation rate:

$$\frac{\Gamma_{flat}}{\Gamma_{round}} = \frac{q'_{flat}}{q'_{round}} = \frac{\bar{h}_L\, 2L\, (T_{sat} - T_w)}{\bar{h}_D\, \pi D\, (T_{sat} - T_w)}$$

$$= \frac{0.943\, k_l\, L^{-1/4}\, C^{1/4}\, 2L}{0.729\, k_l\, D^{-1/4}\, C^{1/4}\, \pi D}$$

$$= \frac{0.943}{0.729}\frac{2}{\pi}\left(\frac{L}{D}\right)^{3/4} = \frac{0.943}{0.729}\left(\frac{2}{\pi}\right)^{1/4}$$

$$= 1.155$$

In conclusion, the flattening of the thin-walled tube promises up to a 15.5 percent increase in the condensation rate.

Problem 8.8. The film temperature of the condensate is (80°C + 100°C)/2 = 90°C, at which the pertinent physical properties have the following values:

$c_{P,l} = 4.205 \frac{kJ}{kg}$ $\mu_l = 3.16 \times 10^{-4} \frac{kg}{m \cdot s}$

$k_l = 0.67 \frac{W}{m \cdot K}$ $\nu_l = 3.27 \times 10^{-7} \frac{m^2}{s}$

$\rho_l = 965.3 \frac{kg}{m^3}$ $Pr_l = 1.98$

Note further that h_{fg} = 2257 kJ/kg, therefore

$$Ja = \frac{c_{P,l}(T_{sat} - T_w)}{h_{fg}} = 0.0373$$

$$h'_{fg} = h_{fg}(1 + 0.68 \times 0.0373) \cong 2314 \frac{kJ}{kg}$$

a) When the surface is tilted at 45° and looks up, we use g cos 45° = 0.707g instead of g in the B parameter of Fig. 8.6:

$$B = L(T_{sat} - T_w)\frac{4 k_l}{\mu_l h'_{fg}}\left(\frac{0.707 g}{\nu_l^2}\right)^{1/3} = 2945$$

By entering this B value and $Pr_l \cong 2$ in Fig. 8.6, we obtain

$$Re_L \cong 730$$

$$\Gamma(L) = 730 \frac{\mu_l}{4} = 0.057 \frac{kg/s}{m}$$

This condensation rate is 10 percent smaller than when the surface is vertical.

b) In the case where the L-wide surface is perfectly horizontal, we use eq. (8.33) in which (note $\rho_l >> \rho_v$),

$$\frac{L^3 h'_{fg}\, g\, \rho_l}{k_l \nu_l (T_{sat} - T_w)} = 5 \times 10^{15}$$

therefore

$$\overline{h}_L = 1.079 \frac{k_l}{L} \left(5 \times 10^{15}\right)^{1/5} = 997.5 \frac{W}{m^2K}$$

$$q' = \overline{h}_L \, L \, (T_{sat} - T_w) = 19\ 949 \ \frac{W}{m}$$

The condensation rate collected over the width L,

$$\dot{m}' = \frac{q'}{h'_{fg}} = \frac{19\ 949 \ \frac{W}{m}}{2314 \ \frac{kJ}{kg}} = 0.0086 \ \frac{kg/s}{m}$$

represents only 13.7 percent of the condensation rate produced by the same surface in the vertical position.

The film Reynolds number at the edge of the horizontal surface can be calculated by first noting the mass flowrate that spills over one edge,

$$\Gamma_{edge} = \frac{\dot{m}'}{2} = 0.0043 \ \frac{kg/s}{m}$$

therefore

$$Re = \frac{4\Gamma}{\mu_l} = \frac{4 \times 0.0043 \ \frac{kg}{s \cdot m}}{3.16 \times 10^{-4} \ \frac{kg}{s \cdot m}} \cong 54$$

This Reynolds number is of the same order as 30, therefore most of the film is in the laminar flow regime.

Problem 8.9. The properties of water at atmospheric pressure and film temperature (100°C + 60°C)/2 = 80°C are (see Appendix C):

$c_{P,l} = 4.196 \ \frac{kJ}{kg\ K}$ $\qquad \mu_l = 3.55 \times 10^{-4} \ \frac{kg}{m \cdot s}$

$k_l = 0.67 \ \frac{W}{m \cdot K}$ $\qquad \nu_l = 3.66 \times 10^{-7} \ \frac{m^2}{s}$

$\rho_l = 971.8 \ \frac{kg}{m^3}$ $\qquad Pr_l = 2.23$

The latent heat of condensation at atmospheric pressure (or at 100°C) is h_{fg} = 2257 kJ/kg. The Jakob number is small,

$$Ja = \frac{c_{P,l}(T_{sat} - T_w)}{h_{fg}} = 0.074$$

therefore h'_{fg} is nearly the same as h_{fg},

$$h'_{fg} = h_{fg}(1 + 0.68 \times 0.074) = 2371 \frac{kJ}{kg}$$

The dimensionless group that appears on the right side of eq. (8.30) is (note $\rho_l >> \rho_v$),

$$\frac{D^3 h'_{fg}\, g\, \rho_l}{k_l\, \nu_l (T_{sat} - T_w)} = \frac{(0.02)^3\, m^3\, 2371 \frac{10^3 J}{kg}\, 9.81 \frac{m}{s^2}\, 971.8 \frac{kg}{m^3}}{0.67 \frac{W}{m \cdot K}(3.66)\, 10^{-7} \frac{m^2}{s}\, 40\, K}$$

$$= 1.844 \times 10^{10}$$

therefore eq. (8.30) yields

$$\overline{Nu_D} = 0.729\,(1.844 \times 10^{10})^{1/4} = 268.6$$

and

$$\bar{h}_D = \frac{k_l}{D}\overline{Nu_D} \cong 9000 \frac{W}{m^2 K}$$

$$q' = \bar{h}_D\, \pi D\, (T_{sat} - T_w) = 2.26 \times 10^4 \frac{W}{m}$$

$$\dot{m}' = \frac{q'}{h'_{fg}} = 0.00954 \frac{kg}{s \cdot m}$$

The mass flowrate of the film of condensate collected on only one side of the tube is

$$\Gamma = \frac{\dot{m}'}{2} = 0.00477 \frac{kg}{s \cdot m}$$

therefore

$$Re = \frac{4}{\mu_l}\Gamma = \frac{(4)\, 0.00477\, kg/(s \cdot m)}{3.55 \times 10^{-4}\, kg/(s \cdot m)}$$

$$\cong 54$$

This Reynolds number is comparable with the laminar range indicated in Fig. 8.5, therefore the laminar film assumption is adequate.

Problem 8.10. The properties of water at atmospheric pressure and film temperature (100°C + 60°C)/2 = 80°C are (see Appendix C):

$$c_{P,l} = 4.196 \frac{kJ}{kg\,K} \qquad \mu_l = 3.55 \times 10^{-4} \frac{kg}{m \cdot s}$$

$$k_l = 0.67 \frac{W}{m \cdot K} \qquad \nu_l = 3.66 \times 10^{-7} \frac{m^2}{s}$$

$$\rho_l = 971.8 \frac{kg}{m^3} \qquad Pr_l = 2.23$$

The latent heat of condensation at atmospheric pressure (or at 100°C) is h_{fg} = 2257 kJ/kg. The Jakob number is small,

$$Ja = \frac{c_{P,l}(T_{sat} - T_w)}{h_{fg}} = 0.074$$

therefore h'_{fg} is nearly the same as h_{fg},

$$h'_{fg} = h_{fg}\,(1 + 0.68 \times 0.074) = 2371 \frac{kJ}{kg}$$

The dimensionless group that appears on the right side of eq. (8.30) is (note $\rho_l >> \rho_v$),

$$\frac{D^3 h'_{fg}\, g\, \rho_l}{k_l\, \nu_l\, (T_{sat} - T_w)} = \frac{(0.02)^3\, m^3\; 2371 \frac{10^3 J}{kg}\; 9.81 \frac{m}{s^2}\; 971.8 \frac{kg}{m^3}}{0.67 \frac{W}{m \cdot K}\; (3.66)\, 10^{-7} \frac{m^2}{s}\; 40\, K}$$

$$= 1.844 \times 10^{10}$$

therefore eq. (8.30) yields

$$\overline{Nu}_D = 0.729\,(1.844 \times 10^{10})^{1/4} = 268.6$$

and

$$\overline{h}_D = \frac{k_l}{D}\, \overline{Nu}_D = 8999 \frac{W}{m^2 K}$$

a) In the first design, we see vertical columns of 3, 4 and 5 tubes. The heat transfer coefficients averaged over each type of column are, eq. (8.32),

$$\bar{h}_{D,3} = \frac{\bar{h}_D}{3^{1/4}} = 6837.6 \frac{W}{m^2K}$$

$$\bar{h}_{D,4} = \frac{\bar{h}_D}{4^{1/4}} = 6363.1 \frac{W}{m^2K}$$

$$\bar{h}_{D,5} = \frac{\bar{h}_D}{5^{1/4}} = 6017.9 \frac{W}{m^2K}$$

There are two 3-tube columns, two 4-tube columns, and one 5-tube column. This means that the heat transfer coefficient averaged over all 19 tubes is

$$\bar{h}_a = \frac{2 \times 3}{19} \bar{h}_{D,3} + \frac{2 \times 4}{19} \bar{h}_{D,4} + \frac{1 \times 5}{19} \bar{h}_{D,5}$$

$$= 6422 \frac{W}{m^2K}$$

and this yields the following heat transfer and condensation rates:

$$q'_a = n\, \pi D\, \bar{h}_a (T_{sat} - T_w)$$

$$= 19\, \pi\, 0.02\text{ m } 6422 \frac{W}{m^2K} 40\text{ K}$$

$$= 3.07 \times 10^5 \frac{W}{m}$$

$$\dot{m}'_a = \frac{q'_a}{h'_{fg}} = \frac{3.07 \times 10^5\text{ W/m}}{2371 \times 10^3\text{ J/kg}}$$

$$= 0.129 \frac{kg}{s{\cdot}m}$$

b) In the second design, there are vertical columns of 1, 2 and 3 tubes, therefore

$$\bar{h}_{D,1} = \frac{\bar{h}_D}{1^{1/4}} = 8999 \frac{W}{m^2K}$$

$$\bar{h}_{D,2} = \frac{\bar{h}_D}{2^{1/4}} = 7567.1 \frac{W}{m^2K}$$

$$\bar{h}_{D,3} = \frac{\bar{h}_D}{3^{1/4}} = 6837.6 \frac{W}{m^2K}$$

$$\bar{h}_b = \frac{2 \times 1}{19} \bar{h}_{D,1} + \frac{4 \times 2}{19} \bar{h}_{D,2} + \frac{3 \times 3}{19} \bar{h}_{D,3}$$

$$= 7372 \frac{W}{m^2K}$$

$$q'_b = n\,\pi D\, \bar{h}_b\,(T_{sat} - T_w) = 3.52 \times 10^5 \frac{W}{m}$$

$$\dot{m}'_b = \frac{q'_b}{h'_{fg}} = 0.148 \frac{kg}{s{\cdot}m}$$

Comparing $\dot{m}'_b$ with $\dot{m}'_a$, we conclude that the 90-degree rotation of the tube bank induces a 15-percent increase in the condensation rate.

Problem 8.11. The characteristic length of the horizontal strip of width L and length Z is

$$L_c = \frac{A}{p} = \frac{LZ}{2Z} = \frac{L}{2} \tag{1}$$

Using this and eq. (8.33), the average heat transfer coefficient can be rewritten as

$$\bar{h}_L = 1.079\, L^{-2/5}\, C$$

$$= 0.818\, L_c^{-2/5}\, C \tag{2}$$

where C is the property group

$$C = k_l \left[\frac{h'_{fg}\, g\,(\rho_l - \rho_v)}{k_l\, v_l\,(T_{sat} - T_w)} \right]^{1/5} \tag{3}$$

The overall Nusselt number based on L_c is therefore

$$\overline{Nu}_{L_c} = \frac{\bar{h}_L\, L_c}{k_l} \cong 0.82 \left[\frac{L_c^3\, h'_{fg}\, g\,(\rho_l - \rho_v)}{k_l\, v_l\,(T_{sat} - T_w)} \right]^{1/5} \tag{4}$$

The same operations can be repeated for the horizontal disc facing upward:

$$L_c = \frac{A}{p} = \frac{\pi D^2/4}{D} = \frac{D}{4}$$

$$\bar{h}_D = 1.368\, D^{-2/5}\, C \qquad (8.34)$$

$$= 0.786\, L_c^{-2/5} C$$

$$\overline{Nu}_{L_c} = \frac{\bar{h}_D L_c}{k_l} \cong 0.79 \left[\frac{L_c^3 h'_{fg}\, g\, (\rho_l - \rho_v)}{k_l \nu_l (T_{sat} - T_w)}\right]^{1/5} \qquad (5)$$

Comparing eqs. (4) and (5), we see that the use of the characteristic length L_c leads to a Nu_{L_c} formula that is almost independent of the shape of the surface. In conclusion, for an upward facing surface whose shape is somewhere between "very long" (eq. 4) and "round" (eq. 5), we can estimate the average heat transfer coefficient with the approximate formula

$$\overline{Nu}_{L_c} = \frac{\bar{h} L_c}{k_l} \cong 0.8 \left[\frac{L_c^3 h'_{fg}\, g\, (\rho_l - \rho_v)}{k_l \nu_l (T_{sat} - T_w)}\right]^{1/5} \qquad (6)$$

Problem 8.12. The film temperature of the condensate is (80°C + 100°C)/2 = 90°C, at which the pertinent physical properties have the following values:

$$c_{P,l} = 4.205 \frac{kJ}{kg \cdot K} \qquad \mu_l = 3.16 \times 10^{-4} \frac{kg}{m \cdot s}$$

$$k_l = 0.67 \frac{W}{m \cdot K} \qquad \nu_l = 3.27 \times 10^{-7} \frac{m^2}{s}$$

$$\rho_l = 965.3 \frac{kg}{m^3} \qquad Pr_l = 1.98$$

Note further that h_{fg} = 2257 kJ/kg, therefore

$$Ja = \frac{c_{P,l}(T_{sat} - T_w)}{h_{fg}} = 0.0373$$

$$h'_{fg} = h_{fg}\,(1 + 0.68 \times 0.0373) \cong 2314 \frac{kJ}{kg}$$

a) In the case of film condensation we use eq. (8.30), which is based on the assumption that the film is laminar

$$\overline{Nu}_D = 0.729 \left[\frac{D^3 h'_{fg}\, g\, (\rho_l - \rho_v)}{k_l\, \nu_l\, (T_{sat} - T_w)} \right]^{1/4}$$

$$= 0.729 \left[\frac{(0.04\text{m})^3\ 2314 \times 10^3\ \frac{\text{J}}{\text{kg}}\ 9.81\ \frac{\text{m}}{\text{s}^2}\ 965.3\ \frac{\text{kg}}{\text{m}^3}}{0.67\ \frac{\text{W}}{\text{m·K}}\ 3.27 \times 10^{-7}\ \frac{\text{m}^2}{\text{s}}\ (100 - 80)\ \text{K}} \right]^{1/4}$$

$$= 548.3$$

$$\bar{h}_D = \overline{Nu}_D \frac{k_l}{D}$$

$$= 548.3 \times 0.67\ \frac{\text{W}}{\text{m·K}}\ \frac{1}{0.04\text{m}} = 8998\ \frac{\text{W}}{\text{m}^2\ \text{K}}$$

$$q' = \bar{h}_D\, \pi D\, (T_{sat} - T_w) = 8998\ \frac{\text{W}}{\text{m}^2\ \text{K}}\ \pi\ 0.04\text{m}\ 20\ \text{K}$$

$$= 2.26 \times 10^4\ \text{W/m}$$

$$\dot{m}' = \frac{q'}{h'_{fg}} = \frac{2.26 \times 10^4\ \text{W/m}}{2314 \times 10^3\ \text{J/kg}} \cong 0.01\ \frac{\text{kg/s}}{\text{m}}$$

We estimate the Reynolds number by noting that the flowrate collected on one side (say, the left side) of the horizontal cylinder is

$$\Gamma = \frac{1}{2}\dot{m}' = 0.005\ \frac{\text{kg/s}}{\text{m}}$$

$$Re = \frac{4\Gamma}{\mu_l} = 4 \times 0.005\ \frac{\text{kg/s}}{\text{m}}\ \frac{\text{m·s}}{3.16 \times 10^{-4}\ \text{kg}} \cong 62$$

At this Reynolds number the film is basically laminar, and eq. (8.30) is adequate.

b) When dropwise condensation occurs on the ouside of the teflon coating, the order of magnitude of the heat transfer coefficient is

$$\bar{h}_{drop} \sim 10\ \bar{h}_D \sim 9 \times 10^4\ \frac{\text{W}}{\text{m}^2\ \text{K}}$$

Unfortunately, the heat transfer must overcome also the thermal resistance of the teflon layer,

$$\frac{1}{\bar{h}} = \frac{1}{\bar{h}_{drop}} + \left(\frac{\delta}{k}\right)_{teflon}$$

$$= \frac{m^2\,K}{9 \times 10^4\,W} + \frac{0.0005m}{0.23\,W/m\cdot K} = 0.00219\,\frac{m^2\,K}{W}$$

$$\bar{h} = 458\,\frac{W}{m^2\,K}$$

where $\bar{h}$ is the "effective" heat transfer coefficient between 80°C and 100°C. It is clear that $\bar{h}$ is controlled (made smaller) by the teflon resistance, as it is almost equal to $(k/\delta)_{teflon}$. Note also that $\bar{h}$ is 20 times smaller than its film-condensation counterpart, and so is the condensation rate:

$$q' = \bar{h}\,\pi D\,(T_{sat} - T_w)$$

$$= 458\,\frac{W}{m^2\,K}\,\pi\,0.04m\,(100 - 80)\,K = 1151\,\frac{W}{m}$$

$$\dot{m}' = \frac{q'}{h'_{fg}} = 0.5\,\frac{g/s}{m}$$

c) In conclusion, it is a bad idea to coat the tube with a 0.5mm layer of teflon. The range of teflon thicknesses that lead to increases in the condensation rate is obtained from the requirement that the teflon resistance must be smaller than the resistance of the laminar film

$$\left(\frac{\delta}{k}\right)_{teflon} < \frac{1}{\bar{h}_D}$$

$$\delta < 0.23\,\frac{W}{m\cdot K}\,\frac{m^2\,K}{9000\,W} = 0.026\text{ mm}$$

Problem 8.13. a) The balance of vertical forces on the hemispherical control volume requires

$$\pi r^2 P_v = \pi r^2 P_l + 2 \pi r \sigma$$

which yields

$$r = \frac{2\sigma}{P_v - P_l} \tag{1}$$

b) Next, by assuming that the vapor is saturated, the pressure difference (P_v - P_l) can be related to the corresponding temperature difference (T_v - T_l) through the Clausius-Clapeyron relation

$$\frac{dP}{dT} = \frac{h_{fg}}{T v_{fg}} \tag{2}$$

in which $T = T_l = T_{sat}$, and

$$v_{fg} = v_g - v_f \cong v_g = \frac{1}{\rho_v}$$

$$\frac{dP}{dT} = \frac{P_v - P_l}{T_v - T_l}, \quad \text{where } T_l = T_{sat}$$

therefore eq. (2) becomes

$$\frac{P_v - P_l}{T_v - T_{sat}} = \frac{h_{fg}\,\rho_v}{T_{sat}} \tag{3}$$

Eliminating (P_v - P_l) between eqs. (1) and (3) leads to

$$r = \frac{2\sigma T_{sat}}{h_{fg}\,\rho_v\,(T_v - T_{sat})}$$

in which T_{sat} is an absolute temperature (i.e. it is expressed in degrees Kelvin).

c) For T_{sat} = 100°C, T_v - T_{sat} = 2°C and the properties listed for water at 100°C in Table 8.2, the bubble radius is

$$r = \frac{2 \times 0.059\,\frac{N}{m}\,373\text{ K}}{2257\,\frac{kJ}{kg}\,0.6\,\frac{kg}{m^3}\,2\text{ K}} = 16.3 \times 10^{-6}\text{m} = 0.016\text{ mm}$$

Problem 8.14. If the excess temperature is left unspecified, $T_w - T_{sat}$, the nucleate boiling heat flux calculated in Example 8.3 is

$$q''_w = \left[\frac{T_w - T_{sat}}{(108 - 100)\,K}\right]^3 7 \times 10^5 \frac{W}{m^2} \tag{1}$$

Equating this with the calculated peak heat flux,

$$q''_{max} = 1.26 \times 10^6 \frac{W}{m^2} \tag{2}$$

we obtain

$$T_w - T_{sat} = \left(\frac{1.26 \times 10^6}{7 \times 10^5}\right)^{1/3} 8\,K \cong 10\,K$$

The actual excess temperature is larger than this value (namely, in the 20-30 K range), because of the "S" shape of the nucleate boiling portion of the boiling curve.

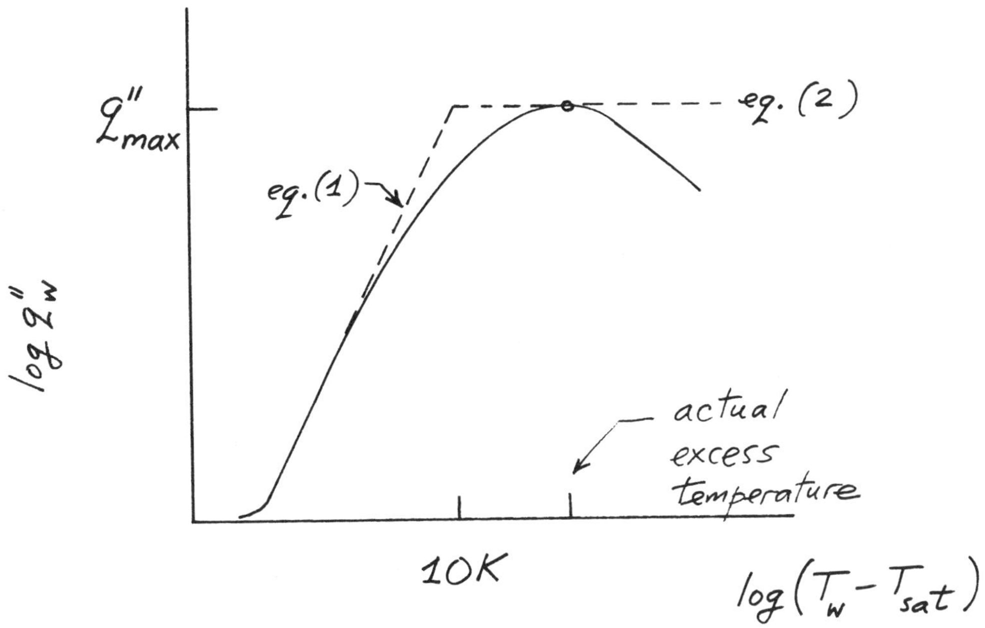

Problem 8.15. The point of maximum slope on the nucleate boiling portion of the curve (Fig. 8.17) is located in the vicinity of the transition from isolated bubbles to columns and slugs,

$$q''_w \cong 10^5 \frac{W}{m^2}$$

$$T_w - T_{sat} \cong 10°C$$

The pertinent properties of water at 100°C are collected from Tables 8.1, 8.2 and Appendix C:

$$\rho_l = 958 \frac{kg}{m^3} \qquad \sigma = 0.059 \frac{N}{m}$$

$$\rho_v = 0.6 \frac{kg}{m^3} \qquad h_{fg} = 2257 \frac{kJ}{kg}$$

$$c_{P,l} = 4.216 \frac{kJ}{kg\,K} \qquad Pr_l = 1.78$$

$$\mu_l = 2.83 \times 10^{-4} \frac{kg}{s{\cdot}m} \qquad s = 1$$

The C_{sf} constant can be calculated using eq. (8.43),

$$C_{sf} = \frac{(T_w - T_{sat})\,c_{P,l}}{h_{fg}\,Pr_l^3} \Big/ \left[\frac{q''_w}{\mu_l\,h_{fg}}\left(\frac{\sigma}{g(\rho_l - \rho_v)}\right)^{1/2}\right]^{1/3}$$

in which we substitute

$$\frac{(T_w - T_{sat})\,c_{P,l}}{h_{fg}\,Pr_l^s} = \frac{10°C\ 4.216 \frac{kJ}{kg\,K}}{2257 \frac{kJ}{kg}\ 1.78} = 0.0187$$

$$\frac{\sigma}{g(\rho_l - \rho_v)} = \frac{0.059 \frac{N}{m}}{9.81 \frac{m}{s^2}(958 - 0.6)\frac{kg}{m^3}} = 6.28 \times 10^{-6}\ m^2$$

$$\frac{q''_w}{\mu_l\,h_{fg}} = \frac{10^5 \frac{W}{m^2}}{2.83 \times 10^{-4} \frac{kg}{s{\cdot}m}\ 2257 \frac{kJ}{kg}} = 156.56 \frac{1}{m}$$

The end result is

$$C_{sf} = \frac{0.0187}{\left[156.56 \frac{1}{m}\left(6.28 \times 10^{-6}\ m2\right)^{1/2}\right]^{1/3}}$$

$$= 0.026$$

This value is compatible with (i.e. of the same order as) the values listed for water-copper in Table 8.1.

Problem 8.16. The properties of saturated helium at atmospheric pressure (T_{sat} = 4.2 K) are listed in Table 8.2 and Appendix C:

$$\rho_l = 125\ \frac{kg}{m3} \qquad \sigma = 10^{-4}\ \frac{N}{m}$$

$$\rho_v = 16.9\ \frac{kg}{m3} \qquad h_{fg} = 20.42\ \frac{kJ}{kg}$$

$$c_{P,l} = 4.98\ \frac{kJ}{kg\ K} \qquad Pr_l = 0.8$$

$$\mu_l = 3.17 \times 10^{-6}\ \frac{kg}{s \cdot m}$$

The excess temperature at $q_w'' = 10^3$ W/m2 can be calculated using eq. (8.43),

$$T_w - T_{sat} = \frac{h_{fg}}{c_{P,l}} Pr_l^s\, C_{sf} \left[\frac{q_w''}{\mu_l h_{fg}} \left(\frac{\sigma}{g\,(\rho_l - \rho_v)}\right)^{1/2}\right]^{1/3} \qquad (8.43)$$

in which

$$\frac{q_w''}{\mu_l h_{fg}} \left(\frac{\sigma}{g\,(\rho_l - \rho_v)}\right)^{1/2} = \frac{10^3\ \frac{W}{m2}}{3.17 \times 10^{-6}\ \frac{kg}{s \cdot m}\ 20.42\ \frac{kJ}{kg}} \left(\frac{10^{-4}\ \frac{N}{m}}{9.81\ \frac{m}{s2}\ (125 - 16.9)\ \frac{kg}{m3}}\right)^{1/2}$$

$$= 4.75$$

Assuming C_{sf} = 0.02 and s = 1.7 in eq. (8.43), we obtain finally

$$T_w - T_{sat} = \frac{20.42\ \frac{kJ}{kg}}{4.98\ \frac{kJ}{kg\ K}}\ 0.8^{1.7}\ 0.02\ (4.75)^{1/3}$$

$$\cong 0.1\ K$$

The peak heat flux can be calculated based on eq. (8.45),

$$q''_{max} = 0.149\, h_{fg}\, \rho_v^{1/2} [\sigma\, g\, (\rho_l - \rho_v)]^{1/4}$$

$$= 0.149 \times 20.42 \frac{kJ}{kg} (16.9)^{1/2} \frac{kg^{1/2}}{m^{3/2}} \left[10^{-4} \frac{N}{m}\, 9.81 \frac{m}{s^2} (125\text{-}16.9) \frac{kg}{m^3}\right]^{1/4} \quad (8.45)$$

$$= 7137 \frac{W}{m^2}$$

to conclude that the heat leak q''_w is roughly 14 percent of the peak nucleate boiling heat flux.

Problem 8.17. The temperature of saturated water at 4.76×10^5 N/m^2 is

$$T_{sat} = 150°C$$

The corresponding properties of water at this temperature are

$$\rho_l = 917 \frac{kg}{m^3} \qquad \sigma = 0.048 \frac{N}{m}$$

$$\rho_v = 2.55 \frac{kg}{m^3} \qquad h_{fg} = 2114 \frac{kJ}{kg}$$

$$c_{P,l} = 4.27 \frac{kJ}{kg\,K} \qquad Pr_l = 1.17$$

$$\mu_l = 1.85 \times 10^{-4} \frac{kg}{s \cdot m}$$

For nucleate boiling of water on a polished copper surface, Table 8.1 suggests $C_{sf} = 0.013$ and $s = 1$, and eq. (8.44) yields

$$\frac{g\,(\rho_l - \rho_v)}{\sigma} = \frac{9.81 \frac{m}{s^2} (917 - 2.55) \frac{kg}{m^3}}{0.048 \frac{N}{m}}$$

$$= 1.869 \times 10^5 \frac{1}{m^2}$$

$$\frac{c_{P,l}(T_w - T_{sat})}{Pr_l^s\, C_{sf}\, h_{fg}} = \frac{4.27 \frac{kJ}{kg\,K} (160 - 150)\,K}{1.17 \times 0.013 \times 2114 \frac{kJ}{kg}}$$

$$= 1.328$$

$$q''_w = 1.85 \times 10^{-4} \frac{kg}{s \cdot m} 2114 \frac{kJ}{kg} \left(1.869 \times 10^5 \frac{1}{m^2}\right)^{1/2} (1.328)^3 \qquad (8.44)$$

$$= 3.96 \times 10^5 \frac{W}{m^2}$$

This heat flux level is only a fraction (17.3 percent) of the peak heat flux, which can be calculated based on eq. (8.45):

$$q''_{max} = 0.149\, h_{fg}\, \rho_v^{1/2} (\sigma\, g\, \rho_l)^{1/4} \qquad (8.45)$$

$$= 0.149 \times 2114 \frac{kJ}{kg} (2.55)^{1/2} \frac{kg^{1/2}}{m^{3/2}} \left(0.048 \frac{N}{m} 9.81 \frac{m}{s^2} 917 \frac{kg}{m^3}\right)^{1/4}$$

$$\cong 2.3 \times 10^6 \frac{W}{m^2}$$

This means that the nucleate-boiling assumption is correct. Finally, we evaluate the total heat transfer rate

$$q_w = \frac{\pi}{4} D^2 q''_w = \frac{\pi}{4} (0.2)^2\, m^2\, 3.96 \times 10^5 \frac{W}{m^2}$$

$$= 1.244 \times 10^4 \text{ W}.$$

Problem 8.18. a) Writing approximately $q_w = \dot{m}\, h_{fg}$, we calculate in order

$$\dot{m} = \frac{q_w}{h_{fg}} = \frac{1.244 \times 10^4 \text{ W}}{2114 \frac{kJ}{kg}} = 0.00588 \frac{kg}{s}$$

$$m = \rho_l \frac{\pi}{4} D^2 H = 917 \frac{kg}{m^3} \frac{\pi}{4} (0.2)^2\, m^2\, 0.05 \text{ m}$$

$$= 1.44 \text{ kg}$$

$$t = \frac{m}{\dot{m}} = \frac{1.44 \text{ kg}}{0.00588 \text{ kg/s}} = 245 \text{ s} = 4.1 \text{ minutes}$$

in which m is the original liquid inventory, and t the time needed to evaporate all the liquid.

b) The vessel defines a control volume whose contents represent a thermodynamic system that operates in unsteady (time dependent) fashion. With reference to the figure attached to the problem statement, we write the first law

$$\frac{dU}{dt} = q_w - \dot{m}\, h_g \tag{1}$$

in which U is the instantaneous internal-energy inventory

$$U = m_f\, u_f + m_g\, u_g \tag{2}$$

and (m_f, m_g) are the (liquid, vapor) mass inventories. Combined, eqs. (1) and (2) yield

$$\dot{m}_f\, u_f + \dot{m}_g\, u_g = q_w - \dot{m}\, h_g \tag{3}$$

The conservation of mass requires

$$\frac{d}{dt}(m_f + m_g) = -\,\dot{m} \tag{4}$$

in other words

$$\dot{m}_f + \dot{m}_g = -\,\dot{m} \tag{5}$$

where $\dot{m}$ is the flowrate of the escaping steam. Finally, we note that the volume of the vessel remains constant,

$$V = m_f\, v_f + m_g\, v_g \tag{6}$$

or, after taking the time derivative,

$$0 = \dot{m}_f\, v_f + \dot{m}_g\, v_g \tag{7}$$

Equations (5) and (7) can be used to express $\dot{m}_f$ and $\dot{m}_g$ in terms of $\dot{m}$ and v_f and v_g:

$$\dot{m}_f = -\,\dot{m}\frac{1}{1 - v_f/v_g} \tag{8}$$

$$\dot{m}_g = \dot{m}\frac{v_f/v_g}{1 - v_f/v_g} \tag{9}$$

Substituting eqs. (8,9) into the first law (3), we obtain the final result

$$\frac{q_w}{\dot{m}} = h_g - \frac{u_f\, v_g - u_g\, v_f}{v_g - v_f} \tag{10}$$

The right hand side can be evaluated numerically using the properties of saturated water. The following results show that this quantity deviates from the h_{fg} value (assumed in part a), as the liquid-vapor mixture approaches the critical point:

T	the right side of eq. (10)	h_{fg}
150°C	$2120.2 \frac{kJ}{kg}$	$2114.3 \frac{kJ}{kg}$
360°C	$990.4 \frac{kJ}{kg}$	$720.5 \frac{kJ}{kg}$

Problem 8.19. According to Fig. 8.17, a temperature difference of 8°C (= $T_w - T_{sat}$) corresponds to nucleate boiling. We will be relying on the Rohsenow correlation listed as eq. (8.44), therefore we collect the following data from the tables:

$C_{sf} = 0.013$ $\quad s = 1$ $\quad$ (Table 8.1)

$c_{P,l} = 4.211 \frac{kJ}{kg \cdot K}$ $\quad Pr_l = 1.75$ $\quad$ (Appendix C, saturated water at 100°C)

$\mu_l = 0.0028 \frac{g}{cm \cdot s}$

$\rho_l = 958 \frac{kg}{m^3}$ $\quad \sigma = 0.059 \frac{N}{m}$ $\quad$ (Table 8.2)

$\rho_v = 0.6 \frac{kg}{m^3}$ $\quad h_{fg} = 2257 \frac{kJ}{kg}$

It is advisable to calculate the right side of eq. (8.44) in two steps, in order to avoid numerical errors and the mix-up of units. We start with the group on the extreme right, which is dimensionless:

$$\frac{c_{P,l}(T_w - T_{sat})}{Pr_l^s\, C_{sf}\, h_{fg}} = \frac{4.211 \frac{kJ}{kg \cdot K}(108 - 100)\,K}{1.75 \times 0.013 \times 2257 \frac{kJ}{kg}}$$

$$= 0.656$$

$$q''_w = \mu_l\, h_{fg}\left[\frac{g(\rho_l - \rho_v)}{\sigma}\right]^{1/2} (0.656)^3$$

$$= 0.0028 \frac{10^{-3}\ \text{kg}}{0.01\ \text{m·s}} 2257 \frac{10^3\ \text{J}}{\text{kg}} \left[\frac{9.81 \frac{\text{m}}{\text{s}^2} 957.4 \frac{\text{kg}}{\text{m}^3}}{0.059 \frac{\text{kg·m}}{\text{s}^2} \frac{1}{\text{m}}} \right]^{1/2} (0.656)^3$$

$$= 7.12 \times 10^4 \frac{\text{W}}{\text{m}^2}$$

This heat flux value agrees well with what can be read off Fig. 8.17. Related quantities are

$$h = \frac{q''_w}{T_w - T_{sat}} = \frac{7.12 \times 10^4\ \text{W/m}^2}{(108 - 100)\ \text{K}}$$

$$= 8.9 \times 10^3 \frac{\text{W}}{\text{m}^2\ \text{K}}$$

$$q = \frac{\pi}{4} D^2 q''_w = \frac{\pi}{4} (0.2\text{m})^2\ 7.12 \times 10^4 \frac{\text{W}}{\text{m}^2}$$

$$= 2.24\ \text{kW}$$

It is worth noting that the h value calculated above falls in the range indicated for boiling water in Fig. 1.12.

Problem 8.20. In order of magnitude terms, the heat flux across the vapor film can be expressed as

$$q''_w = \bar{h} (T_w - T_{sat}) \sim k_v \frac{T_w - T_{sat}}{\delta}$$

where δ is the scale of the vapor film thickness. Using the numerical data of Example 8.4, we conclude that

$$\delta \sim \frac{k_v}{\bar{h}} = \frac{0.0334\ \text{W}}{\text{m·K}} \frac{\text{m}^2\ \text{K}}{224\ \text{W}}$$

$$= 0.15\ \text{mm}$$

Problem 8.21. Assuming that the dominant mode of heat transfer across the film is radiation,

$$\bar{h} \cong \bar{h}_{rad}$$

we write

$$q''_w = \bar{h}_{rad} (T_w - T_{sat})$$

$$= \frac{\sigma \, \varepsilon_w \left(T_w^4 - T_{sat}^4\right)}{T_w - T_{sat}} (T_w - T_{sat})$$

$$\cong \sigma \, T_w^4$$

because $T_w^4 << T_{sat}^4$, and $\varepsilon_w = 1$. Numerically, this means

$$T_w = \left(\frac{q''_w}{\sigma}\right)^{1/4} = \left(\frac{10^6 \frac{W}{m^2}}{5.669 \times 10^{-8} \frac{W}{m^2 \, K^4}}\right)^{1/4}$$

$$= 2049 \text{ K} = 1776°\text{C}$$

or an excess temperature of 1776°C - 100°C = 1676°C. This excess temperature is compatible with the values suggested in Fig. 8.17.

The actual excess temperature will be somewhat lower than the value calculated above, because the convection effect will contribute to an $\bar{h}$ value that is a bit higher than $\bar{h}_{rad}$.

Problem 8.22. a) With T_{sat} = 100°C and an excess temperature of 200°C, the wall and average film temperatures are T_w = 300°C and (300°C + 100°C)/2 = 200°C. The relevant properties are

$$\rho_v = 0.46\ \frac{kg}{m^3}, \qquad \text{Appendix D, steam at 200°C}$$

$$\left.\begin{aligned} \rho_l &= 958\ \frac{kg}{m^3} \\ h_{fg} &= 2257\ \frac{kJ}{kg} \\ \sigma &= 0.059\ \frac{N}{m} \end{aligned}\right\} \qquad \text{Table 8.2, water at 100°C}$$

and the minimum heat flux can be calculated using eq. (8.46):

$$q''_{min} = 0.09 \times 2257\ \frac{kJ}{kg} \times 0.46\ \frac{kg}{m^3} \left[\frac{0.059\ \frac{N}{m}\ 9.81\ \frac{m}{s^2}\ (9.58 - 0.46)\ \frac{kg}{m^3}}{(958 + 0.46)^2\ \frac{kg^2}{m^6}} \right]^{1/4}$$

$$= 1.46 \times 10^4\ \frac{W}{m^2}$$

This minimum heat flux is compatible with the level indicated in Fig. 8.17.

b) The same calculation can be repeated for T_{sat} = 300°C and an excess temperature of 200°C, which means T_w = 500°C and an average film temperature of 400°C.

$$\rho_v = 0.323\ \frac{kg}{m^3}, \qquad \text{Appendix D, steam at 400°C}$$

$$\left.\begin{aligned} \rho_l &= 712\ \frac{kg}{m^3} \\ h_{fg} &= 1405\ \frac{kJ}{kg} \\ \sigma &= 0.014\ \frac{N}{m} \end{aligned}\right\} \qquad \text{Table 8.2, water at 300°C}$$

$$q''_{min} = 0.09 \times 1405 \frac{kJ}{kg} \times 0.323 \frac{kg}{m^3} \left[\frac{0.014 \frac{N}{m} 9.81 \frac{m}{s^2} (712 - 0.323) \frac{kg}{m^3}}{(712 + 0.323)^2 \frac{kg^2}{m^6}} \right]^{1/4}$$

$$= 4.8 \times 10^3 \frac{W}{m^2}$$

In conclusion, the minimum heat flux decreases by a factor of 3 as the pool temperature increases from 100°C to 300°C, while the excess temperature remains constant.

Project 8.1. After examining Fig. 1.12, on the condensing side we expect a heat transfer coefficient that is two or three orders of magnitude greater than the heat transfer coefficient on the air side. This means that the air natural convection thermal resistance (or the difference $T_w - T_\infty$) would be much greater than the thermal resistance posed by the film of condensate (or $T_{sat} - T_w$).

Therefore, as a starting guess we assume that the wall temperature is nearly the same as the saturated vapor temperature,

$$T_w \cong 100°C$$

On the air side, the natural convection "film" temperature is (100°C + 20°C)/2 = 60°C, therefore

$$Ra_L = \left(\frac{g\beta}{\alpha\nu}\right)_{\substack{air \\ 60°C}} H^3 \left(T_w - T_\infty\right)$$

$$= \frac{57.1}{cm^3 K} (50)^3\ cm^3\ (100\text{-}20)\ K = 5.71 \times 10^8$$

$$k_a = 0.028 \frac{W}{m \cdot K} \qquad \text{(air, 60°C)}$$

$$\overline{Nu_L} = 0.68 + 0.515\ Ra_L^{1/4} \qquad \text{(eq. 7.62')}$$

$$= 80.29$$

$$\bar{h}_a = \overline{Nu_L} \frac{k_a}{L} = 80.29 \frac{0.028\ W}{m \cdot K} \frac{1}{0.5 m}$$

$$= 4.5 \frac{W}{m^2 K}$$

$$q' = \bar{h}_a L \left(T_w - T_\infty\right) = 4.5 \frac{W}{m^2 K} 0.5 m\ 80 K$$

$$= 179.8 \frac{W}{m}$$

The same heat transfer rate q' enters the vertical surface through the water side, therefore the total condensation rate is

$$\Gamma(L) = \frac{q'}{h_{fg}} = \frac{179.8\ W/m}{2257\ kJ/kg}$$

$$= 7.97 \times 10^{-5} \frac{kg}{s \cdot m}$$

Note the use of h_{fg} instead of h'_{fg}, because when $T_w \cong 100°C$ the condensate experiences negligible subcooling.

We can now verify that the $T_w \cong 100°C$ assumption was correct. This can be done by using the answer q' = 179.8 W/m and eq. (8.15) in order to estimate the "true" temperature difference across the liquid film ($T_{sat} - T_w$):

$$q' = \overline{h}_L \, L \, (T_{sat} - T_w) = \overline{Nu}_L \, k_l \, (T_{sat} - T_w)$$

$$= 0.943 \left[\frac{L^3 \, h_{fg} \, g \, \rho_l}{k_l \, \nu_l \, (T_{sat} - T_w)} \right]^{1/4} k_l \, (T_{sat} - T_w)$$

This will be easier to manipulate if we use the known temperature difference $\Delta T = T_{sat} - T_\infty = 80°C$, therefore we write

$$\frac{q'}{k_l \, \Delta T} = 0.943 \left(\frac{L^3 \, h_{fg} \, g \, \rho_l}{k_l \, \nu_l \, \Delta T} \right)^{1/4} \left(\frac{T_{sat} - T_w}{\Delta T} \right)^{3/4} \qquad \text{(a)}$$

in which we evaluate all the water properties at 100°C:

$$\frac{q'}{k_l \, \Delta T} = \frac{179.8 \text{ W/m}}{0.68 \frac{W}{m \cdot K} \, 80K} = 3.305$$

$$\frac{L^3 \, h_{fg} \, g \, \rho_l}{k_l \, \nu_l \, \Delta T} = \frac{(0.5)^3 \, m^3 \; 2257 \, \frac{10^3 J}{kg} \; 9.81 \, \frac{m}{s^2} \; 958.4 \, \frac{kg}{m^3}}{0.68 \, \frac{W}{m \cdot K} \, (2.95) \, 10^{-7} \, \frac{m^2}{s} \, 80 \text{ K}}$$

$$= 1.65 \times 10^{14}$$

Equation (a) yields in order

$$\left(\frac{T_{sat} - T_w}{\Delta T} \right)^{3/4} = 0.000977$$

$$T_{sat} - T_w \cong 0.01°C$$

proving in this way that the assumption $T_w \cong 100°C$ was justified.

Project 8.2. a) The steam properties (Appendix D) are evaluated at the average temperature of the vapor film, (354°C + 100°C)/2 = 227°C:

$$\rho_v = 0.435\ \frac{kg}{m^3} \qquad c_{P,v} = 1.983\ \frac{kJ}{kg\cdot K}$$

$$\nu_v = 4 \times 10^{-5}\ \frac{m^2}{s} \qquad k_v = 0.0358\ \frac{W}{m\cdot K}$$

The relevant properties of saturated 100°C water are (Table 8.2),

$$\rho_l = 958\ \frac{kg}{m^3} \qquad \sigma = 0.059\ \frac{N}{m}$$

$$h_{fg} = 2257\ \frac{kJ}{kg}$$

and, using eq. (8.49),

$$h'_{fg} = h_{fg} + 0.4\ c_{P,v}\,(T_w - T_{sat}) \tag{8.49}$$

$$= 2257\ \frac{kJ}{kg} + 0.4 \times 1.983\ \frac{kJ}{kg\ K}\ (354 - 100)\ K$$

$$= 2458.5\ \frac{kJ}{kg}$$

The convection heat transfer coefficient for film boiling on the sphere is

$$\frac{\bar{h}_D\, D}{k_v} = 0.67 \left[\frac{D^3\, h'_{fg}\, g\, (\rho_l - \rho_v)}{k_v\, \nu_v\, (T_w - T_{sat})}\right]^{1/4} \tag{8.48}$$

$$= 0.67 \left(\frac{0.02^3\ m^3\ 2458.5\ \frac{kJ}{kg}\ 9.81\ \frac{m}{s^2}\ (958 - 0.435)\ \frac{kg}{m^3}}{0.0358\ \frac{W}{m\cdot K}\ 4 \times 10^{-5}\ \frac{m^2}{s}\ (354 - 100)\ K}\right)$$

$$= 0.67 \left(5.08 \times 10^8\right)^{1/4} = 100.6$$

$$\bar{h}_D = 100.6\ \frac{k_v}{D} = 100.6\ \frac{0.0358\ W}{m\cdot K}\ \frac{1}{0.02\ m}$$

$$= 180\ \frac{W}{m^2\ K}$$

This value is augmented somewhat by a contribution due to radiation ($\varepsilon_w = 0.05$),

$$\bar{h}_{rad} = \frac{\sigma\, \varepsilon_w \left(T_w^4 - T_{sat}^4\right)}{T_w - T_{sat}}$$

$$= \frac{5.669 \times 10^{-8}\ \text{W}}{\text{m}^2\ \text{K}^4}\, 0.05\, \frac{\left(627^4 - 373^4\right) \text{K}^4}{(627 - 373)\ \text{K}}$$

$$= 1.51\ \frac{\text{W}}{\text{m}^2\ \text{K}}$$

$$\bar{h} = \bar{h}_D + \frac{3}{4}\bar{h}_{rad} = 181.1\ \frac{\text{W}}{\text{m}^2\ \text{K}}$$

leading to the following heat transfer rate for the entire sphere:

$$q_w = 4\pi \left(\frac{D}{2}\right)^2 q_w'' = \pi D^2\, \bar{h}\, (T_w - T_{sat})$$

$$= \pi\ 0.02^2\ \text{m}^2\ 181.1\ \frac{\text{W}}{\text{m}^2\ \text{K}} (354 - 100)\ \text{K}$$

$$= 57.8\ \text{W}$$

b) The first law for the copper ball as a lumped capacitance is

$$-\rho\, c_p V \frac{dT_w}{dt} = q_w$$

for which the approximate property values are (Appendix B):

$$\rho = 8954\ \frac{\text{kg}}{\text{m}^3} \qquad k = 381\ \frac{\text{W}}{\text{m}\cdot\text{K}}$$

$$c_p = 0.384\ \frac{\text{kJ}}{\text{kg}\cdot\text{K}}$$

Calculating first the volume of the ball,

$$V = \frac{4}{3}\pi \left(\frac{D}{2}\right)^3 = \frac{\pi}{6} D^3 = \frac{\pi}{6}\, 0.02^3\ \text{m}^3 = 4.19 \times 10^{-6}\ \text{m}^3$$

we obtain the instantaneous rate of cooling

$$-\frac{dT}{dt} = \frac{57.8\ \frac{\text{J}}{\text{s}}}{8954\ \frac{\text{kg}}{\text{m}^3}\ 0.384\ \frac{10^3\ \text{J}}{\text{kg}\cdot\text{K}}\ 4.19 \times 10^{-6}\ \text{m}^3}$$

$$= 4.01\ \frac{\text{K}}{\text{s}}$$

According to Fig. 4.5, the lumped capacitance model is valid when

$$\frac{hr_o}{k} << 1, \text{ and} \qquad \text{(Bi)}$$

$$\frac{\alpha t}{r_o^2} >> 1 \qquad \text{(Fo)}$$

The (Bi) criterion is clearly satisfied:

$$Bi = \frac{hr_o}{k} = \frac{181.1 \text{ W}}{m^2 \text{ K}} \frac{0.01 \text{ m}}{381 \frac{W}{m \cdot K}} = 0.0047 << 1$$

The (Fo) requirement is met when

$$t > \frac{r_o^2}{\alpha} = \frac{10^{-4} \text{ m}^2}{1.11 \times 10^{-4} \frac{m^2}{s}} = 0.9 \text{ s}$$

because the thermal diffusivity is approximately

$$\alpha = \frac{k}{\rho c_P} \cong \frac{381 \text{ W}}{m \cdot K} \frac{m^3}{8954 \text{ kg}} \frac{kg \cdot K}{384 \text{ J}}$$

$$\cong 1.11 \times 10^{-4} \frac{m^2}{s}$$

The time interval of interest $\Delta t = 10$ s, is therefore long enough to satisfy the (Fo) criterion, and justify the use of the lumped capacitance model. The temperature change after 10 seconds is

$$\Delta T_w = \frac{dT_w}{dt} \Delta t = -4.01 \frac{K}{s} 10 \text{ s}$$

$$\cong -40 \text{ K}$$

which corresponds to a new ball temperature of

$$T_w = (354 - 40)°C = 314°C$$

Chapter 9

HEAT EXCHANGERS

Problem 9.1. The inverse of r_S is the scale "heat transfer coefficient" that bridges the gap from the wall surface (T_W) to the wetted surface (T_S):

$$h_S = \frac{1}{r_S}$$

The analysis contained between eqs. (9.1) and (9.3) in the text can be repeated for h_S (in place of h) and (T_W - T_S) (in place of T_W - T_∞):

$$\begin{aligned} q &= \eta\, h_S A_f (T_W - T_S) + h_S A_u (T_W - T_S) \\ &= \left(\eta \frac{A_f}{A} + \frac{A_u}{A}\right) h_S A (T_W - T_S) \\ &= \varepsilon\, h_S A (T_W - T_S) \qquad (1) \end{aligned}$$

Under the same circumstances, eq. (9.3) reads

$$q = \varepsilon\, hA \left(T_S - T_\infty\right) \qquad (2)$$

Rewriting eqs. (1) and (2) as

$$T_W - T_S = \frac{q}{\varepsilon\, h_S A}$$

$$T_S - T_\infty = \frac{q}{\varepsilon\, hA}$$

and adding them side by side, we obtain

$$\begin{aligned} T_W - T_\infty &= \frac{q}{\varepsilon A}\left(\frac{1}{h_S} + \frac{1}{h}\right) \\ &= \frac{q}{\varepsilon\, h_e A} \end{aligned}$$

where

$$\frac{1}{h_e} = \frac{1}{h_S} + \frac{1}{h} = r_S + \frac{1}{h}$$

Problem 9.2. a) According to Table 9.1, the fouling factors are

$$r_{s,h} = 0.0002 \frac{m^2 K}{W} \quad \text{(city water)} \qquad r_{s,c} = 0.0002 \frac{m^2 K}{W} \quad \text{(compressed air)}$$

The effective heat transfer coefficients for both sides of the wall are given by eq. (9.7):

$$h_{e,h} = \left(r_s + \frac{1}{h}\right)_h^{-1} = \left(0.0002 + \frac{1}{100}\right)^{-1} \frac{W}{m^2 K}$$

$$= 98 \frac{W}{m^2 K}$$

$$h_{e,c} = \left(r_s + \frac{1}{h}\right)_c^{-1} = \left(0.0002 + \frac{1}{10}\right)^{-1} \frac{W}{m^2 K}$$

$$\cong 10 \frac{W}{m^2 K}$$

In conclusion, fouling has an effect (a minor one) only on the water side of the surface.

The geometry of the heat exchanger surface is described by

$$A_h = 1 \text{ m}^2 \qquad \text{(on the hot side)}$$

and, on the cold side by

one fin per $(1.4 \text{ cm})^2$

$$n = \frac{1 \text{ m}^2}{(1.4 \text{ cm})^2/\text{fin}} = 5102 \text{ fins}$$

$$A_u = 1 \text{ m}^2 - n\,\pi \left(\frac{D}{2}\right)^2$$

$$= 1 \text{ m}^2 - 5102\,\pi \left(\frac{0.003 \text{ m}}{2}\right)^2 = 0.964 \text{ m}^2$$

The effective (corrected) length of each fin is

$$L_c = L + \frac{D}{4} = 1.5 \text{ cm} + \frac{0.3 \text{ cm}}{4} = 1.575 \text{ cm}$$

therefore

$$A_f = n\,\pi\,D\,L_c = 5102\,\pi\,0.003\text{ m } 0.01575\text{ m}$$

$$= 0.757\text{ m}^2$$

$$A_c = A_f + A_u = 1.721\text{ m2}$$

For the overall efficiency of the cold side, we calculate in order

$$mL_c = \left(\frac{h_{e,c}\,p}{kA}\right)^{1/2} L_c = 2\left(\frac{h_{e,c}}{kD}\right)^{1/2} L_c$$

$$= 2\left(\frac{10\text{ W}}{\text{m}^2\text{ K}}\;\frac{\text{m}\cdot\text{K}}{50\text{ W}}\;\frac{1}{0.003\text{ m}}\right)^{1/2} 0.01575\text{ m}$$

$$= 0.257$$

$$\eta = \frac{\tanh\,(m\,L_c)}{m\,L_c} = 0.979$$

$$\varepsilon_c = \eta\,\frac{A_f}{A_c} + \frac{A_u}{A_c}$$

$$= 0.979\,\frac{0.757\text{ m2}}{1.721\text{ m}^2} + \frac{0.964\text{ m2}}{1.721\text{ m}^2} = 0.991$$

We now have all the information for calculating the overall thermal resistance and the heat transfer rate:

$$\frac{1}{U_h\,A_h} = \frac{1}{h_{e,h}\,A_h} + \frac{t}{k\,A_h} + \frac{1}{\varepsilon_c\,h_{e,c}\,A_c}$$

$$= \left(\frac{1}{98\times 1} + \frac{0.004}{50\times 1} + \frac{1}{0.991\times 10\times 1.721}\right)\frac{\text{K}}{\text{W}}$$

$$= 0.0689\,\frac{\text{K}}{\text{W}}$$

$$q_a = U_h\,A_h\,(T_h - T_c) = \frac{T_h - T_c}{\dfrac{1}{U_h\,A_h}}$$

$$= \frac{(80 - 30)\text{ K}}{0.0689\text{ K/W}} = 726\text{ W}$$

b) In the absence of fins, the total area on the cold side is $A_c = 1\text{ m}^2$, therefore

$$\frac{1}{U_h A_h} = \frac{1}{h_{e,h} A_h} + \frac{t}{k A_h} + \frac{1}{h_{e,c} A_c}$$

$$= \left(0.0102 + 0.00008 + \frac{1}{10 \times 1}\right) \frac{K}{W}$$

$$= 0.1103 \, \frac{K}{W}$$

$$q_b = \frac{(80 - 30) \text{ K}}{0.1103 \text{ K/W}} = 453 \text{ W}$$

Comparing this heat transfer rate with the previous estimate (726 W), we conclude that the finning of the cold side leads to a 60-percent increase in the overall heat transfer rate.

Problem 9.3. To calculate the total number of plate fins mounted on a wall of 0.4 m^2, we note that the most elementary building block in the staggered pattern is a square with the side of 1 cm. Inside this square, the wall area is occupied by the facing one-fourths of two closest fins. In conclusion, 1 cm^2 of wall area holds half of one fin, or one fin corresponds to a wall area of 2 cm^2. The total number of fins is therefore

$$n = \frac{0.4 \text{ m}^2}{2 \text{ cm}^2\text{/fin}} = \frac{4000 \text{ cm}^2}{2 \text{ cm}^2\text{/fin}} = 2000 \text{ fins}$$

The unfinned portion of the wall is

$$A_u = 0.4 \text{ m}^2 - n \cdot z \cdot y$$

$$= 0.4 \text{ m}^2 - 2000 \times 0.0025 \text{ m} \times 0.01 \text{ m} = 0.35 \text{ m}^2$$

The effective length of each plate fin is, cf. eq. (2.105)

$$L_c = L + \frac{z}{2} = 1.5 \text{ cm} + \frac{0.25 \text{ cm}}{2} = 1.625 \text{ cm}$$

therefore, for the total finned area we write

$$A_f = n \, L_c \, 2 \, (z + y)$$

$$= 2000 \times 0.01625 \text{ m} \times 2 \, (0.0025 + 0.01) \text{ m} = 0.8125 \text{ m}^2$$

The total heat transfer area on the hot side is

$$A_h = A_f + A_u = 0.8125 \text{ m}^2 + 0.35 \text{ m}^2 =$$

$$= 1.1625 \text{ m}^2$$

For the overall efficiency of the hot surface (ε_h) we must calculate in order

$$m\,L_c = \left(\frac{h_h\,p}{kA}\right)^{1/2} L_c = \left(\frac{h_h\,2\,(z+y)}{k\,z\,y}\right)^{1/2} L_c$$

$$= \left[\frac{40\ \text{W}}{\text{m}^2\,\text{K}}\,\frac{2\,(0.0025+0.01)\ \text{m}}{20\,\frac{\text{W}}{\text{mK}}\,0.0025\ \text{m}\ 0.01\ \text{m}}\right]^{1/2} 0.01625\ \text{m}$$

$$= 0.7267$$

$$\eta = \frac{\tanh\,(m\,L_c)}{m\,L_c} = 0.855$$

$$\varepsilon_h = \eta\,\frac{A_f}{A_h} + \frac{A_u}{A_h}$$

$$= 0.855\,\frac{0.8125\ \text{m}^2}{1.1625\ \text{m}^2} + \frac{0.35\ \text{m}^2}{1.1625\ \text{m}^2} = 0.898$$

The overall thermal resistance and heat transfer rate are

$$\frac{1}{U_c\,A_c} = \frac{1}{h_c\,A_c} + \frac{t}{k\,A_c} + \frac{1}{\varepsilon_h\,h_h\,A_h}$$

$$= \left(\frac{1}{150\times 0.4} + \frac{0.005}{20\times 0.4} + \frac{1}{0.898\times 40\times 1.1625}\right)\frac{\text{K}}{\text{W}}$$

$$= (0.0167 + 0.00063 + 0.02395)\,\frac{\text{K}}{\text{W}}$$

$$= 0.0413\,\frac{\text{K}}{\text{W}}$$

$$q = U_c\,A_c\,(T_h - T_c) = \frac{T_h - T_c}{\frac{1}{U_c\,A_c}}$$

$$= \frac{(200 - 20)\ \text{K}}{0.0413\ \text{K/W}} = 4361\ \text{W}$$

The one-dimensional conduction model is applicable to the plate fin when the inequality (2.129) is satisfied,

$$\left(\frac{h_h t}{k}\right)^{1/2} << 1$$

In the present case, we have

$$\left(\frac{h_h t}{k}\right)^{1/2} = \left(\frac{40 \text{ W}}{\text{m}^2 \text{ K}} 0.0025 \text{ m} \frac{\text{m·K}}{20 \text{ W}}\right)^{1/2} = 0.071$$

i.e. a number smaller than 1, therefore the use of the one-dimensional conduction model was justified.

Problem 9.4. We begin with eq. (9.8), in which air occupies the hot side:

$$\frac{1}{U_c A_c} = \frac{1}{\varepsilon_h h_{e,h} A_h} + R_{t,w} + \frac{1}{\varepsilon_c h_{e,c} A_c} \tag{9.8}$$

In this equation we know:

$$h_{e,c} = h_c \quad \text{(no scale)}$$
$$= 1000 \frac{\text{W}}{\text{m}^2 \text{ K}}$$
$$\varepsilon_c = 1 \quad \text{(no fins on the cold side)}$$
$$h_{e,h} = h_h \quad \text{(no scale)}$$
$$= 80 \frac{\text{W}}{\text{m}^2 \text{ K}}$$

Multiplied by A_c, eq. (9.8) reads now

$$\frac{1}{U_c} = \frac{1}{\varepsilon_h h_h A_h/A_c} + A_c R_{t,w} + \frac{1}{h_c} \tag{1}$$

The ratio A_h/A_c can be evaluated by assuming that the fin thickness is negligible:

$$A_c = \pi D_i L \tag{2}$$

$$A_h = \pi D_o L + A_{f,h} \tag{3}$$

In eq. (3) $\pi D_o L$ is the outer surface of all the tubes (total length L), while $A_{f,h}$ represents the area contributed by the fins. We divide eqs. (2) and (3) and obtain

$$\frac{A_c}{A_h} = \frac{D_i}{D_o} \frac{1}{1 + \dfrac{A_{f,h}}{\pi D_o L}} = \frac{D_i}{D_o} \frac{1}{1 + \dfrac{A_{f,h}}{A_h - A_{f,h}}}$$

$$= \frac{D_i}{D_o}\left(1 - \frac{A_{f,h}}{A_h}\right) \tag{4}$$

In our case the numerical values are

$D_o = 10.7$ mm (Fig. 9.39)

$D_i = 8.3$ mm

$\left(\dfrac{A_f}{A}\right)_h = 0.876$ (Fig. 9.39)

Equation (4) yields:

$$\frac{A_c}{A_h} = \frac{8.3}{10.7}(1 - 0.876) = 0.0962$$

$$\frac{A_h}{A_c} = 10.4$$

The resistance of the cylindrical wall is furnished by eq. (2.33),

$$R_{t,w} = \frac{\ln(D_o/D_i)}{2\pi k_w L} \tag{5}$$

where $k_w = 215$ W/m·K. Multiplied by A_c, eq. (5) yields

$$A_c R_{t,w} = \frac{\ln(D_o/D_i)}{2\pi k_w L} \pi D_i L$$

$$= \frac{\ln(10.7/8.3)}{2 \times 215 \dfrac{W}{m \cdot K}} 0.0083m = 4.9 \times 10^{-6} \frac{m^2 K}{W}$$

The overall efficiency of the hot surface can be calculated based on eq. (9.4),

$$\varepsilon_h = 1 - \frac{A_{f,h}}{A_h}(1 - \eta)$$

$$= 1 - 0.876(1 - \eta) \tag{6}$$

for which η is provided by Fig. 2.16:

$$\frac{r_o}{r_i} = \frac{D_o}{D_i} = \frac{10.7}{8.3} \cong 1.3$$

$$L_c \cong L + \frac{t}{2} = \frac{21.9 - 10.7}{2}\text{ mm} + \frac{0.48}{2}\text{ mm}$$
$$= 5.84\text{ mm}$$

$$L_c\left(\frac{2h_h}{k_w t}\right)^{1/2} = 0.00584\text{m}\left(\frac{2 \times 80 \dfrac{\text{W}}{\text{m}^2\text{ K}}}{215 \dfrac{\text{W}}{\text{m·K}} 0.00048\text{m}}\right)^{1/2}$$
$$= 0.23$$

$$\eta \cong 0.975 \quad \text{(Fig. 2.16)}$$

Equation (6) yields finally

$$\varepsilon_h = 1 - 0.876\ (1 - 0.975) = 0.978$$

Now we have all the ingredients to substitute in eq. (1):

$$\frac{1}{U_c} = \frac{1}{0.978 \times 80 \dfrac{\text{W}}{\text{m}^2\text{ K}} 10.4} + 4.9 \times 10^{-6} \frac{\text{m}^2\text{ K}}{\text{W}} + \frac{1}{1000 \dfrac{\text{W}}{\text{m}^2\text{ K}}}$$
$$= \left(1.23 \times 10^{-3} + 4.9 \times 10^{-6} + 10^{-3}\right)\frac{\text{m}^2\text{ K}}{\text{W}}$$
$$= 0.00223 \frac{\text{m}^2\text{ K}}{\text{W}}$$
$$U_c = 448 \frac{\text{W}}{\text{m}^2\text{ K}}$$

The overall heat transfer coefficient based on the hot side is

$$U_h = \frac{A_c}{A_h} U_c = 0.0962 \times 448 \frac{\text{W}}{\text{m}^2\text{ K}}$$
$$= 43 \frac{\text{W}}{\text{m}^2\text{ K}}$$

Problem 9.5. The outlet temperature of the cold stream can be determined by invoking the first law of thermodynamics for the parallel flow heat exchanger (modelled as an open system, in the steady state)

$$C_h \left(T_{h,in} - T_{h,out}\right) = C_c \left(T_{c,out} - T_{c,in}\right)$$

where

$$\frac{C_h}{C_c} = \frac{(\dot{m}\, c_P)_h}{(\dot{m}\, c_P)_c} \cong \frac{\dot{m}_h}{\dot{m}_c} = \frac{1}{2}$$

therefore,

$$T_{c,out} = T_{c,in} + \frac{C_h}{C_c}\left(T_{h,in} - T_{h,out}\right)$$

$$= 40°C + \frac{1}{2}(90 - 60)°C = 55°C$$

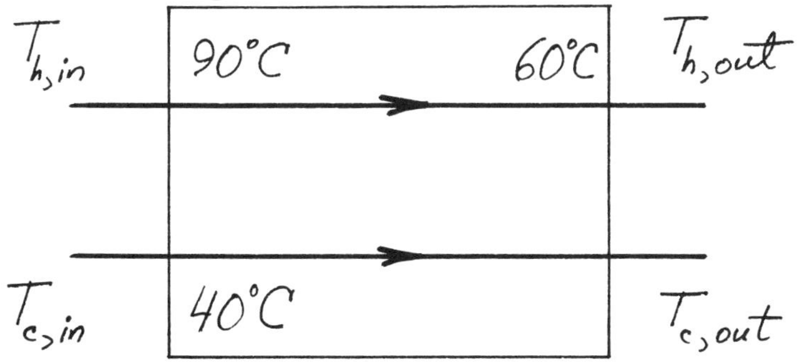

Following the steps summarized at the end of Example 9.2, we calculate in order

$$\Delta T_1 = T_{h,in} - T_{c,in} = 50°C$$

$$\Delta T_2 = T_{h,out} - T_{c,out} = 5°C$$

$$\Delta T_{lm} = \frac{(50 - 5)°C}{\ln \frac{50}{5}} = 19.54°C$$

$$A = \frac{q}{U\,\Delta T_{lm}} = \frac{\dot{m}_h\, c_{P,h}\left(T_{h,in} - T_{h,out}\right)}{U\,\Delta T_{lm}}$$

$$= \frac{1\,\frac{kg}{s}\,4.19\,\frac{10^3\,J}{kg\,K}\,(90 - 60)°C}{10^3\,\frac{W}{m^2\,K}\,19.54°C} = 6.43\ m^2$$

This area requirement is 37 percent larger than in the first part of Example 9.2, where the two water streams were running in counterflow. In conclusion, the counterflow arrangement is more cost-effective, in the sense that it does the same job (it cools the hot stream from 90°C to 60°C) while requiring the smallest heat exchanger surface.

Problem 9.6. The first steps are identical to what we saw in Example 9.2:

$$C_h\left(T_{h,in} - T_{h,out}\right) = C_c\left(T_{c,out} - T_{c,in}\right)$$

$$T_{c,out} = T_{c,in} + \frac{C_h}{C_c}\left(T_{h,in} - T_{h,out}\right)$$

$$= 40°C + \frac{1}{2}\,(90°C - 60°C) = 55°C$$

$$\Delta T_1 = 35°C \qquad \Delta T_2 = 20°C$$

$$\Delta T_{lm} = \frac{(35 - 20)°C}{\ln\frac{35}{20}} = 26.8°C$$

Different is the correction factor F, which is determined from Fig. 9.16,

$$P = \frac{t_2 - t_1}{T_1 - t_1} = \frac{55 - 40}{90 - 40} = 0.3$$

$$R = \frac{T_1 - T_2}{t_2 - t_1} = \frac{90 - 60}{55 - 40} = 2$$

$F \cong 0.93$ (Fig. 9.16, with P = 0.3 and R = 2)

therefore

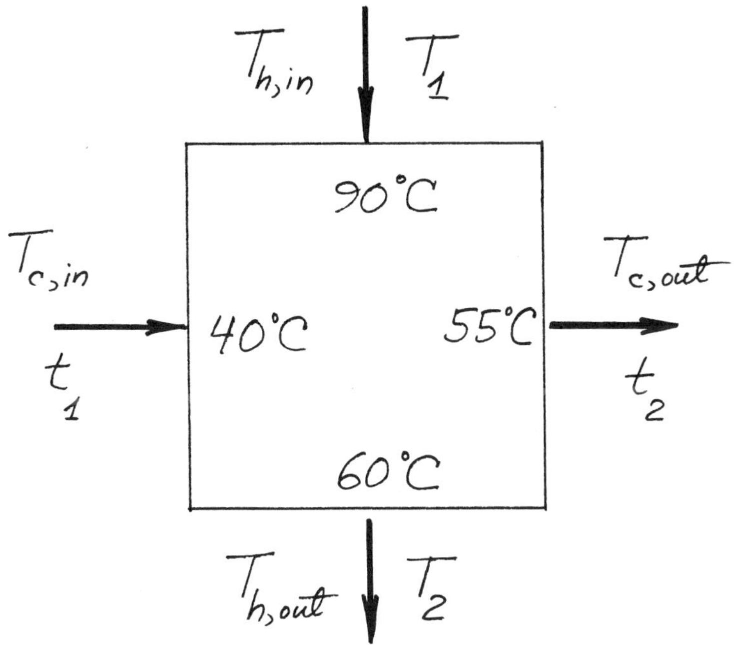

$$A = \frac{q}{U\,\Delta T_{lm}\,F} = \frac{\dot{m}_h\, c_{P,h}\left(T_{h,in} - T_{h,out}\right)}{U\,\Delta T_{lm}\,F}$$

$$= \frac{1\,\frac{kg}{s}\,4.19\,\frac{10^3 J}{kg\,K}\,(90 - 60)°C}{1000\,\frac{W}{m^2\,K}\,26.8°C\;0.93} = 5.04\ m^2$$

This area is 5 percent smaller than the area required by the cross-flow arrangement with both fluids mixed ($A = 5.3\ m^2$, Example 9.2). In conclusion, if we strive to prevent the mixing of each stream, we can save at most 5 percent in the size of the heat exchanger surface.

Problem 9.7. With reference to the counterflow temperature distribution shown in the upper-right quadrant of Fig. 9.15, we recognize first the stream-to-stream heat transfer rate through the heat exchanger area dA,

$$dq = (T_h - T_c)\,U\,dA \qquad (1)$$

and the first law for each stream (control volume) of "length" dA,

$$dq = C_h\,dT_h \qquad (2)$$

$$dq = C_c\,dT_c \qquad (3)$$

The area A increases toward the right on the abscissa of Fig. 9.15.

After these preliminary steps, we follow the course of the analysis listed between eqs. (9.13) and (9.22) in the text:

$$d\left(T_h - T_c\right) = dT_h - dT_c = \frac{dq}{C_h} - \frac{dq}{C_c}$$

$$= \left(\frac{1}{C_h} - \frac{1}{C_c}\right) dq$$

$$= \left(\frac{1}{C_h} - \frac{1}{C_c}\right)\left(T_h - T_c\right) U\, dA \qquad (4)$$

Integrating from the left side (A = 0) all the way to the right side (A = A), we obtain

$$\ln \frac{\Delta T_2}{\Delta T_1} = \left(\frac{1}{C_h} - \frac{1}{C_c}\right) UA \qquad (5)$$

where

$$\Delta T_1 = T_{h,out} - T_{c,in} \qquad \text{and} \qquad \Delta T_2 = T_{h,in} - T_{c,out} \qquad (6)$$

Next, we integrate eqs. (2)-(3) from the left side to the right side of the heat exchanger,

$$q = C_h \left(T_{h,in} - T_{h,out}\right) \qquad (7)$$

$$q = C_c \left(T_{c,out} - T_{c,in}\right) \qquad (8)$$

therefore

$$\frac{1}{C_h} - \frac{1}{C_c} = \frac{1}{q}\left(T_{h,in} - T_{h,out} - T_{c,out} + T_{c,in}\right)$$

$$= \frac{1}{q}\left(\Delta T_2 - \Delta T_1\right) \qquad (9)$$

Combined, eqs. (5) and (9) lead to

$$\ln \frac{\Delta T_2}{\Delta T_1} = \frac{1}{q}\left(\Delta T_2 - \Delta T_1\right) UA \qquad (10)$$

or

$$q = UA\, \Delta T_{lm} \qquad (11)$$

in which ΔT_{lm} has the form given in eq. (9.22) in the text.

Problem 9.8. a) The capacity rates of the two streams are

$$C_h = \dot{m}_h c_{P,h} = 1.5 \frac{kg}{s} 2.25 \frac{kJ}{kg\,K} = 3.38 \frac{kW}{K}$$

$$C_c = \dot{m}_c c_{P,c} = 0.5 \frac{kg}{s} 4.18 \frac{kJ}{kg\,K} = 2.09 \frac{kW}{K}$$

We follow the method of Example 9.3, in order to calculate the outlet temperatures and the stream-to-stream heat transfer rate.

I) Assume $T_{h,out} = 70°C$

$$C_h (T_{h,in} - T_{h,out}) = C_c (T_{c,out} - T_{c,in})$$

$$T_{c,out} = T_{c,in} + \frac{C_h}{C_c}\left(T_{h,in} - T_{h,out}\right)$$

$$= 15°C + \frac{3.38}{2.09}(110 - 70)°C$$

$$= 79.69°C$$

This cold-stream outlet temperature is higher than the adjacent hot-stream temperature (70°C)! This is a violation of the second law (see subsection 9.4.1), and it means that the assumed value of $T_{h,out} = 70°C$ is too low. We must restart the trial-and-error procedure with a higher guess for $T_{h,out}$.

I) Assume $T_{h,out} = 80°C$

$$T_{c,out} = 15°C + \frac{3.38}{2.09}(110 - 80)°C$$

$$= 63.52°C$$

II) $\Delta T_1 = (110 - 15)°C = 95°C$

$$\Delta T_2 = (80 - 63.52)°C = 16.48°C$$

$$\Delta T_{lm} = \frac{(95 - 16.48)°C}{\ln \frac{95}{16.48}} = 44.83°C$$

III) $q = UA\,\Delta T_{lm}$

$$= 500 \frac{W}{m^2\,K} 10\, m^2\, 44.83°C + 2.24 \times 10^5\, W$$

$$T_{h,out} = T_{h,in} - \frac{q}{C_h}$$

$$= 110°C - \frac{2.24 \times 10^5 \text{ W}}{3.38 \times 10^3 \text{ W/K}}$$

$$= 43.69°C$$

Comparing this $T_{h,out}$ estimate (43.69°C) with the assumed value (80°C), we conclude that the next $T_{h,out}$ guess must be lower.

I) $T_{h,out} = 75°C$ (assumed)

$T_{c,out} = 71.6°C$

II) $\Delta T_1 = 95°C$

$\Delta T_2 = 3.4°C$

$\Delta T_{lm} = 27.51°C$

III) $q = 1.375 \times 10^5$ W

IV) $T_{h,out} = 69.31°C$ (calculated)

The calculated value (69.31°C) is still below the assumed value (75°C), therefore, in the next try we assume an even lower value for $T_{h,out}$.

I) $T_{h,out} = 74°C$ (assumed)

$T_{c,out} = 73.22°C$

II) $\Delta T_1 = 95°C$

$\Delta T_2 = 0.78°C$

$\Delta T_{lm} = 19.62°C$

III) $q = 0.981 \times 10^5$ W

IV) $T_{h,out} = 81°C$ (calculated)

This time the calculated value is (barely) greater than the assumed value. In conclusion, the correct guess is somewhere between the last two guesses, for example:

I) $T_{h,out} = 74.5°C$ (assumed)

$T_{c,out} = 72.41°C$

II) $\Delta T_1 = 95°C$

$\Delta T_2 = 2.09°C$

$\Delta T_{lm} = 24.34°C$

III) $q = 1.217 \times 10^5$ W

IV $T_{h,out} = 74°C$ (calculated)

b) The total heat transfer rate in the case of the counterflow arrangement would have been $q = 1.6 \times 10^5$ W (see the end of Example 9.3). The present heat transfer rate is only 76 percent of the rate accommodated by the counterflow arrangement.

Another way of seeing the relative inefficiency of the parallel flow arrangement, is by comparing the outlet temperature of the oil stream (~74°C) with the corresponding value in the counterflow heat exchanger (62.6°C, Example 9.3). In conclusion, with the same A, U and flowrates, the counterflow heat exchanger lowers the temperature of the oil stream by an additional 11.4°C below the final temperature of the oil delivered by the parallel flow heat exchanger.

Problem 9.9. With reference to Fig. 9.14 we assume that

$$C_h = C_{min} \qquad \text{and} \qquad C_c = C_{max} \tag{1}$$

The effectiveness is therefore defined as

$$\varepsilon = \frac{T_{h,1} - T_{h,2}}{T_{h,1} - T_{c,1}} = \frac{C_{max}}{C_{min}} \frac{T_{c,2} - T_{c,1}}{T_{h,1} - T_{c,1}} \tag{2}$$

Instead of eq. (9.20) we obtain

$$\ln \frac{\Delta T_2}{\Delta T_1} = - \frac{1}{q} (\Delta T_1 - \Delta T_2) \frac{UA}{C_{min}} C_{min}$$

$$= - \frac{T_{h,1} - T_{c,1} - T_{h,2} + T_{c,2}}{C_{min} (T_{h,1} - T_{h,2})} NTU\, C_{min}$$

$$= - NTU \left[1 + \underbrace{\left(\frac{T_{c,2} - T_{c,1}}{T_{h,1} - T_{h,2}} \cdot \frac{C_c}{C_h} \right)}_{=1} \frac{C_h}{C_c} \right]$$

$$= - NTU\left(1 + \frac{C_{min}}{C_{max}}\right) \tag{3}$$

For the ratio $\Delta T_2/\Delta T_1$ we write sequentially

$$\frac{\Delta T_2}{\Delta T_1} = \frac{T_{h,2} - T_{c,2}}{T_{h,1} - T_{c,1}} = 1 + \frac{T_{h,2} - T_{c,2} - T_{h,1} + T_{c,1}}{T_{h,1} - T_{c,1}}$$

$$= 1 + \frac{T_{h,2} - T_{c,2} - T_{h,1} + T_{c,1}}{\frac{1}{\varepsilon}\left(T_{h,1} - T_{h,2}\right)}$$

$$= 1 + \varepsilon\left[- 1 + \underbrace{\left(\frac{T_{c,1} - T_{c,2}}{T_{h,1} - T_{h,2}} \cdot \frac{C_c}{C_h}\right)}_{= -1}\frac{C_h}{C_c}\right]$$

$$= 1 + \varepsilon\left(- 1 - \frac{C_{min}}{C_{max}}\right) \tag{4}$$

Equations (3) and (4) are the same as eqs. (9.35) and (9.36) in the text. In conclusion, the assumption (1) leads to the same ε-NTU expression as in eq. (9.37).

<u>Problem 9.10</u>. a) First, we identify C_{max} and C_{min},

$$C_h = (\dot{m}\, c_P)_{oil}$$

$$= 2\,\frac{kg}{s}\,2.25\,\frac{kJ}{kg\ K} = 4.5\,\frac{kW}{K}$$

$$C_c = (\dot{m}\, c_P)_{water}$$

$$= 1\,\frac{kg}{s}\,4.18\,\frac{kJ}{kg\ K} = 4.18\,\frac{kW}{K}$$

Conclusion:

$$C_{max} = C_h \qquad C_{min} = C_c$$

$$\frac{C_{min}}{C_{max}} = \frac{4.18}{4.5} = 0.929$$

Next, we calculate sequentially ε, NTU and A:

$$\varepsilon = \frac{C_h}{C_{min}} \frac{T_{h,in} - T_{h,out}}{T_{h,in} - T_{c,in}}$$

$$= \frac{4.5}{4.18} \frac{110 - 50}{110 - 20} = 0.718$$

$$NTU = \frac{\ln\left(\dfrac{1 - 0.718 \times 0.929}{1 - 0.718}\right)}{1 - 0.929} = 2.352$$

$$A = NTU \frac{C_{min}}{U}$$

$$= 2.352 \frac{4.18 \dfrac{10^3\ W}{K}}{400 \dfrac{W}{m^2\ K}} = 24.6\ m^2$$

b) The total heat transfer rate can be calculated by drawing a control volume around the stream for which we know both the inlet and outlet temperatures,

$$q = C_h \left(T_{h,in} - T_{h,out}\right)$$

$$= 4.5 \frac{10^3\ W}{K} (110 - 50)\ K = 2.7 \times 10^5\ W$$

The remaining outlet temperature, $T_{c,out}$, can be calculated by writing the first law for the control volume that contains the cold stream,

$$q = C_c \left(T_{c,out} - T_{c,in}\right)$$

$$T_{c,out} = T_{c,in} + \frac{q}{C_c}$$

$$= 20°C + \frac{2.7 \times 10^5\ W}{4.18 \times 10^3\ W/K} = 84.6°C$$

<u>Problem 9.11</u>. Specified in this problem are the surface characteristics (U, A), the flowrates, and the two inlet temperatures. We can calculate q and the outlet temperatures if we first determine the NTU and ε. Here are the main steps:

I) Identify C_{min}, C_{max} and NTU:

$$C_{min} = C_h = \left(\dot{m}\ c_P\right)_h =$$

$$= 1 \frac{kg}{s} 4.19 \frac{kJ}{kg\ K} = 4.19 \frac{kW}{K}$$

$$C_{max} = C_c = (\dot{m}\, c_P)_c$$

$$= 4\,\frac{kg}{s}\; 4.19\,\frac{kJ}{kg\,K} = 16.76\,\frac{kW}{K}$$

$$\frac{C_{min}}{C_{max}} = \frac{1}{4}$$

$$NTU = \frac{AU}{C_{min}} = 10\ m^2\ 600\,\frac{W}{m^2\,K}\;\frac{1}{4.19}\;\frac{K}{10^3\,W}$$

$$= 1.432$$

II) Calculate the effectiveness using eq. (9.42):

$$\varepsilon = \frac{1 - \exp\,[-1.432\,(1 - 0.25)]}{1 - 0.25\,\exp\,[-1.432\,(1 - 0.25)]} = 0.72$$

III) Calculate the total heat transfer rate using eqs. (9.29) and (9.30):

$$q = \varepsilon\, C_{min}\left(T_{h,in} - T_{c,in}\right)$$

$$= 0.72 \times 4.19\,\frac{kW}{K}\,(90 - 20)°C$$

$$= 2.11 \times 10^5\ W$$

IV) Calculate the outlet temperatures:

$$C_c\left(T_{c,out} - T_{c,in}\right) = q$$

$$T_{c,out} = T_{c,in} + \frac{q}{C_c}$$

$$= 20°C + \frac{2.11 \times 10^5\ W}{16.76 \times 10^3\ W/K}$$

$$= 32.6°C$$

$$C_h\left(T_{h,in} - T_{h,out}\right) = q$$

$$T_{h,out} = T_{h,in} - \frac{q}{C_h}$$

$$= 90°C - \frac{2.11 \times 10^5\ W}{4.19 \times 10^3\ W/K}$$

$$= 39.6°C$$

Problem 9.12. a) The water flowrates are equal, therefore, if we assume $c_{P,h} = c_{P,c} = 4.19$ kJ/kg K, we have a "balanced" parallel flow heat exchanger,

$$C_h = C_c = 1 \frac{kg}{s} 4.19 \frac{kJ}{kg\ K} = 4.19 \frac{kW}{K}$$

The effectiveness is

$$\varepsilon = \frac{T_{h,in} - T_{h,out}}{T_{h,in} - T_{c,in}} = \frac{80 - 40}{80 - 20} = \frac{2}{3}$$

In order to calculate the NTU, we turn eq. (9.39) inside-out,

$$NTU = -\frac{1}{2} \ln (1 - 2\varepsilon)$$

$$= -\frac{1}{2} \ln\left(1 - 2\frac{2}{3}\right) = -\frac{1}{2} \underbrace{\ln\left(-\frac{1}{3}\right)}_{???}$$

and conclude that there is no real NTU (!?) for the specified design conditions. This mathematical impossibility has a physical basis, of course. Forget about the NTU, and calculate the outlet temperature of the cold stream that "would have to exist" according to this design. Written for the entire heat exchanger, the first law dictates

$$C_h \left(T_{h,in} - T_{h,out}\right) = C_c \left(T_{c,out} - T_{c,in}\right)$$

$$T_{c,out} = T_{c,in} + \left(T_{h,in} - T_{h,out}\right)$$

$$= 20°C + (80°C - 40°C) = 60°C$$

This outlet temperature of the cold stream (60°C) would be higher than the temperature of the (facing) hot stream outlet, 40°C. This can only happen if the temperature distributions cross inside the heat exchanger, i.e. if the second law of thermodynamics is violated (cf. A. Bejan, Advanced Engineering Thermodynamics, Wiley, 1988, p. 535).

b) One way out of difficulty is to arrange the streams in counterflow. In this case, the numerical work begins with turning eq. (9.44) inside-out, in order to calculate the NTU:

$$NTU = \frac{\varepsilon}{1 - \varepsilon} = \frac{2/3}{1 - 2/3} = 2$$

$$A = NTU \frac{C_{min}}{U}$$

$$= 2 \times 4.19 \frac{10^3\ W}{K} \frac{1}{800} \frac{m^2\ K}{W} = 10.48\ m^2$$

Problem 9.13. a) With reference to the parallel-plate model for the water space, we write

$$NTU >> 1 \tag{1}$$

as the requirement for good water-solid thermal contact. For only when the NTU is much larger than 1 the water leaves the channel with a temperature that approaches the temperature of the wall (grain). In the NTU definition

$$NTU = \frac{hA}{\dot{m}\, c_P} \tag{2}$$

we substitute the following order-of-magnitude quantities (W is the width of the channel):

$$A \sim DW \tag{3}$$

$$\dot{m} \sim \rho U \delta W \tag{4}$$

$$h \sim \frac{k}{\delta} \tag{5}$$

The last of these, $h \sim k/\delta$, is based on the assumption that the Reynolds number $U\delta/\nu$ is so small that the heat transfer regime is laminar and fully developed. In this regime, the Nusselt number based on δ is a constant (see Table 6.1),

$$\frac{h\delta}{k} = \text{constant} \tag{6}$$

By substituting the scales (3)-(5) into the NTU definition (2), the inequality (1) becomes

$$\frac{U\delta}{\alpha} << \frac{D}{\delta} \tag{7}$$

or

$$\frac{U\delta^2}{\alpha D} << 1 \tag{8}$$

On the left side of eq. (7), we recognize the Peclet number based on δ. When the grain size D is comparable with the pore size δ, the condition for local thermal equilibrium is simply

$$\frac{UD}{\alpha} << 1 \tag{9}$$

b) In the numerical case represented by

$$U = 0.1\ \frac{\text{cm}}{\text{s}} \qquad D = 0.4\ \text{cm}$$

$$\alpha = 0.00144\ \frac{\text{cm}^2}{\text{s}} \qquad \delta = 0.1\ \text{cm}$$

the left side of criterion (8) has the value

$$\frac{U\delta^2}{\alpha D} = \frac{0.1\ \text{cm}}{\text{s}}\ \frac{(0.1)^2\ \text{cm}^2}{0.00144\ \text{cm}^2/\text{s}}\ \frac{1}{0.4\ \text{cm}}$$

$$= 1.74$$

This value is not much smaller than 1, therefore, one cannot neglect the local temperature difference between the sphere and the surrounding water.

Problem 9.14. a) The metal sheet and the gas stream form a counterflow heat exchanger. The gas temperature (300°C at inlet) drops in the direction of flow. The gas outlet temperature is one of the unknowns, however, for the purpose of estimating the gas properties we assume (as a first try) that the outlet gas temperature drops to the level of the metal inlet temperature (100°C). Therefore we evaluate the gas properties at the mean temperature (300°C + 100°C)/2 = 200°C:

$$\nu = 0.346\ \frac{\text{cm}^2}{\text{s}} \qquad Pr = 0.68$$

$$\rho = 0.746\ \frac{\text{kg}}{\text{m}^3} \qquad c_P = 1.025\ \frac{\text{kJ}}{\text{kg}\cdot\text{K}}$$

Next, we note that $D_h = 2D = 0.3\text{m}$, and the Reynolds number

$$Re_{D_h} = \frac{V\ D_h}{\nu} = 2\ \frac{\text{m}}{\text{s}}\ \frac{0.3\text{m}}{0.346 \times 10^{-4}\ \text{m}^2/\text{s}} = 1.73 \times 10^4$$

shows that the flow is turbulent. The entrance length is about ten times D_h, namely $10 \times 0.3\text{m} = 3\text{m}$, i.e. about one tenth of the overall length of the oven. Consequently, we treat the gas flow as fully developed, and starting with eq. (6.89) calculate, in order,

$$St = \frac{\frac{1}{2}f}{Pr^{2/3}} = \frac{h}{\rho\ c_P\ V} \qquad (6.89)$$

Moody chart, Fig. 6.14,

$$k_s = 0.05\ \text{mm} \qquad \text{(commercial steel)}$$

$$\frac{k_s}{D_h} = \frac{0.05\ \text{mm}}{300\ \text{mm}} = 0.00017 \qquad \left(\text{i.e. nearly "smooth" in our } Re_{D_h} \text{ range}\right)$$

$$4f \cong 0.0265 \quad \text{(Fig. 6.14, smooth, } Re_{D_h} = 1.73 \times 10^4\text{)}$$

$$f \cong 0.00663$$

$$\frac{h}{\rho\, c_P\, V} = \frac{\frac{1}{2}\, 0.00663}{(0.68)^{2/3}} = 0.00428$$

$$h = 0.00428 \times 0.746\, \frac{kg}{m^3}\, 1.025\, \frac{10^3\ J}{kg{\cdot}K}\, 2\, \frac{m}{s}$$

$$= 6.55\, \frac{W}{m^2\, K}$$

$$C'_{gas} = \rho\, VD\, c_P$$

$$= 0.746\, \frac{kg}{m^3}\, 2\, \frac{m}{s}\, 0.15m\ 1.025\, \frac{10^3\ J}{kg{\cdot}K}$$

$$= 229.4\, \frac{W}{m{\cdot}K} \quad \text{(the prime in C' means "per unit length")}$$

Metal velocity:

$$U = \frac{\text{oven length}}{\text{residence time}} = \frac{30m}{2 \times 3600s} = 4.17\, \frac{mm}{s}$$

$$C'_{metal} = \rho_s\, Ut\, c_s \quad \text{(s = steel, t = thickness)}$$

$$= 7817\, \frac{kg}{m^3}\, 4.17\, \frac{mm}{s}\, 0.1m\ 0.46\, \frac{10^3\ J}{kg{\cdot}K}$$

$$= 1499.5\, \frac{W}{m{\cdot}K}$$

$$C'_{min} = C'_{gas} \quad \text{and} \quad C'_{max} = C'_{metal}$$

$$NTU = \frac{hLW}{C'_{min}W} \quad \text{(W = width, normal to the plane of the figure)}$$

$$= \frac{6.55\ W}{m^2\ K}\, \frac{30m}{229.4\ W/m{\cdot}K} = 0.857$$

$$\frac{C'_{min}}{C'_{max}} = \frac{229.4}{1499.5} = 0.153$$

$$1 - \frac{C'_{min}}{C'_{max}} = 0.847$$

Equation (9.42):

$$\varepsilon = \frac{1 - \exp(-0.857 \times 0.847)}{1 - 0.153 \exp(-0.857 \times 0.847)} = 0.557$$

Equation (9.41):

$$\varepsilon = \frac{C'_{gas}}{C'_{min}} \frac{T_{gas,in} - T_{gas,out}}{T_{gas,in} - T_{metal,in}}$$

$$T_{gas,out} = T_{gas,in} - \varepsilon\left(T_{gas,in} - T_{metal,in}\right)$$

$$= 300°C - 0.557\ (300°C - 100°C)$$

$$= 188.5°C$$

This intermediate conclusion shows that the gas properties should have been evaluated at (188.5°C + 300°C)/2 = 244°C, instead of 200°C. For greater accuracy, the preceding calculations can be repeated using the new property estimates, but the results will not differ much from the following, which are based on the 200°C properties used until now:

$$C'_{gas}\left(T_{in} - T_{out}\right)_{gas} = C'_{metal}\left(T_{out} - T_{in}\right)_{metal}$$

$$T_{metal,out} = T_{metal,in} + \frac{C'_{gas}}{C'_{metal}}\left(T_{in} - T_{out}\right)_{gas}$$

$$= 100°C + 0.153\ (300 - 188.5)°C$$

$$= 117°C$$

b) If the metal final temperature is to be higher (200°C), the residence time must be greater than 2 hours. We assume a residence time of 10 hours, and repeat the calculations outlined in part (a), using the same properties. Here are the more important results:

$$C'_{metal} = 300 \frac{W}{m \cdot K} = C'_{max}$$

$$\frac{C'_{min}}{C'_{max}} = 0.765$$

$$\varepsilon = 0.487 \qquad (NTU = 0.857)$$

$$T_{gas,out} = 300°C - 0.487\,(300 - 100)°C$$

$$= 202.6°C$$

$$T_{metal,out} = 100°C + 0.765\,(300 - 202.6)°C$$

$$= 174.5°C$$

The outlet metal temperature is less than the desired 200°C, therefore the residence time must be longer than 10 hours. If we assume a residence time of 20 hours, the preceding calculations yield

$$C'_{metal} = 150 \frac{W}{m \cdot K} = C'_{min}$$ (note that C'_{metal} is now the smaller capacity rate)

$$C'_{gas} = C'_{max} = 229.4 \frac{W}{m \cdot K}$$

$$\frac{C'_{min}}{C'_{max}} = 0.654$$

$$NTU = \frac{hL}{C'_{min}} = 1.31$$

$$\varepsilon = 0.624$$

$$\varepsilon = \frac{C'_{metal}}{C'_{min}} \frac{(T_{out} - T_{in})_{metal}}{T_{gas,in} - T_{metal,in}} \tag{9.41}$$

$$T_{metal,out} = T_{metal,in} + \varepsilon\left(T_{gas,in} - T_{metal,in}\right)$$

$$= 100°C + 0.624\,(300 - 100)°C$$

$$= 224°C$$

By interpolating linearly between the results for 10 and 20 hours, we find that the metal outlet temperature reaches 200°C during a residence time of approximately 15 hours.

Problem 9.15. We begin with the first law of thermodynamics for a control-volume slice of thickness dx,

$$-\dot{m}\, c_P \frac{dT}{dx} = hp\,(T - T_c) + \frac{kp}{t}(T - T_0) \tag{1}$$

where p is the perimeter of contact between the wall (inner surface) and the stream. In the second term on the right-hand side we have assumed that the temperature drop across the wall thickness t is nearly the same as the difference between the local bulk temperature T(x) and the temperature of the wall outer surface T_0. Equation (1) leads sequentially to

$$-\dot{m}\, c_P \frac{dT}{dx} = hp\left(1 + \frac{k}{ht}\right)(T - T_0) - hp\,(T_c - T_0) \tag{2}$$

and

$$-\frac{dT}{d\xi} = T - a \tag{3}$$

where

$$\xi = \frac{x}{L}\,\mathrm{NTU}\left(1 + \frac{1}{Bi}\right) \tag{4}$$

$$Bi = \frac{ht}{k}, \qquad \mathrm{NTU} = \frac{hp\,L}{\dot{m}\, c_P} \tag{5}$$

$$a = \frac{Bi\, T_c + T_0}{Bi + 1} \tag{6}$$

By integrating eq. (3) from the inlet, with $T = T_h$ at $\xi = 0$, we obtain the temperature distribution along the stream

$$T = a + (T_h - a)\exp\left[-\frac{x}{L}\,\mathrm{NTU}\left(1 + \frac{1}{Bi}\right)\right] \tag{7}$$

and the outlet temperature

$$T_{out} = a + (T_h - a)\exp\left[-\mathrm{NTU}\left(1 + \frac{1}{Bi}\right)\right] \tag{8}$$

The total enthalpy drop experienced by the stream is

$$q = \dot{m}\, c_P (T_h - T_{out})$$
$$= \dot{m}\, c_P (T_h - a)\left\{1 - \exp\left[- NTU\left(1 + \frac{1}{Bi}\right)\right]\right\} \qquad (9)$$

while the heat transfer (through the wall) to the ambient is

$$q_0 = \int_0^L \frac{kp}{t}(T - T_0)\, dx = \cdots$$
$$= \frac{\dot{m}\, c_P}{Bi + 1}\left\{(a - T_0)\, NTU\left(1 + \frac{1}{Bi}\right) + (T_h - a)\left[1 - \exp\left(- NTU\left(1 + \frac{1}{Bi}\right)\right)\right]\right\} \qquad (10)$$

The heat transfer rate received by the T_c stream can be obtained by writing $q_c = q - q_0$ and using eqs. (9) and (10).

In relative terms, the heat loss to the ambient is represented by the fraction

$$\frac{q_0}{q} = \frac{1}{Bi + 1}\left\{1 + \frac{a - T_0}{T_h - a} \cdot \frac{NTU\left(1 + \frac{1}{Bi}\right)}{1 - \exp\left[- NTU\left(1 + \frac{1}{Bi}\right)\right]}\right\} \qquad (11)$$

The perfectly insulated heat exchanger corresponds to the limit $Bi \rightarrow \infty$, where eq. (11) reduces to $q_0/q \rightarrow 0$, and $a \rightarrow T_c$. In the same limit eq. (8) assumes the known form

$$T_{out} = T_c + (T_h - T_c) \exp(- NTU) \qquad (12)$$

Problem 9.16. As shown in Fig. 9.34, we label with a, b, c, and d the planes before and after the step change in cross-sectional area. The properties of air at 1 atm are

$$\rho_{a,b} = \rho\,(100°C) = 0.946 \frac{kg}{m^3}$$
$$\rho_{c,d} = \rho\,(300°C) = 0.616 \frac{kg}{m^3}$$

For the 5m-long straight tube, we evaluate the air properties at (100°C + 300°C)/2 = 200°C:

$$\rho = 0.746 \frac{kg}{m^3} \qquad \nu = 0.346 \frac{cm^2}{s}$$

In the straight section the pressure drop is due to friction and acceleration from b to c:

$$A_{flow} = 50 \frac{\pi}{4} (0.02\text{m})^2 = 0.0157 \text{ m}^2$$

$$G = \frac{\dot{m}}{A_{flow}} = \frac{500 \text{ kg/h}}{0.0157 \text{ m}^2} = 8.84 \frac{\text{kg}}{\text{m}^2\text{s}}$$

$$V = \frac{G}{\rho} = 8.84 \frac{\text{kg}}{\text{m}^2\text{s}} \frac{\text{m}^3}{0.746 \text{ kg}} = 11.85 \frac{\text{m}}{\text{s}}$$

$$Re_D = \frac{VD}{\nu} = 11.85 \frac{\text{m}}{\text{s}} 0.02\text{m} \frac{1}{0.346 \times 10^{-4} \text{ m}^2/\text{s}}$$

$$= 6850 \quad \text{(turbulent)}$$

$$\Delta P_{acc} = G^2 \left(\frac{1}{\rho_{c,d}} - \frac{1}{\rho_{a,b}} \right)$$

$$= \left(8.84 \frac{\text{kg}}{\text{m}^2\text{s}} \right)^2 \left(\frac{1}{0.616} - \frac{1}{0.946} \right) \frac{\text{m}^3}{\text{kg}}$$

$$= 44.24 \frac{\text{N}}{\text{m}^2} \quad \text{(acceleration)}$$

$$f \cong 0.0791 \, Re_D^{-0.25} = 0.0087$$

$$\Delta P_{friction} = f \frac{4L}{D} \frac{1}{2} \rho V^2$$

$$= 0.0087 \frac{4 \times 5\text{m}}{0.02\text{m}} \frac{1}{2} 0.746 \frac{\text{kg}}{\text{m}^3} \left(11.85 \frac{\text{m}}{\text{s}} \right)^2$$

$$= 455.7 \frac{\text{N}}{\text{m}^2}$$

$$\Delta P_s = P_b - P_c = \Delta P_{friction} + \Delta P_{acc}$$

$$= (455.7 + 44.24) \frac{\text{N}}{\text{m}^2} = 499.94 \frac{\text{N}}{\text{m}^2}$$

For the pressure drop due to contraction we use eq. (9.51):

$$A_{header} = \frac{\pi}{4} (0.3\text{m})^2 = 0.0707 \text{ m}^2$$

$$\sigma = \frac{A_{flow}}{A_{header}} = \frac{0.0157 \text{ m}^2}{0.0707 \text{ m}^2} = 0.222$$

$K_c \cong 0.42$ (Fig. 9.30, at $\sigma = 0.222$ and $Re_D \cong 7000$)

$$V_b = V\frac{\rho}{\rho_b}$$

$$= 11.85\ \frac{m}{s}\ \frac{0.746}{0.946} = 9.34\ \frac{m}{s}$$

$$P_a - P_b = \left(1 - \sigma^2 + K_c\right)\frac{1}{2}\rho_b V_b^2$$

$$= \left(1 - 0.222^2 + 0.42\right)\frac{1}{2}\,0.946\ \frac{kg}{m^3}\left(9.34\ \frac{m}{s}\right)^2$$

$$= 56.57\ \frac{N}{m^2}$$

For the pressure rise through the enlargement we use eq. (9.54):

$K_e \cong 0.575$ (Fig. 9.30, at $\sigma = 0.222$ and $Re_D \cong 7000$)

$$V_c = V\frac{\rho}{\rho_c} = 11.85\ \frac{m}{s}\ \frac{0.746}{0.616} = 14.35\ \frac{m}{s}$$

$$P_d - P_c = \left(1 - \sigma^2 - K_e\right)\frac{1}{2}\rho_c V_c^2$$

$$= \left(1 - 0.222^2 - 0.575\right)\frac{1}{2}\,0.616\ \frac{kg}{m^3}\left(14.35\ \frac{m}{s}\right)^2$$

$$= 23.85\ \frac{N}{m^2}$$

Finally, we calculate the overall pressure drop by writing

$$\Delta P = P_a - P_d = \left(P_a - P_b\right) + \left(P_b - P_c\right) - \left(P_d - P_c\right)$$

$$= (56.57 + 499.94 - 23.85)\ \frac{N}{m^2}$$

$$= 532.76\ \frac{N}{m^2}$$

The frictional pressure drop over the straight section contributes 94 percent to the total pressure drop.

Problem 9.17. For 25°C water we have

$$\nu = 0.00894 \frac{cm^2}{s}$$

From the comparison indicated in the problem statement,

$$\left(1 - \sigma^2 + K_c\right)\frac{1}{2}\rho V^2 = f\frac{4L}{D_h}\frac{1}{2}\rho V^2$$

we deduce

$$L = \frac{D_h}{4f}\left(1 - \sigma^2 + K_c\right) \qquad (1)$$

where

$$D_h = 2 \times 0.4m = 0.8m$$

$$Re = \frac{V D_h}{\nu} = 0.15 \frac{m}{s} 0.8m \frac{1}{0.00894 \times 10^{-4} \frac{m^2}{s}}$$

$$= 1.342 \times 10^5$$

$$f = 0.046\, Re^{-0.2} = 0.00434$$

$\sigma \cong 0.464$ (measured on Fig. 9.33)

$K_c \cong 0.35$ (Fig. 9.31)

Equation (1) yields

$$L = \frac{0.8m}{4 \times 0.00434}\left(1 - 0.464^2 + 0.35\right) = 52.3m$$

The pressure drop through the abrupt contraction is given by eq. (9.51)

$$P_a - P_b = \left(1 - \sigma^2 + K_c\right)\frac{1}{2}\rho V^2$$

$$= \left(1 - 0.464^2 + 0.35\right)\frac{1}{2} 997 \frac{kg}{m^3}\left(0.15 \frac{m}{s}\right)^2$$

$$= 12.7 \frac{N}{m^2}$$

Problem 9.18. According to the notation employed in Fig. 9.34, we label with a, b, c, and d the planes before and after the step change in cross-sectional area. The properties of air at 1 atm are

$$\rho_{a,b} = \rho\ (500°C) = 0.456\ \frac{kg}{m^3}$$

$$\rho_{c,d} = \rho\ (100°C) = 0.946\ \frac{kg}{m^3}$$

For each 3m-long tube we evaluate the air properties at (500°C + 100°C)/2 = 300°C:

$$\rho = 0.616\ \frac{kg}{m^3} \qquad\qquad \nu = 0.481\ \frac{cm^2}{s}$$

We evaluate sequentially the pressure changes associated with the straight section, contraction and enlargement.

Straight section (deceleration):

$$A_{flow} = 300\ \frac{\pi}{4}\ (0.015m)^2 = 0.053m^2$$

$$G = \frac{\dot{m}}{A_{flow}} = \frac{5000\ kg/h}{0.053\ m^2} = 26.2\ \frac{kg}{m^2 s}$$

$$V = \frac{G}{\rho} = 26.2\ \frac{kg}{m^2 s}\ \frac{1}{0.616\ kg/m^3} = 42.54\ \frac{m}{s}$$

$$Re_D = \frac{VD}{\nu} = 42.54\ \frac{m}{s}\ 0.015m\ \frac{1}{0.481 \times 10^{-4}\ m^2/s}$$

$$= 1.33 \times 10^4 \quad \text{(turbulent)}$$

$$\Delta P_{acc} = G^2 \left(\frac{1}{\rho_{c,d}} - \frac{1}{\rho_{a,b}}\right)$$

$$= \left(26.2\ \frac{kg}{m^2 s}\right)^2 \left(\frac{1}{0.946} - \frac{1}{0.456}\right) \frac{m^3}{kg}$$

$$= -\ 779.7\ \frac{N}{m^2} \quad \text{(negative sign means deceleration)}$$

Straight section (friction):

$$f \cong 0.0791\ Re_D^{-0.25} = 0.00737$$

$$\Delta P_{friction} = f\,\frac{4L}{D}\,\frac{1}{2}\,\rho\,V^2$$

$$= 0.00737\,\frac{4 \times 3m}{0.015m}\,\frac{1}{2}\,0.616\,\frac{kg}{m^3}\left(42.54\,\frac{m}{s}\right)^2$$

$$= 3286.3\,\frac{N}{m^2}$$

Straight section (total):

$$\Delta P_s = -\,779.7\,\frac{N}{m^2} + 3286.3\,\frac{N}{m^2} = 2506.6\,\frac{N}{m^2}$$

Contraction:

$\sigma = 0.6$ (given)

$K_c \cong 0.25$ (Fig. 9.30, at $\sigma = 0.6$ and $Re \cong 1.3 \times 10^4$)

$$V_b = \frac{G}{\rho_b} = \frac{26.2\ kg/m^2s}{0.456\ kg/m^3} = 57.46\,\frac{m}{s}$$

$$P_a - P_b = \left(1 - \sigma^2 + K_c\right)\frac{1}{2}\,\rho_b\,V_b^2$$

$$= \left(1 - 0.6^2 + 0.25\right)\frac{1}{2}\,0.456\,\frac{kg}{m^3}\left(57.46\,\frac{m}{s}\right)^2$$

$$= 669.97\,\frac{N}{m^2}$$

Enlargement:

$K_e \cong 0.11$ (Fig. 9.30, at $\sigma = 0.6$ and $Re \cong 1.3 \times 10^4$)

$$V_c = \frac{G}{\rho_c} = \frac{26.2\ kg/m^2s}{0.946\ kg/m^3} = 27.7\,\frac{m}{s}$$

$$P_d - P_c = \left(1 - \sigma^2 - K_e\right)\frac{1}{2}\rho_c V_c^2$$

$$= \left(1 - 0.6^2 - 0.11\right)\frac{1}{2}0.946\,\frac{kg}{m^3}\left(27.7\,\frac{m}{s}\right)^2$$

$$= 192.35\,\frac{N}{m^2}$$

Total pressure drop:

$$\Delta P = \left(P_a - P_b\right) + \Delta P_s - \left(P_d - P_c\right)$$

$$= \left(669.97 + 2506.6 - 192.35\right)\frac{N}{m^2}$$

$$= 2984.2\,\frac{N}{m^2} = 0.029 \text{ atm}$$

The pressure drop due to tube friction is larger than the total pressure drop,

$$\frac{\Delta P_{friction}}{\Delta P} = \frac{3286.3}{2984.2} = 1.10$$

because the deceleration of the stream contributes to "pushing" the stream through the heat exchanger.

Problem 9.19. The number of tube rows in the longitudinal direction is

$$n_l = \frac{L}{X_l} = \frac{0.5\text{m}}{0.0203\text{m}} \cong 25$$

The tube array is characterized also by

$$X_t^* = \frac{X_t}{D} = \frac{24.8 \text{ mm}}{10.7 \text{ mm}} = 2.32$$

$$X_l^* = \frac{X_l}{D} = \frac{20.3 \text{ mm}}{10.7 \text{ mm}} = 1.90$$

$$X_t^*/X_l^* = 1.22$$

$$\chi \cong 1 \quad \text{(Fig. 9.38 insert)}$$

The number of tubes in the transverse direction is

$$n_t = \frac{0.5\text{m}}{0.0248\text{m}} \cong 20$$

In order to calculate V_{max}, we must first determine the minimum free-flow area A_c. The latter depends on the spacings between two adjacent tubes:

transverse (vertical) spacing

$$S_t = (24.8 - 10.7)\text{ mm} = 14.10\text{ mm}$$

diagonal spacing

$$S_d = \left[20.3^2 + \left(\frac{1}{2}\,24.8\right)^2\right]^{1/2}\text{mm} - 10.7\text{ mm} = 13.09\text{ mm}$$

The flow blade that passes through S_t must pass through $2S_d$. Since $S_t < 2\,S_d$, we conclude that the transverse spacing pinches the flow the most,

$$A_c = S_t\cdot(\text{width})\cdot n_t$$
$$= 0.0141\text{m} \times 0.5\text{m} \times 20 = 0.141\text{ m}^2$$

$$V_{max} = \frac{\dot{m}}{\rho\, A_c} = \frac{1500\text{ kg}}{3600\text{s}}\,\frac{\text{m}^3}{0.746\text{ kg}}\,\frac{1}{0.141\text{ m}^2}$$
$$=3.96\,\frac{\text{m}}{\text{s}}$$

$$Re_D = \frac{V_{max}\,D}{\nu} = 3.96\,\frac{\text{m}}{\text{s}}\,0.0107\text{m}\,\frac{1}{0.346 \times 10^{-4}\text{ m}^2/\text{s}}$$
$$=1225$$

$f \cong 0.42$ (Fig. 9.38)

Now we have all we need in order to use eq. (9.62):

$$\Delta P = n_l\, f\, \chi\, \frac{1}{2}\, \rho\, V_{max}^2$$
$$= 25 \times 0.42 \times 1 \times \frac{1}{2} \times 0.746\,\frac{\text{kg}}{\text{m}^3}\left(3.96\,\frac{\text{m}}{\text{s}}\right)^2$$
$$= 61.4\,\frac{\text{N}}{\text{m}^2}$$

This pressure drop is only 30 percent of the pressure drop registered when the tubes are finned.

Problem 9.20. a) We rely on the definitions (9.64)-(9.67) and write sequentially

$$D_h = 4\frac{A_c L}{A} = 4\frac{\sigma A_{fr} L}{A}$$

$$= 4\frac{\sigma \mathcal{V}}{A} = 4\frac{\sigma}{\alpha}$$

b) Figure 9.12 shows that D_h = 6mm corresponds to $\alpha \cong 550\ m^2/m^3$. The assumed value of σ was therefore

$$\sigma = D_h \frac{\alpha}{4} \cong 0.006\text{m}\ \frac{550/\text{m}}{4} = 0.83$$

c) Equation (9.74) follows from eqs. (9.64) and (9.65):

$$\frac{A}{A_c} = \frac{\alpha\,\mathcal{V}}{\sigma A_{fr}} = \frac{\alpha}{\sigma} L = \frac{4}{D_h} L$$

Problem 9.21. a) The properties of air at 100°C are

$$\rho = 0.946\ \frac{\text{kg}}{\text{m}^3} \qquad \nu = 0.23\ \frac{\text{cm}^2}{\text{s}}$$

The pressure drop along the square array can be calculated by using eq. (9.62) and Fig. 9.37:

$$V_{max} = V_\infty \frac{X_t}{X_t - D} = 3\ \frac{\text{m}}{\text{s}}\ \frac{9\ \text{cm}}{(9-5)\ \text{cm}} = 6.75\ \frac{\text{m}}{\text{s}}$$

$$Re_D = V_{max}\frac{D}{\nu} = 6.75\ \frac{\text{m}}{\text{s}}\ \frac{0.05\text{m}}{0.23 \times 10^{-4}\ \text{m}^2/\text{s}} = 1.47 \times 10^4$$

$$X_l^* = \frac{9\ \text{cm}}{5\ \text{cm}} = 1.8$$

$$f \cong 0.25 \qquad \text{(Fig. 9.37)}$$

$$\frac{X_t^* - 1}{X_l^* - 1} = 1$$

$$\chi = 1 \qquad \text{(Fig. 9.37 insert)}$$

$$\Delta P = n_l\, f\, \chi \frac{1}{2} \rho V_{max}^2 \qquad (9.62)$$

$$= 21 \times 0.25 \times 1 \times \frac{1}{2} \times 0.946 \frac{kg}{m^3} \left(6.75 \frac{m}{s}\right)^2$$

$$= 113 \frac{N}{m^2}$$

b) The staggered array with the same X_t and X_l will have the same V_{max}, Re_D and X_t^* as in part (a):

$$V_{max} = 6.75 \frac{m}{s}, \qquad Re_D = 1.47 \times 10^4, \qquad X_t^* = 1.8$$

The f and χ factors provided by Fig. 9.38 will have different values:

$$f \cong 0.33$$

$$\chi \cong 1.02, \text{ because } \frac{X_t^*}{X_l^*} = 1$$

Equation (9.62) yields

$$\Delta P = 21 \times 0.33 \times 1.02 \times \frac{1}{2} \times 0.946 \frac{kg}{m^3} \left(6.75 \frac{m}{s}\right)^2$$

$$= 152 \frac{N}{m^2}$$

In conclusion, the pressure drop along the staggered tubes is 35 percent greater than the pressure drop along the aligned tubes.

Project 9.1. a) We draw a control volume around the water tank, and write the laws that govern the conservation of mass and energy

$$\frac{dm}{dt} = \dot{m} \qquad (1)$$

$$\frac{d}{dt}(mc\,T) = \dot{m}c\,T_h \qquad (2)$$

Equation (2) represents the first law of thermodynamics, in which water was modelled as an incompressible substance at constant pressure. Integrated from t = 0 to any t, eqs. (1) and (2) read

$$\dot{m} = m_0 + \dot{m}t \qquad (3)$$

$$mc\,T - m_0\,c\,T_0 = c\,T_h \underbrace{(m - m_0)}_{\dot{m}t} \qquad (4)$$

Finally, eq. (4) may be written as

$$T = \frac{T_0 + \tau T_h}{1 + \tau} \qquad (5)$$

in which τ is the dimensionless time

$$\tau = \frac{\dot{m}t}{m_0} \qquad (6)$$

A more useful rewriting of eq. (5) is the dimensionless version

$$\frac{T_h - T}{T_h - T_0} = \frac{1}{1 + \tau} \qquad (7)$$

which is plotted as curve (a) on the attached figure.

b) The analysis of the second arrangement follows the same steps. The first steps are the mass conservation principle and the first law of thermodynamics (incompressible liquid, constant pressure):

$$m = m_0 \text{ (constant)} \qquad (8)$$

$$\frac{d}{dt}(m_0\,c\,T) = \dot{m}c\,T_h - \dot{m}\,c\,T \qquad (9)$$

Equation (9) yields in order

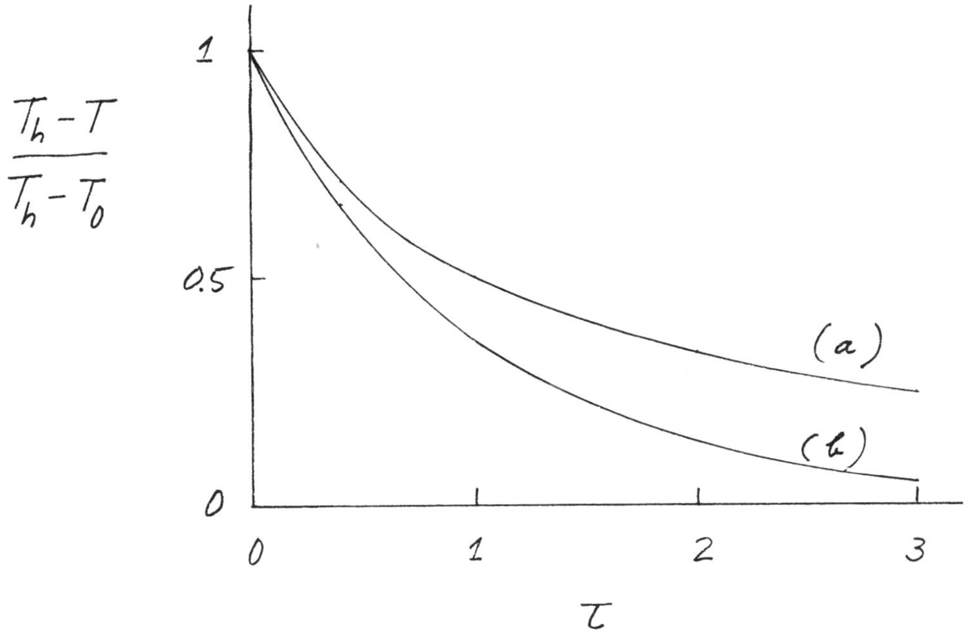

$$\frac{dT}{T_h - T} = \frac{\dot{m}}{m_0}\, dt \tag{10}$$

$$-\ln (T_h - T) + \ln (T_h - T_0) = \tau \tag{11}$$

$$\ln \frac{T_h - T_0}{T_h - T} = \tau \tag{12}$$

$$\frac{T_h - T}{T_h - T_0} = \exp(-\tau) \tag{13}$$

Equation (13) is represented by curve (b) on the same figure. At a fixed τ, a lower curve indicates a warmer batch (i.e. a T value closer to T_h). In conclusion, method (b) is preferable when the objective is to raise the batch temperature the fastest, subject to a set of fixed parameters ($\dot{m}$, T_0, T_h) and, of course, the assumption that $T_{in} \cong T_h$.

Other directions in which this project may be expanded:

c) The effect of the size of the external heat exchanger, when the NTU is of the order of 1 or smaller. This effect can be included in the preceding analysis by using eq. (9.40).

d) The relative effect of imperfect mixing (spatially nonuniform T in the batch) on scheme (a) vs. scheme (b). For example, if in scheme (b) the batch (pool) is shallow, would you position the drain right under the hot jet that falls from above?

Project 9. 2. According to the model suggested in the text, the longitudinal variation of the water and rock temperature is neglected by assigning a single instantaneous temperature $T_{out}(t)$ to the water inventory in the crack. This is equivalent to the assumption that the instantaneous water inventory is well mixed to a single temperature that matches the temperature of the outflowing stream, T_{out}. The water inlet temperature is T_{in}. Treating the control volume swept by the $\dot{m}/2$ stream as an open thermodynamic system, we note that the first law requires

$$\frac{\dot{m}}{2} c_P \left(T_{out} - T_{in}\right) = hA \left(T_0 - T_{out}\right) \tag{1}$$

In this equation, A is the area swept by the $\dot{m}/2$ stream, and h is the average heat transfer coefficient based on A. The rock-stream thermal conductance hA is assumed constant. The temperature of the rock-stream interface is T_0. On the lower side of the rock-stream interface (at y = 0) the convective heat flux $h(T_0 - T_{out})$ is matched by the conduction heat flux that arrives from the semiinfinite rock medium,

$$h \left(T_0 - T_{out}\right) = k \left(\frac{\partial T}{\partial y}\right)_{y=0} \tag{2}$$

The time-dependent temperature distribution inside the rock, T(y,t), must satisfy the conduction equation

$$\frac{\partial T}{\partial t} = \alpha \frac{\partial^2 T}{\partial y^2} \tag{3}$$

in which α is the thermal diffusivity of the rock. The rock temperature distribution T(y,t) must satisfy also the interface condition (2), the far-field condition

$$T = T_\infty \quad \text{at} \quad y = \infty \tag{4}$$

and the initial condition

$$T = T_\infty \quad \text{at} \quad t = 0 \tag{5}$$

It is useful to nondimensionalize the problem defined between equations (1)-(5), by introducing the dimensionless variables

$$\theta = \frac{T - T_{in}}{T_\infty - T_{in}}, \qquad \theta_{out} = \frac{T_{out} - T_{in}}{T_\infty - T_{in}}, \qquad \theta_0 = \frac{T_0 - T_{in}}{T_\infty - T_{in}} \tag{6}$$

$$\eta = \frac{y}{k/h}, \qquad \tau = \frac{t}{k^2/\left(h^2 \alpha\right)} \tag{7}$$

The dimensionless counterparts of equations (1)-(5) are, in order,

$$(1 + M)\, \theta_{out} = \theta_0 \qquad (8)$$

$$\theta_0 - \theta_{out} = \left(\frac{\partial \theta}{\partial \eta}\right)_{\eta=0} \qquad (9)$$

$$\frac{\partial \theta}{\partial \tau} = \frac{\partial^2 \theta}{\partial \eta^2} \qquad (10)$$

$$\theta = 1 \qquad \text{at} \qquad \eta = \infty \qquad (11)$$

$$\theta = 1 \qquad \text{at} \qquad \tau = 0 \qquad (12)$$

where M is the dimensionless mass flowrate (or the inverse NTU)

$$M = \frac{(\dot{m}/2)\, c_P}{hA} \qquad (13)$$

<u>Integral Solution</u>

The approximate behavior of the temperature distribution in the water and rock system can be determined analytically by assuming a rock temperature profile that resembles the one sketched on the right side of the figure shown in the project statement,

$$\theta = (\theta_0 - 1) \exp\left(-\frac{\eta}{\delta}\right) + 1 \qquad (14)$$

The function $\delta(\tau)$ is the unknown dimensionless thickness of the conduction boundary layer that forms between the T_0 interface and the T_∞ far field. Substituting the temperature profile (14) in equations (9) and (10), and integrating from $\eta = 0$ to $\eta = \infty$ we obtain

$$\theta_0 - \theta_{out} = \frac{1 - \theta_0}{\delta} \qquad (15)$$

$$\frac{d}{dt}\left[(1 - \theta_0)\, \delta\right] = \frac{1 - \theta_0}{\delta} \qquad (16)$$

Equations (8), (15) and (16) can be solved numerically or analytically for θ_0, θ_{out} and δ as functions of time (τ) and flowrate (M). The analytical solution can be obtained by first eliminating θ_{out} and δ between equations (8), (15) and (16). What remains is a first-order ordinary differential equation for $\theta_0(\tau)$, which can be integrated by separation of variables. After invoking the initial condition $\theta_0 = 1$ at $\tau = 0$, the complete solution is given implicitly by the following sequence,

$$\frac{1}{2}\left(\theta_0^{-2} - 1\right) + \ln \theta_0 = \left(\frac{M}{1 + M}\right)^2 \tau \qquad (17)$$

$$\theta_{out} = \frac{\theta_0}{1 + M} \qquad (18)$$

$$\delta = \left(\frac{1}{\theta_0} - 1\right)\left(\frac{1}{M} + 1\right) \qquad (19)$$

This solution is presented graphically in the attached figures. The interface temperature θ_0 decreases monotonically as both τ and M increase. The temperature of the crack fluid θ_{out} exhibits a similar behavior, although it is generally lower than the interface temperature θ_0. The last figure shows that the thickness of the conduction boundary layer increases approximately as $\tau^{1/2}$, which is the expected behavior in unidirectional time-dependent conduction. The mass flowrate number (or the NTU) has a negligible effect on $\delta(t)$ in the M range 1-10.

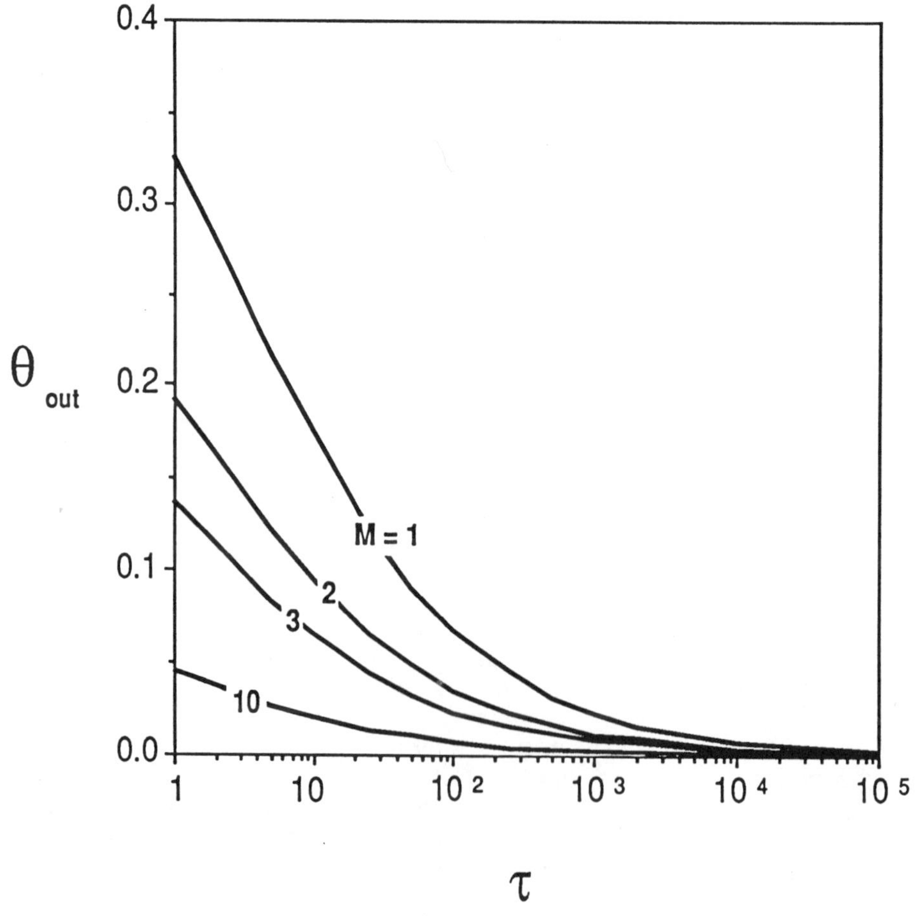

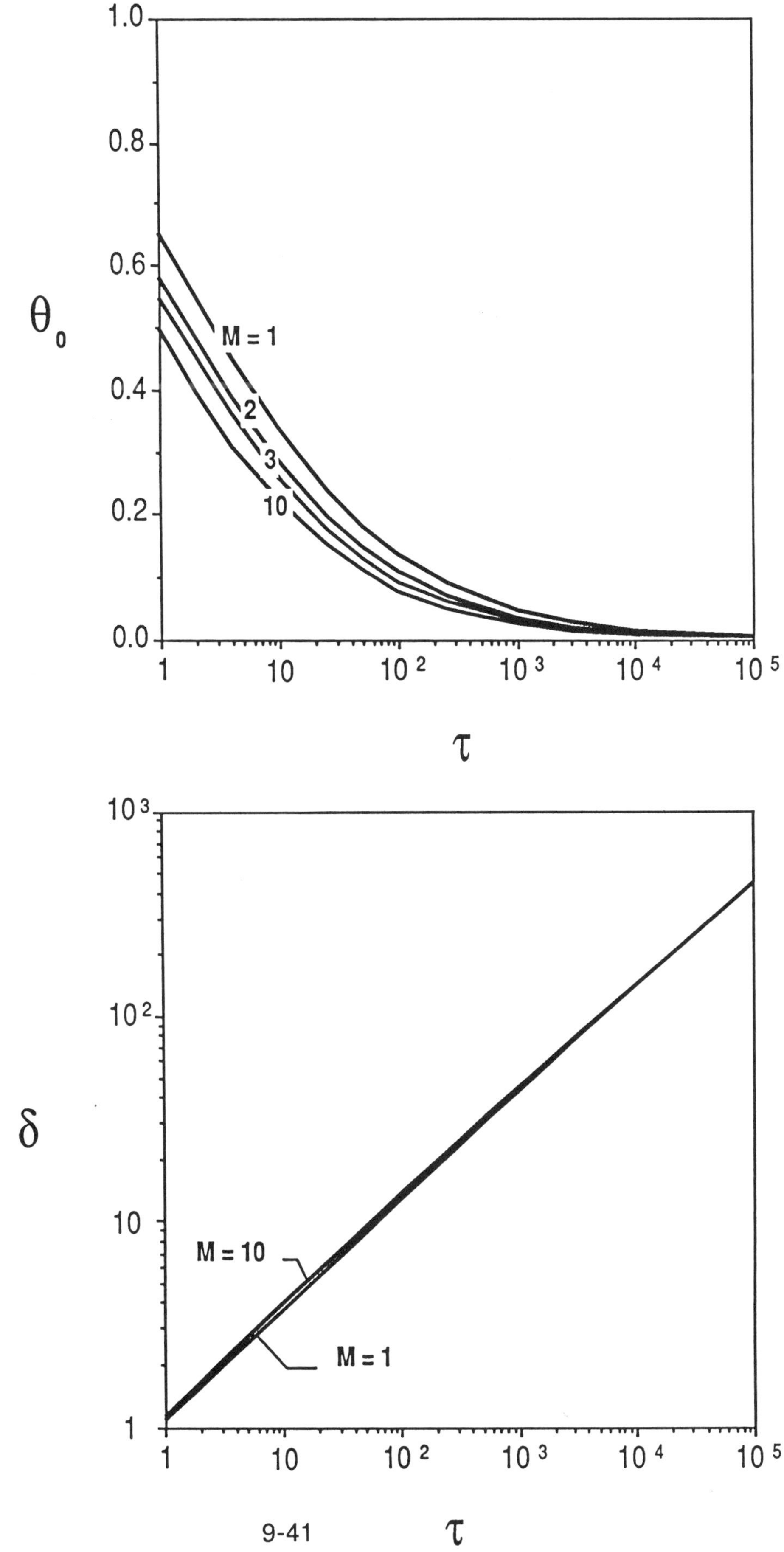

1.0
0.8
0.6
0.4
0.2
0.0
θ₀
M = 1
2
3
10
1
10
10 2
10 3
10 4
10 5
τ
10 3
10 2
10
1
δ
M = 10
M = 1
1
10
10 2
10 3
10 4
10 5
τ

9-41

Project 9.3. a) First, we assume that the fluid flows, i.e. that the D/2 spacing is not so narrow that the fluid is "stuck" in it. We will check this assumption later.

Second, the assumed flow will form a balanced counterflow, with cold fluid coming down on one side (say, the left side), and a warm branch rising through the right-hand channel. The counterflow is balanced because the net mass flow between the top and bottom reservoirs can only be zero.

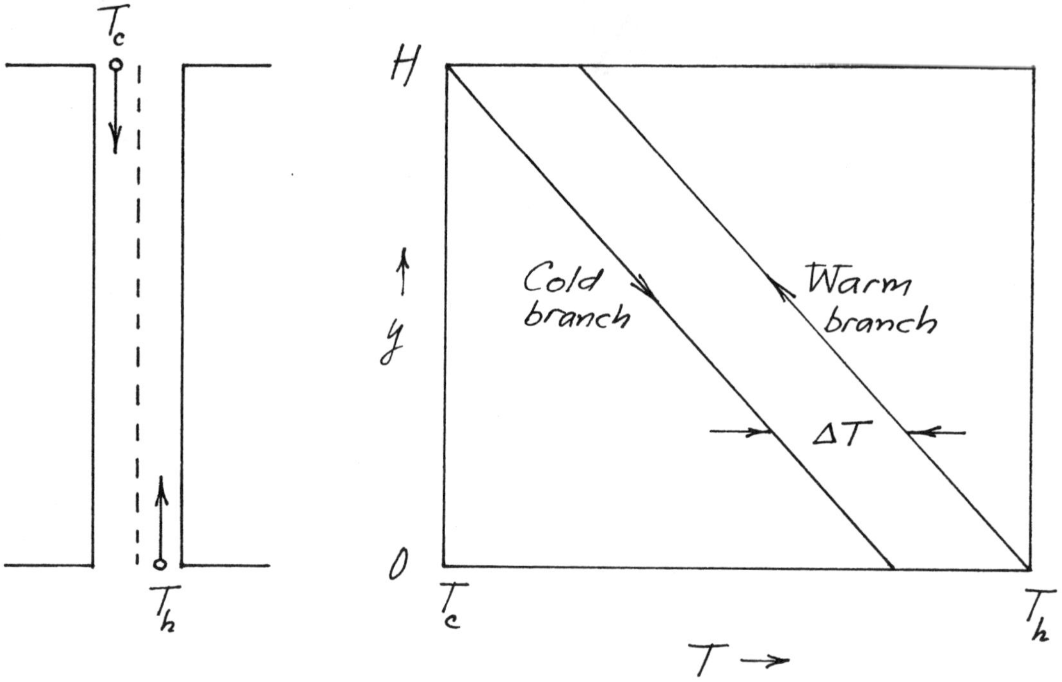

We know that in a balanced counterflow heat exchanger the stream-to-stream temperature difference (in any cut perpendicular to the streams) is constant. This feature is illustrated in the attached sketch. Let ΔT be the stream-to-stream temperature difference. To it corresponds a stream-to-stream density difference

$$\Delta\rho = \rho\beta\,\Delta T$$

where ρ is some representative (average) density of the fluid that inhabits the entire counterflow. As in chapter 7, this analysis is valid when $\Delta\rho \ll \rho$, or when $\beta\,\Delta T \ll 1$.

It follows that the left (cold) column of fluid creates a hydrostatic pressure rise (over the height H) of order $(\rho + \Delta\rho)gH$. The hydrostatic pressure generated by the right (warm) column of fluid is ρgH. The difference (mismatch) between these two pressure rises is $\Delta\rho \cdot gH$. It can be said that this pressure difference drives the counterflow.

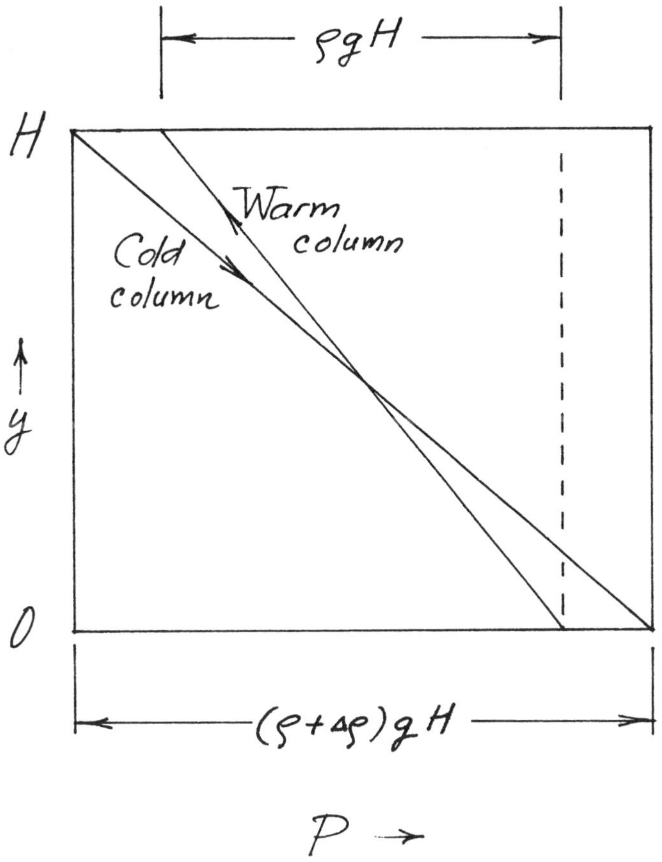

Now, in order for the pressures of the two streams to match everywhere along H, they both must flow fast enough so that the frictional pressure drop along each stream eliminates the pressure difference estimated in the preceding paragraph. Specifically, one stream must experience the frictional pressure drop

$$\Delta P = \frac{\Delta\rho}{2} gH = \frac{1}{2} \rho \beta \, \Delta T \, gH \tag{1}$$

The mean velocity through one D/2-wide channel is indicated by eq. (6.22),

$$U = \frac{(D/2)^2}{12\mu} \frac{\Delta P}{H} = \frac{g\beta \, \Delta T \, D^2}{96\nu} \tag{2}$$

The mass flowrate of one stream is (per unit length normal to the plane of the figure),

$$\dot{m}' = \rho U \frac{D}{2} = \frac{\rho g \beta \, \Delta T \, D^3}{192\nu} \tag{3}$$

The net energy convected upward by the counterflow (hot fluid upward, cold fluid downward) is

$$q' = \dot{m}' c_P \, \Delta T = \frac{k \, \Delta T}{192} \cdot \frac{g\beta \, \Delta T \, D^3}{\alpha\nu} \quad (4)$$

In this expression for the vertical heat transfer rate q' (per unit length normal to the plane $H \times D$) the only unknown is ΔT. The additional equation that is needed is provided by the ε-NTU relation for a balanced counterflow heat exchanger,

$$\varepsilon = \frac{NTU}{1 + NTU} \quad (5)$$

where

$$\varepsilon = \frac{T_h - (T_c + \Delta T)}{T_h - T_c} = 1 - \frac{\Delta T}{T_h - T_c} \quad (6)$$

and

$$NTU = \frac{hH}{\dot{m}' c_P} \quad (7)$$

Assuming that each stream is laminar and fully-developed, we find the heat transfer coefficient h by reading the last line of Table 6.1,

$$\frac{h \, D_h}{k} = 5.385 \quad (8)$$

Note that each D/2-wide parallel-plate channel has one insulated wall, and one "wall" (the midplane of the D-wide channel) with uniform heat flux. Combining eqs. (3), (7) and (8) we obtain

$$NTU = \underbrace{1033.92}_{\cong 1034} \frac{H}{D} \frac{\alpha\nu}{g\beta \, (T_h - T_c) \, D^3} \cdot \frac{T_h - T_c}{\Delta T} \quad (9)$$

On the other hand, eqs. (5) and (6) yield

$$\frac{\Delta T}{T_h - T_c} = \frac{1}{1 + NTU} \quad (10)$$

Finally, eqs. (9) and (10) deliver the needed answer for the stream-to-stream temperature difference,

$$\frac{\Delta T}{T_h - T_c} = 1 - 1034 \frac{H/D}{Ra} \quad (11)$$

in such a way that Ra is the Rayleigh number based on D and the (given) reservoir-to-reservoir temperature difference,

$$Ra = \frac{g\beta \, (T_h - T_c) \, D^3}{\alpha\nu} \quad (12)$$

The formula for the net vertical heat transfer rate, eq. (4), becomes

$$\frac{q'}{k(T_h - T_c)} = \frac{1}{192}\, Ra \left(1 - 1034\, \frac{H/D}{Ra}\right)^2 \qquad (13)$$

b) The onset of convection is marked by q' = 0 in eq. (13). This condition yields immediately the Rayleigh number (based on D, and top-bottom temperature difference) needed for convection,

$$Ra > 1034\, \frac{H}{D} \qquad (14)$$

It is instructive to interpret this result as something related to (i.e., another manifestation of) the onset of Bénard convection in a horizontal fluid layer of depth H. Note that near the onset of Bénard convection (Fig. 7.22) each roll is approximately square, i.e. as in the present counterflow if D happens to be comparable with H. In this case, eq. (14) can be written approximately as

$$\frac{g\beta\,(T_h - T_c)\, H^3}{\alpha\nu} \sim 10^3 \qquad (15)$$

This agrees in an order of magnitude sense with the onset criterion listed in eq. (7.104).

c) The transition to turbulence occurs when the local Reynolds number (Appendix F) of each branch of the counterflow exceeds the order of magnitude 10^2. Therefore the laminar regime prevails when

$$\frac{\begin{pmatrix}\text{longitudinal}\\ \text{velocity}\end{pmatrix} \times \begin{pmatrix}\text{transversal}\\ \text{length}\end{pmatrix}}{\nu} < 10^2 \qquad (16)$$

This yields in order

$$\frac{U\,\frac{D}{2}}{\nu} < 10^2 \qquad (17)$$

$$\frac{\Delta T}{T_h - T_c}\, Ra < 2 \times 10^4\, Pr \qquad (18)$$

$$Ra - 1034\, \frac{H}{D} < 2 \times 10^4\, Pr \qquad (19)$$

In summary, conditions (14) and (19) pinpoint the Ra range in which: a) the assumed flow is present, and b) the flow is laminar (purely vertical),

$$0 < Ra - 1034\, \frac{H}{D} < 2 \times 10^4\, Pr \qquad (20)$$

Project 9.4. The constraint that the amount of insulation material (volume V) is fixed means that the L-averaged insulation thickness t_{avg} is fixed when the duct length L and wrapped perimeter p are fixed,

$$V = \int_0^L p\, t(x)\, dx = p\, t_{avg}\, L \tag{1}$$

In the pursuit of the best taper for the insulation layer, we choose the thickness function

$$t(x) = t_{avg}\left[1 - b\left(\frac{x}{L} - \frac{1}{2}\right)\right] \tag{2}$$

where b is the dimensionless "taper parameter". Note that b can only be a number in the range

$$-2 < b < 2 \tag{3}$$

because t(x) must be positive over the entire length L.

If we write the first law of thermodynamics for a large enough control volume that contains the entire duct and its insulation, and we write q_0 for the total rate of heat transfer (loss) to the ambient, we have

$$q_0 = \dot{m}\, c_P\, (T_h - T_{out}) \tag{4}$$

This shows that the task of minimizing q_0 is equivalent to maximizing the outlet bulk temperature of the stream. We can determine T_{out} by deriving first the temperature distribution along the stream. This begins with writing the first law for a control volume of longitudinal length dx,

$$-\dot{m}\, c_P\, dT = kp\,\frac{T - T_0}{t(x)}\, dx \tag{5}$$

where it has been assumed that the duct wall temperature T(x) is nearly the same as the local bulk temperature of the stream ("nearly the same" when both are compared with the ambient temperature T_0). This assumption is equivalent to saying that the heat transfer coefficient on the internal side of the wall, h, is sufficiently greater than k/t, or that the internal convective resistance 1/h is negligible relative to the thermal resistance of the layer of insulation, t/k.

It has been assumed also that the temperature drop across the duct wall (metallic) is negligible with respect to $T - T_0$, and that the outer surface of the insulation layer has a temperature that is nearly the same as T_0 (i.e. constant). With all these assumptions in mind, we integrate eq. (5) with the t(x) function of eq. (2), from x = 0 (where $T = T_h$) to x = L (where $T = T_{out}$), and obtain

$$\ln \frac{T_h - T_0}{T_{out} - T_0} = \frac{kp\,L}{\dot{m}\,c_P\,t_{avg}} \int_0^1 \frac{dm}{1 - b\left(m - \frac{1}{2}\right)}, \qquad \left(\text{note: } m = \frac{x}{L}\right)$$

$$= \frac{kp\,L}{\dot{m}\,c_P\,t_{avg}} \frac{1}{b} \ln\left[1 - b\left(m - \frac{1}{2}\right)\right]\Bigg|_1^0$$

$$= \frac{kp\,L}{\dot{m}\,c_P\,t_{avg}} \frac{1}{b} \ln \frac{1 + b/2}{1 - b/2} \tag{6}$$

The wall temperature decreases exponentially in the direction of flow: this can be seen by replacing L with x, and T_{out} with T(x) in eq. (6). In conclusion, the outlet temperature is

$$\frac{T_{out} - T_0}{T_h - T_0} = \exp\left(- \frac{kp\,L}{\dot{m}\,c_P\,t_{avg}} \frac{1}{b} \ln \frac{1 + b/2}{1 - b/2}\right) \tag{7}$$

The outlet temperature depends on two dimensionless numbers, the taper parameter b, and the number of heat transfer units

$$\text{NTU} = \frac{(k/t_{avg})\,p\,L}{\dot{m}\,c_P} \tag{8}$$

in which k/t_{avg} plays the role of the overall heat transfer coefficient of the usual NTU definition. The most instructive way of examining the effect of taper on T_{out} (or heat loss) is by comparing eq. (7) with the no-taper limit of the same result,

$$\frac{(T_{out} - T_0)_{b=0}}{T_h - T_0} = \exp(-\,\text{NTU}) \tag{9}$$

Equation (9) can be obtained by repeating the analysis (5)-(7) using $t = t_{avg}$ (constant), or by setting $b \rightarrow 0$ in eq. (7) and calculating the limit

$$\lim_{b\to 0} \frac{1}{b} \ln \frac{1 + b/2}{1 - b/2} = \lim_{b\to 0} \frac{1}{b}\left[\ln\left(1 + \frac{b}{2}\right) - \ln\left(1 - \frac{b}{2}\right)\right]$$

$$= \frac{1}{b}\left[\frac{b}{2} \cdots - \left(-\frac{b}{2} \cdots\right)\right]$$

$$= 1 \tag{10}$$

Finally, we divide eq. (7) by eq. (9),

$$\frac{T_{out} - T_0}{(T_{out} - T_0)_{b=0}} = \frac{\exp\left(-\frac{\text{NTU}}{b} \ln \frac{1 + b/2}{1 - b/2}\right)}{\exp(-\,\text{NTU})} \tag{11}$$

This quantity is plotted in the attached graph, as a function of NTU and b. The highest outlet temperature occurs when the insulation thickness is uniform (b = 0): this is the best design, especially when the supply of insulation material (or t_{avg}) is so small that NTU is larger than approximately 0.1.

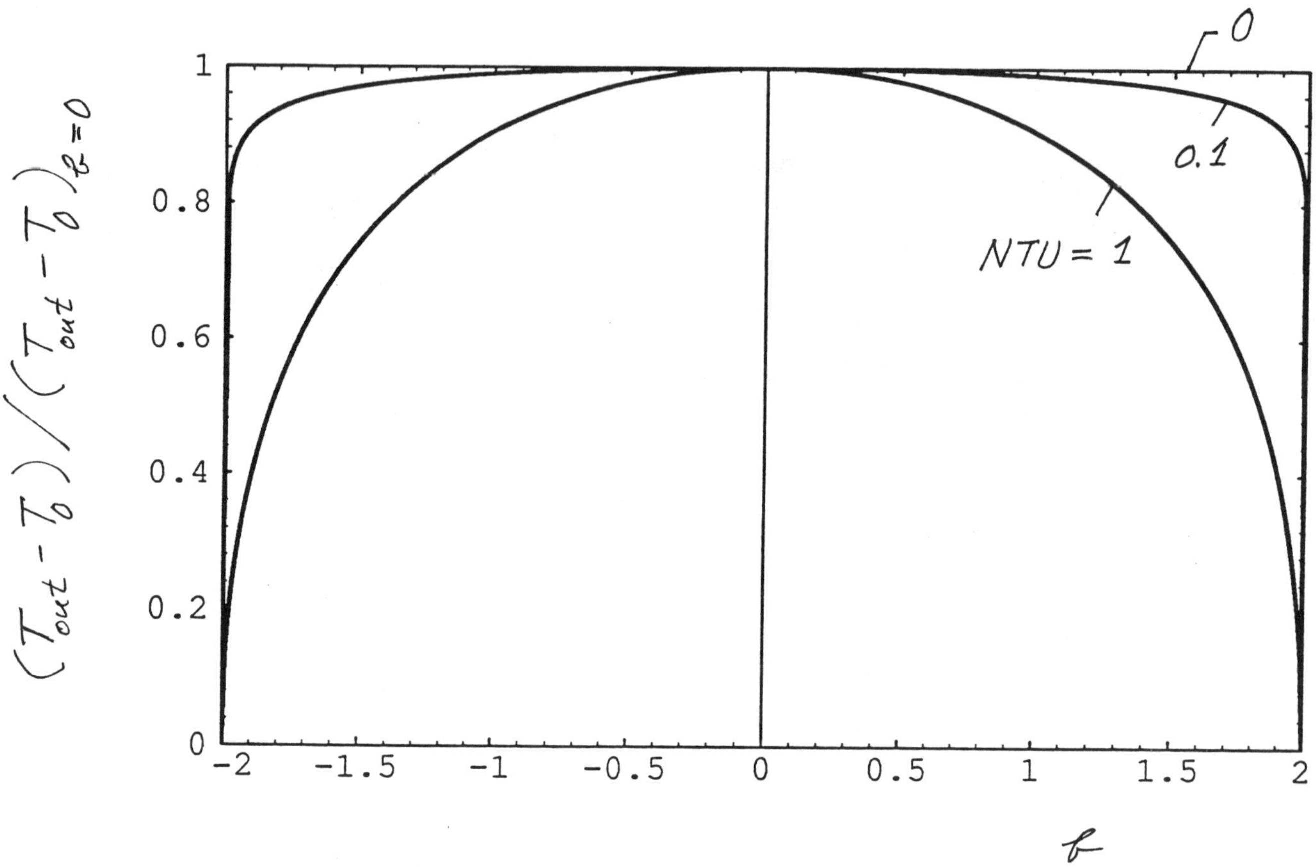

It is indeed fascinating that what engineers must have been doing all along, for ease of installation, or expediency (e.g. wrapping insulation uniformly over a pipe carrying a hot fluid) turns out to be the best way of spreading a limited supply of insulation material.

<u>Proof that t = constant is the best design</u>

It turns out that t = constant is the optimal way of distributing the insulation material even when t(x) is not constrained to vary linearly, and when all the simplifying assumptions made in Project 9.4 are abandoned. The following development is outside the scope of this course. I must report it here (i.e. somewhere*), because the present topic and its punchline are original.

Assume that the pipe wall has the outer radius r, and the insulation layer has the outer radius r_o. The insulation thickness $t(x) = r_o(x) - r$ is not necessarily small when compared with r. The overall heat transfer coefficient U between the stream T(x) and the environment T_0 is given by

$$\frac{1}{U\,2\pi r} = \frac{1}{h_o\,2\pi r_o} + \frac{\ln\left[1 + \frac{t(x)}{r}\right]}{2\pi k} + \frac{t_w}{k_w\,2\pi r} + \frac{1}{h\,2\pi r} \qquad (12)$$

The four resistances on the right-hand side represent, in order, convection outside the insulation, conduction through the insulation (now a thick cylindrical shell), conduction through the pipe wall (thickness $t_w << r$, conductivity k_w), and convection inside the pipe.

The group $U\,2\pi r$ replaces kp/t in eq. (5),

$$-\dot{m}\,c_P\,dT = U\,2\pi r\,(T - T_0)\,dx \qquad (13)$$

Integrating from x = 0 (where $T = T_h$) to x = L (where $T = T_{out}$), we arrive at

$$\ln\frac{T_h - T_0}{T_{out} - T_0} = \int_0^L \frac{U\,2\pi r\,dx}{\dot{m}\,c_P} \qquad (14)$$

This integral must be maximized (because T_{out} must be maximized) subject to the volume constraint

$$V = \int_0^L \left(\pi r_0^2 - \pi r^2\right) dx$$

$$= \int_0^L \pi r^2 \left[\left(1 + \frac{t}{r}\right)^2 - 1\right] dx \qquad (15)$$

The variational calculus problem of maximizing the integral (14) subject to the volume constraint (15) is equivalent to seeking the extremum of the aggregate integral (see pp. 722-723 in A. Bejan, Advanced Engineering Thermodynamics, Wiley, New York, 1988)

---

* I am actually forced into pressing claims of this sort in this solutions manual because of my experience with writing problems for my previous books. Several of those problems were "original" in the sense that they could have been developed into articles in archival journals. Because I did not emphasize that they were being formulated (and solved, of course) for the first time, they were overlooked in subsequent publications.

(see pp. 722-723 in A. Bejan, Advanced Engineering Thermodynamics, Wiley, New York, 1988)

$$\Phi = \int_0^L \underbrace{\left\{ \frac{U\, 2\pi r}{\dot{m}\, c_P} + \lambda\, \pi r^2 \left[\left(1 + \frac{t}{r}\right)^2 - 1 \right] \right\}}_{F} dx \tag{16}$$

Note that the integrand of Φ, namely { } = F, is a linear combination of the integrands of (14) and (15), and that λ is a Lagrange multiplier. Note further that U depends on x through the function t(x), in accordance with eq. (12).

The optimal function t(x) is the solution to the Euler equation

$$\frac{\partial F}{\partial t} = 0 \tag{17}$$

which can be written down by using eq. (12) for U(t). This last step is not necessary if we notice that U decreases when t increases, while the group multiplied by λ in F increases when t increases. In this way, it is clear that F has a minimum with respect to t(x). That minimum can be pinpointed by solving eq. (17), however, since all the other parameters that will be present in that equation are x-independent, the t(x) solution of equation (17) is

$$t_{opt} = K, \text{ constant} \tag{18}$$

The constant K is evaluated finally by substituting eq. (18) back in the volume constraint (15), and the result is

$$t_{opt} = r\left[\left(\frac{V}{\pi r^2 L} + 1\right)^{1/2} - 1\right], \text{ constant} \tag{19}$$

This is then a general rule for designing duct insulation for minimum heat loss to the ambient, when the amount of insulation is fixed. The same conclusion is reached (in a completely analogous analysis) when the amount of insulation is minimized subject to a fixed rate of heat transfer to the ambient.

Chapter 10

RADIATION

Problem 10.1. The projected image of sun on the sphere with earth as center and r as radius is $A_n = \pi D^2/4$. This projected area is the area of the solar disc. The solid angle definition (10.10) yields

$$\omega = \frac{A_n}{r^2} = \frac{\pi}{4}\left(\frac{D}{r}\right)^2 = \frac{\pi}{4}\left(\frac{1.392 \times 10^6 \text{ km}}{1.447 \times 10^8 \text{ km}}\right)^2$$

$$= 0.73 \times 10^{-4} \text{sr}$$

Problem 10.2. The maximum of the monochromatic hemispherical emissive power can be located by solving $\partial E_{b,\lambda}/\partial\lambda = 0$, for which $E_{b,\lambda}$ is listed in eq. (10.16). We obtain in this manner

$$- 5\lambda^{-6}\left[\exp\left(\frac{C_2}{\lambda T}\right) - 1\right] - \lambda^{-5}\exp\left(\frac{C_2}{\lambda T}\right)\frac{(- C_2)}{\lambda^2 T} = 0 \qquad (1)$$

By calling $C_2/\lambda T = u$, eq. (1) can be rewritten as

$$u = 5(1 - e^{-u}) \qquad (2)$$

The trial-and-error solution of eq. (2) is u = 4.965, which means also that at the $E_{b,\lambda}$ maximum

$$\lambda T = \frac{C_2}{4.965} = 0.0029 \text{ m K} \qquad (3)$$

The corresponding peak value of $E_{b,\lambda}$ is obtained by substituting the λ value of eq. (3) into the $E_{b,\lambda}$ expression (10.16):

$$E_{b,\lambda,max} = E_{b,\lambda}(\lambda = C_2/uT) = \frac{C_1}{C_2^5}\frac{u^5}{e^u - 1}T^5$$

$$= \frac{12.87 \times 10^{-6} \text{W}}{\text{m}^3 \text{ K}^5} T^5 \qquad (4)$$

The last step, eq. (4), was executed by setting $C_1 = 3.742 \times 10^{-16} \text{W m}^2$, $C_2 = 1.439 \times 10^{-2} \text{m K}$ and u = 4.965.

Problem 10.3. Dividing the radiation function (10.24) by σT^4 we obtain, in order,

$$\frac{E_b(0 - \lambda_1 T)}{\sigma T^4} = \frac{1}{\sigma T^4}\int_0^{\lambda_1} \frac{C_1 \lambda^{-5}}{\exp(C_2/\lambda T) - 1}\, d\lambda$$

$$= \int_0^{\lambda_1} \frac{C_1/\sigma}{\exp(C_2/\lambda T) - 1} \frac{d(\lambda T)}{(\lambda T)^5}$$

$$= \int_0^{\lambda_1 T} \frac{C_1/\sigma}{\exp(C_2/m) - 1} \frac{dm}{m^5} \qquad (1)$$

in which the dummy variable is $m = \lambda T$. Equation (1) proves that the dimensionless radiation function depends only on the value of the group $\lambda_1 T$.

Problem 10.4. Guided by the nomenclature employed in Table 10.1, we calculate in order

$$\lambda_1 T = 0.4 \times 10^{-6} \mathrm{m}\ 5800\ \mathrm{K} = 0.00232$$

$$\lambda_2 T = 1.0 \times 10^{-6} \mathrm{m}\ 5800\ \mathrm{K} = 0.0058$$

In Table 10.1 we use linear interpolation to calculate

$$\frac{E_b\left(0 - \lambda_1 T\right)}{\sigma T^4} \cong 0.128$$

$$\frac{E_b\left(0 - \lambda_2 T\right)}{\sigma T^4} \cong 0.719$$

and, based on these values,

$$\frac{E_b\left(\lambda_1 T - \lambda_2 T\right)}{\sigma T^4} = 0.719 - 0.128 = 0.591$$

$$\frac{E_b\left(\lambda_2 T - \infty\right)}{\sigma T^4} = 1 - 0.719 = 0.281$$

The absorbed fraction of the incident solar (T = 5800 K) radiation is

$$0.8 \times 0.128 + 0.1 \times 0.591 + 1 \times 0.281 = 0.443$$

Problem 10.5. a) With reference to the drawing attached to the problem statement, we calculate in order:

$$\tan \phi_1 = \frac{0.1\text{m}}{1\text{m}} = 0.1$$

$$\phi_1 = \tan^{-1}(0.1) = 0.009967 \text{ rad} = 5.7°$$

$$A_{n,1} = A_1 \cos \phi_1 = 1 \text{ cm}^2\, 0.995$$

$$= 0.995 \text{ cm}^2$$

$$\beta = 90° - 5.7° = 84.3°$$

$$\phi_2 + (\beta - 15°) = 90°$$

$$\phi_2 = 90° + 15 - 84.3 = 20.7°$$

$$\cos \phi_2 = 0.9354$$

$$A_{n,2} = A_2 \cos \phi_2 = 2 \text{ cm}^2\, 0.9354$$

$$= 1.871 \text{ cm}^2$$

$$r = [1 \text{ m}^2 + (0.1\text{m})^2]^{1/2} = 1.005 \text{ m}$$

The solid angle subtended by A_2 is

$$\omega_{1\text{-}2} = \frac{A_{n,2}}{r^2} = \frac{1.871 \text{ cm}^2}{(1.005 \text{ m})^2} = 1.852 \times 10^{-4} \text{ sr}$$

and the heat current emitted by A_1 and intercepted by A_2 is

$$q_{1\to 2} = I_{b,1}\, A_{n,1}\, \omega_{1\text{-}2}$$

$$= 3000 \frac{\text{W}}{\text{m}^2 \text{ sr}}\, 0.995 \text{ cm}^2\, 1.852 \times 10^{-4} \text{ sr}$$

$$= 5.53 \times 10^{-5} \text{ W}$$

b) When A_1 and A_2 are parallel and opposite one another, $\phi_1 = \phi_2 = 0$, $A_{n,1} = A_1$, $r = 1$ m, and

$$\omega_{1\text{-}2} = \frac{A_2}{r^2} = \frac{2\ \text{cm}^2}{1\ \text{m}^2} = 2 \times 10^{-4}\ \text{sr}$$

$$q_{1\text{-}2} = I_{b,1}\, A_1\, \omega_{1\text{-}2}$$

$$= 3000\ \frac{\text{W}}{\text{m}^2\cdot\text{sr}}\ 1\ \text{cm}^2\ 2 \times 10^{-4}\ \text{sr}$$

$$= 6 \times 10^{-5}\ \text{W}$$

By comparing this result with the answer to part (a) we see that the proper alignment of A_1 and A_2 leads to a 8.5-percent increase in the one-way heat current from A_1 to A_2.

Problem 10.6. In the small-λT limit, the radiation function definition (10.24) yields, in order,

$$E_b(0 - \lambda_1 T) = \int_0^{\lambda_1} \frac{C_1 \lambda^{-5} d\lambda}{\exp(C_2/\lambda T) - 1} \cong \int_0^{\lambda_1} \frac{C_1 \lambda^{-5} d\lambda}{\exp(C_2/\lambda T)}$$

$$\cong \frac{C_1 T^4}{C_2^4} \int_{C_2/\lambda_1 T}^{\infty} \beta^3\, e^{-\beta}\, d\beta \qquad (1)$$

where the dummy variable is $\beta = C_2/\lambda T$. The integral can be evaluated by first consulting a set of mathematical tables, and learning that

$$\int x^m e^{ax} dx = e^{ax} \sum_{r=0}^{m} (-1)^r \frac{m!\, x^{m-r}}{(m-r)!\, a^{r+1}} \qquad (2)$$

In the present case, the integral listed in eq. (1) yields

$$\int_{C_2/\lambda_1 T}^{\infty} \beta^3 e^{-\beta} d\beta = (u^3 + 3u^2 + 6u + 6)\, e^{-u} \qquad (3)$$

in which we are using the notation $u = C_2/\lambda_1 T$. Dividing eq. (1) by σT^4, we obtain the small-λT asymptote of the dimensionless radiation function,

$$\frac{E_b(0 - \lambda_1 T)}{\sigma T^4} \cong \frac{15}{\pi^4}(u^3 + 3u^2 + 6u + 6)\, e^{-u} \qquad (4)$$

The values calculated with this expression can be compared with those listed in Table 10.1, in order to show that eq. (4) is accurate within 1 percent if $\lambda_1 T < 0.0042$ m K:

$\lambda_1 T$ (m K)	u	$E_b(0 - \lambda_1 T)/\sigma T^4$ Table 10.1	Eq. (4)	Error (percent)
0.003	4.797	0.273	0.2723	0.32
0.0042	3.426	0.516	0.5106	1.05
0.005	2.878	0.634	0.6232	1.65

In the large-λT limit, the corresponding asymptotic analysis is somewhat simpler:

$$E_b(0 - \lambda_1 T) = \int_0^{\lambda_1} \frac{C_1 \lambda^{-5} d\lambda}{\exp(C_2/\lambda T) - 1}$$

$$= \int_0^{\infty} \frac{C_1 \lambda^{-5} d\lambda}{\exp(C_2/\lambda T) - 1} - \int_0^{\lambda_1} \frac{C_1 \lambda^{-5} d\lambda}{\exp(C_2/\lambda T) - 1}$$

$$\cong \sigma T^4 - \frac{C_1}{C_2} T \int_0^{\lambda_1} \lambda^{-4} d\lambda$$

$$\cong \sigma T^4 - \frac{C_1}{C_2} \frac{T}{3\lambda_1^3} \qquad (5)$$

The dimensionless version of eq. (5) is

$$\frac{E_b(0 - \lambda_1 T)}{\sigma T^4} \cong 1 - \left(\frac{0.00535 \text{ m K}}{\lambda_1 T}\right)^3 \qquad (6)$$

Equation (6) approximates within 1 percent the Table 10.1 data when $\lambda_1 T > 0.016$ m K:

$\lambda_1 T$ (m K)	$E_b(0 - \lambda_1 T)/\sigma T^4$		Error (percent)
	Table 10.1	Eq. (6)	
0.01	0.914	0.847	7.36
0.016	0.974	0.9626	1.16
0.02	0.986	0.9809	0.49

Problem 10.7. With reference to the definition (10.33) and the sketch shown below, we must calculate

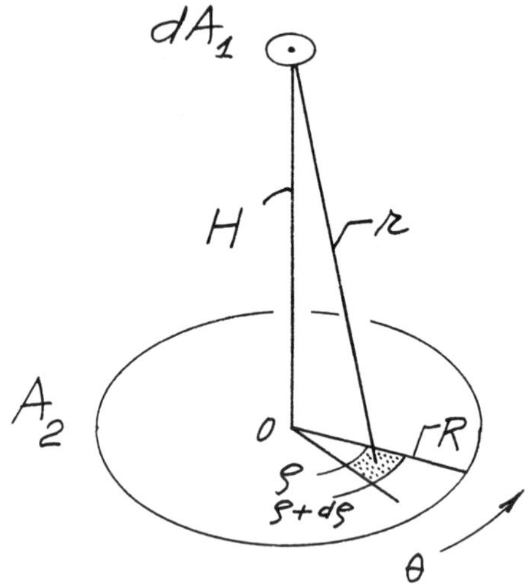

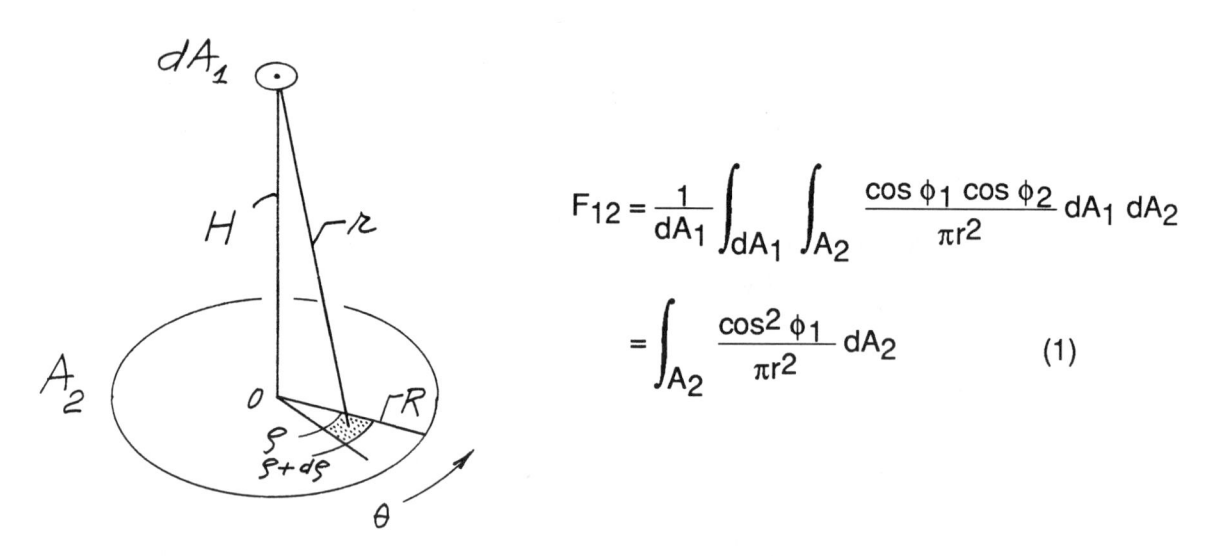

$$F_{12} = \frac{1}{dA_1}\int_{dA_1}\int_{A_2} \frac{\cos\phi_1 \cos\phi_2}{\pi r^2}\, dA_1\, dA_2$$

$$= \int_{A_2} \frac{\cos^2\phi_1}{\pi r^2}\, dA_2 \qquad (1)$$

where $\cos\phi_1 = \cos\phi_2 = H/r$, $r^2 = H^2 + \rho^2$, and $dA_2 = (\rho d\theta)d\rho$. The integral (1) can be evaluated in closed form:

$$F_{12} = \int_0^{2\pi} d\theta \int_0^R \frac{H^2}{\pi(H^2+\rho^2)^2}\, \rho\, d\rho$$

$$= 2\pi\frac{H^2}{2\pi}\int_{H^2}^{H^2+R^2} \frac{dm}{m^2} = H^2\left(-\frac{1}{H^2+R^2} + \frac{1}{H^2}\right)$$

$$= \frac{R^2}{H^2+R^2} \qquad (2)$$

One interesting limit of this configuration is $H/R \to 0$, that is the case of an infinitesimal element dA_1 positioned infinitesimally close to the disc. In this limit eq. (2) yields $F_{12} = 1$, which means that all the blackbody radiation emitted hemispherically by dA_1 (downward, toward A_2) is intercepted entirely by A_2.

Problem 10.8. The enclosure with triangular cross-section (Table 10.2, the third entry) has three surfaces, therefore the system (10.43) becomes

$$
\begin{aligned}
F_{11} + F_{12} + F_{13} &= 1 \\
F_{21} + F_{22} + F_{23} &= 1 \\
F_{31} + F_{32} + F_{33} &= 1
\end{aligned}
\qquad \text{(a)}
$$

Each surface is plane, therefore we can write

$$F_{11} = F_{22} = F_{33} = 0 \qquad \text{(b)}$$

Three more relations are recommended by the reciprocity property,

$$
\begin{aligned}
L_1 F_{12} &= L_2 F_{21} \\
L_1 F_{13} &= L_3 F_{31} \\
L_2 F_{23} &= L_3 F_{32}
\end{aligned}
\qquad \text{(c)}
$$

Using the zero values (b), and eliminating F_{21}, F_{31} and F_{32} between eqs. (c) and (a), leads to the new system of three equations:

$$F_{12} + F_{13} = 1 \qquad \text{(d)}$$

$$L_1 F_{12} + L_2 F_{23} = L_2 \qquad \text{(e)}$$

$$L_1 F_{13} + L_2 F_{23} = L_3 \qquad \text{(f)}$$

Equations (d) and (f) state that

$$F_{13} = 1 - F_{12}, \qquad \text{and} \qquad F_{23} = \frac{L_3}{L_2} - \frac{L_1}{L_2} + \frac{L_1}{L_2} F_{12} \qquad \text{(g)}$$

The substitution of these F_{13} and F_{23} expressions into eq. (e) yields finally

$$F_{12} = \frac{L_1 + L_2 - L_3}{2L_1} \qquad \text{(h)}$$

Problem 10.9. a) With reference to the sphere-disc geometry of Table 10.2 (the last entry), we recognize the following dimensions

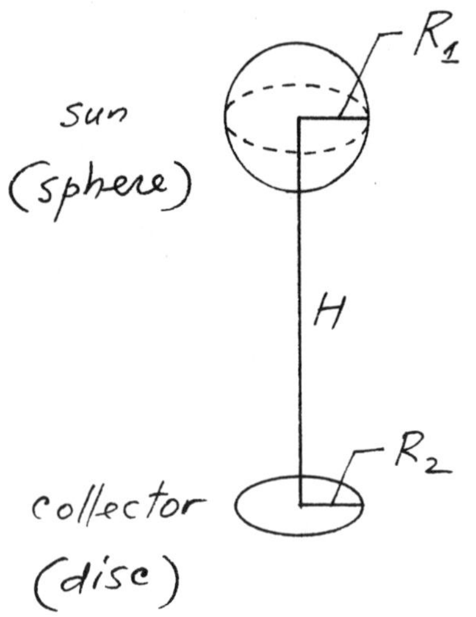

$$R_1 = \frac{1.392}{2} 10^6 \text{ km} = 0.696 \times 10^6 \text{ km}$$

$$H = 1.447 \times 10^8 \text{ km}$$

We are asked to calculate F_{21}, however, the table allows us to calculate only F_{12} in which

$$x = \frac{R_2}{H} << 1$$

i.e. "small" because the collector (R_2, unknown) is man-made. In the $x \rightarrow 0$ limit, the F_{12} expression becomes

$$F_{12} = \frac{1}{2}\left[1 - \left(1 - \frac{1}{2} x^2\right)\right] = \frac{1}{4} x^2$$

$$= \frac{1}{4}\left(\frac{R_2}{H}\right)^2$$

The relationship between F_{21} and F_{12} is

$$A_2 F_{21} = A_1 F_{12}$$

therefore

$$F_{21} = F_{12}\frac{A_1}{A_2} = \frac{1}{4}\frac{R_2^2}{H^2}\frac{4\pi R_1^2}{\pi R_2^2}$$

$$= \left(\frac{R_1}{H}\right)^2 = \left(\frac{0.696 \times 10^6}{1.447 \times 10^8}\right)^2 = 2.314 \times 10^{-5}$$

b) The view factor from one disc (1) to another disc (2), which is parallel and coaxial, is given by the fifth entry in Table 10.2,

$$F_{12} = \frac{1}{2}\left\{X - \left[X^2 - 4\left(\frac{x_2}{x_1}\right)^2\right]^{1/2}\right\}$$

where

$$x_1 = \frac{R_1}{H} << 1 \qquad \text{and} \qquad x_2 = \frac{R_2}{H} << 1$$

therefore

$$X = 1 + \frac{1 + x_2^2}{x_1^2} \cong \frac{1}{x_1^2}$$

In the small-(x_1,x_2) limit, the F_{12} formula reduces to

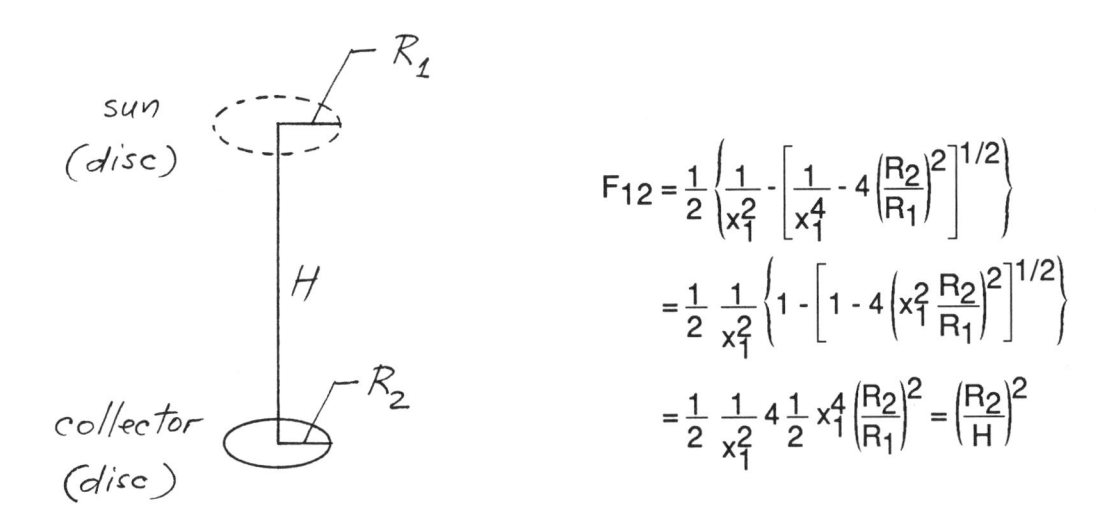

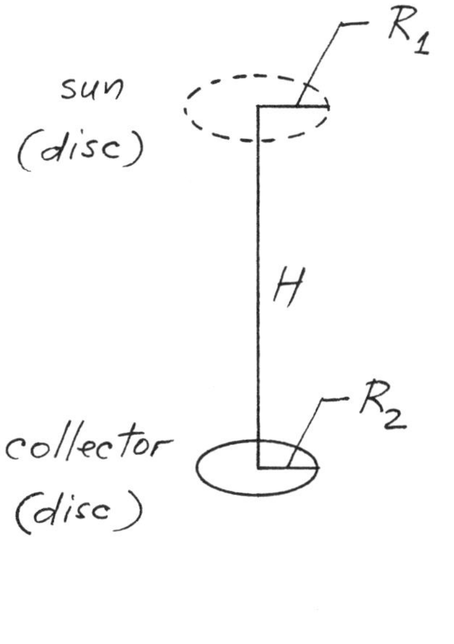

$$F_{12} = \frac{1}{2}\left\{\frac{1}{x_1^2} - \left[\frac{1}{x_1^4} - 4\left(\frac{R_2}{R_1}\right)^2\right]^{1/2}\right\}$$

$$= \frac{1}{2}\frac{1}{x_1^2}\left\{1 - \left[1 - 4\left(x_1^2\frac{R_2}{R_1}\right)^2\right]^{1/2}\right\}$$

$$= \frac{1}{2}\frac{1}{x_1^2}4\frac{1}{2}x_1^4\left(\frac{R_2}{R_1}\right)^2 = \left(\frac{R_2}{H}\right)^2$$

The reciprocity relation requires

$$A_1F_{12} = A_2F_{21}$$

and this yields

$$F_{21} = F_{12}\frac{A_1}{A_2} = F_{12}\frac{\pi R_1^2}{\pi R_2^2} = \frac{R_2^2}{H^2}\frac{R_1^2}{R_2^2}$$

$$= \left(\frac{R_1}{H}\right)^2 = 2.314 \times 10^{-5}$$

In general, of course, the F_{21} values of the disc-sphere and disc-disc arrangements are not equal (after all, the formulas listed in Table 10.2 are quite different). The answers to parts (a) and (b) happen to be identical because of the limit $R_1/H \ll 1$, $R_2/H \ll 1$. In this limit the solar sphere is so far away from earth (and subtends such a small solid angle) that the spherical shape of the solar ball matters extremely little to the observer who travels from side to side on the flat disc of the collector. The collector is so small (narrow) and the sun is so distant that the observer sees always the same solar disc regardless of his/her position on the collector. Said another way, by moving to the edge of the collector you cannot see "behind" the solar-disc image that you see from the center of the collector.

Problem 10.10. Consider the sphere of radius $R = (H^2 + R_2^2)^{1/2}$ drawn concentrically around the R_1 sphere. On this larger sphere, the disc of radius R_2 serves as base for the spherical segment of height $h = (H^2 + R_2^2)^{1/2} - H$. The spherical area of this segment – the spherical "zone" – has the following area (consult a set of mathematical tables):

$$A = 2\pi R h = 2\pi[H^2 + R_2^2 - H\,(H^2 + R_2^2)^{1/2}] \qquad \text{(a)}$$

We are interested in the view factor F_{12}, that is from the inner sphere (A_1) to the disc (R_2). The same F_{12} value holds between A_1 and the spherical zone A subtended by the disc,

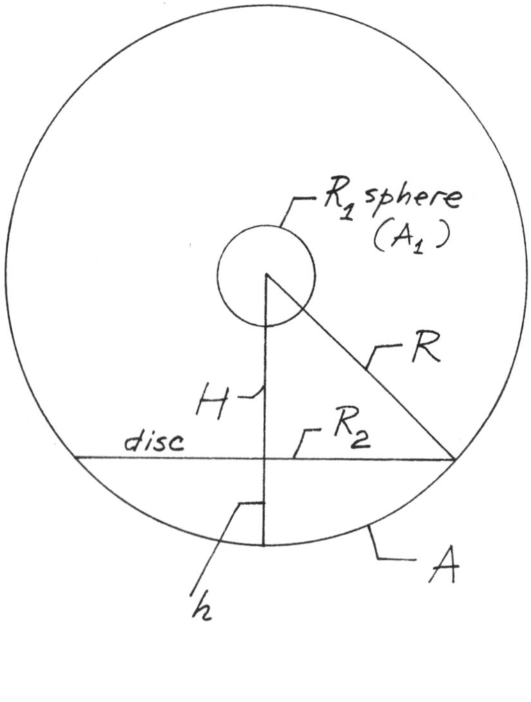

$$F_{12} = F_{A_1 - A} \qquad \text{(b)}$$

We also know that the view factor from the inner sphere to the entire spherical surface of radius R is equal to 1 (see the two-sphere configuration of Fig. 10.15),

$$F_{A_1 - R} = 1 \qquad \text{(c)}$$

The spherical surface of radius R (area $4\pi R^2$) is a certain multiple m of the spherical zone A,

$$4\pi R^2 = m\,A \qquad \text{(d)}$$

In view of the additivity rule (10.40), the view factor from A_1 to the whole R sphere must be equal to the same multiple of the view factor from A_1 to the spherical zone A,

$$F_{A_1 - R} = m\, F_{A_1 - A} \tag{e}$$

Equations (a) - (e) can be used in reverse order to calculate F_{12}:

$$F_{A_1 - A} = \frac{1}{m} = \frac{4\pi R^2}{A}$$

$$= \frac{1}{2}\left[1 - \frac{H}{(H^2 + R_2{}^2)^{1/2}}\right] = F_{12} \tag{f}$$

Equation (f) validates the formula listed in Table 10.2: indeed, F_{12} does not depend on the radius of the A_1 sphere.

Problem 10.11. The surfaces L_1, c, L_2 and d form an enclosure, for which eq. (10.43) states

$$1 = F_{1c} + F_{12} + F_{1d} \tag{1}$$

This intermediate result shows that we must first estimate F_{1c} and F_{1d}, in order to determine F_{12}. For the triangular enclosure (L_1, b, d), the third configuration of Table 10.2 lists

$$F_{1d} = \frac{L_1 + d - b}{2L_1} \tag{2}$$

Similarly, for the triangular enclosure (L_1, c, a) we can write

$$F_{1c} = \frac{L_1 + c - a}{2L_1} \tag{3}$$

Eliminating F_{1d} and F_{1c} between eqs. (1) - (3) leads to the wanted result:

$$F_{12} = \frac{a + b - (c + d)}{2L_1} \tag{4}$$

Problem 10.12. To apply the crossed-strings method, we note that

$$a = b = (H^2 + X^2)^{1/2}$$

$$c = d = H$$

$$L_1 = X$$

The view factor is therefore

$$F_{12} = \frac{a + b - (c + d)}{2L_1} = \frac{2(H^2 + X^2)^{1/2} - 2H}{2X}$$

$$= \left[\left(\frac{H}{X}\right)^2 + 1\right]^{1/2} - \frac{H}{X}$$

This formula reproduces the values indicated by the long-strips limit ($Y/H \to \infty$) in Fig. 10.11.

Problem 10.13. a) Consider the thermal resistance associated with the first gap. The temperature gap from T_H to T_1 is bridged by three resistances in series, therefore, according to the sketch,

$$R_{one\ gap} = \frac{1 - \varepsilon}{\varepsilon A} + \frac{1}{A} + \frac{1 - \varepsilon}{\varepsilon A} = \frac{1}{A}\left(\frac{2}{\varepsilon} - 1\right) \qquad (1)$$

Since there are $(n + 1)$ gaps of this kind, the total thermal resistance from T_H to T_L is

$$R_{total} = (n + 1)\, R_{one\ gap} = \frac{n + 1}{A}\left(\frac{2}{\varepsilon} - 1\right) \qquad (2)$$

The heat transfer rate from T_H to T_L is therefore

$$q_{H-L}^{(a)} = \frac{E_{bH} - E_{bL}}{R_{total}} = \frac{1}{n + 1}\,\frac{\sigma A(T_H^4 - T_L^4)}{\frac{2}{\varepsilon} - 1} \qquad (3)$$

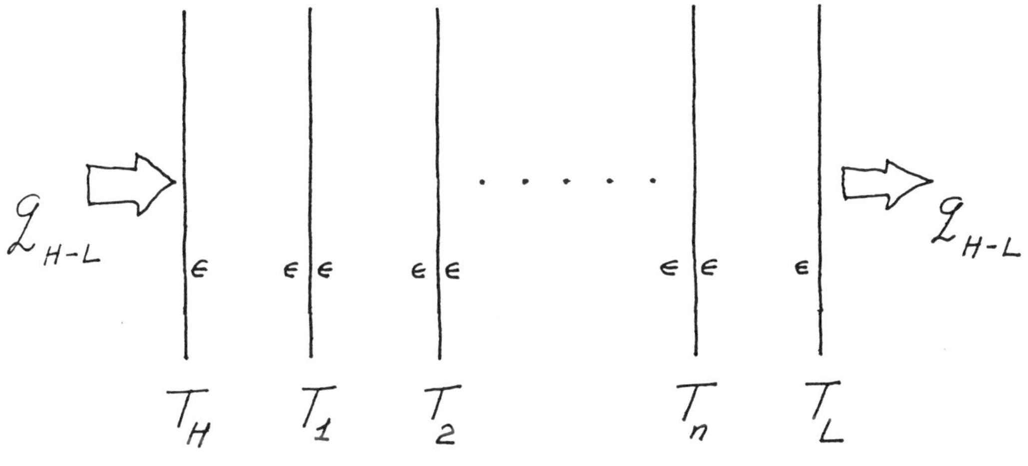

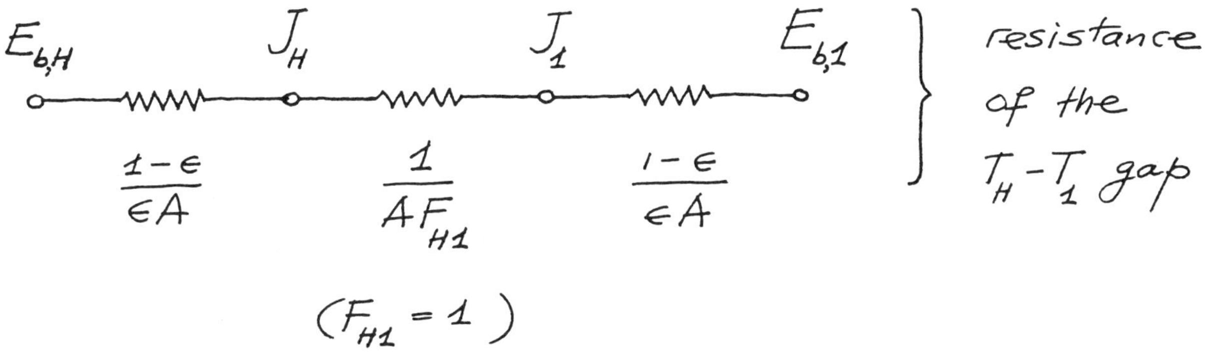

b) In the absence of radiation shields, the total heat transfer rate from T_H to T_L is

$$q_{H-L}^{(b)} = \frac{AF_{HL}\sigma(T_H{}^4 - T_L{}^4)}{\frac{1}{\varepsilon} + \frac{1}{\varepsilon} - 1} = \frac{\sigma A(T_H{}^4 - T_L{}^4)}{\frac{2}{\varepsilon} - 1} \qquad (4)$$

This result shows that $q_{H-L}^{(a)} = q_{H-L}^{(b)}/(n + 1)$, in other words, that the use of n identical shields reduces the heat transfer rate by the factor $(1 + n)^{-1}$.

Problem 10.14. a) The path of the heat transferred from T_1 to T_3 is shown in the attached network, in which

$$A_1 = 2 \text{ sides} \times \pi R_1^2 = 2\pi\,(0.1)^2 \text{ m}^2 = 0.063 \text{ m}^2$$

$$A_2 = 4\pi R_2^2 = 4\pi\,(0.3)^2 \text{ m}^2 = 1.131 \text{ m}^2$$

$$F_{12} = 1$$

$$F_{23} = 1$$

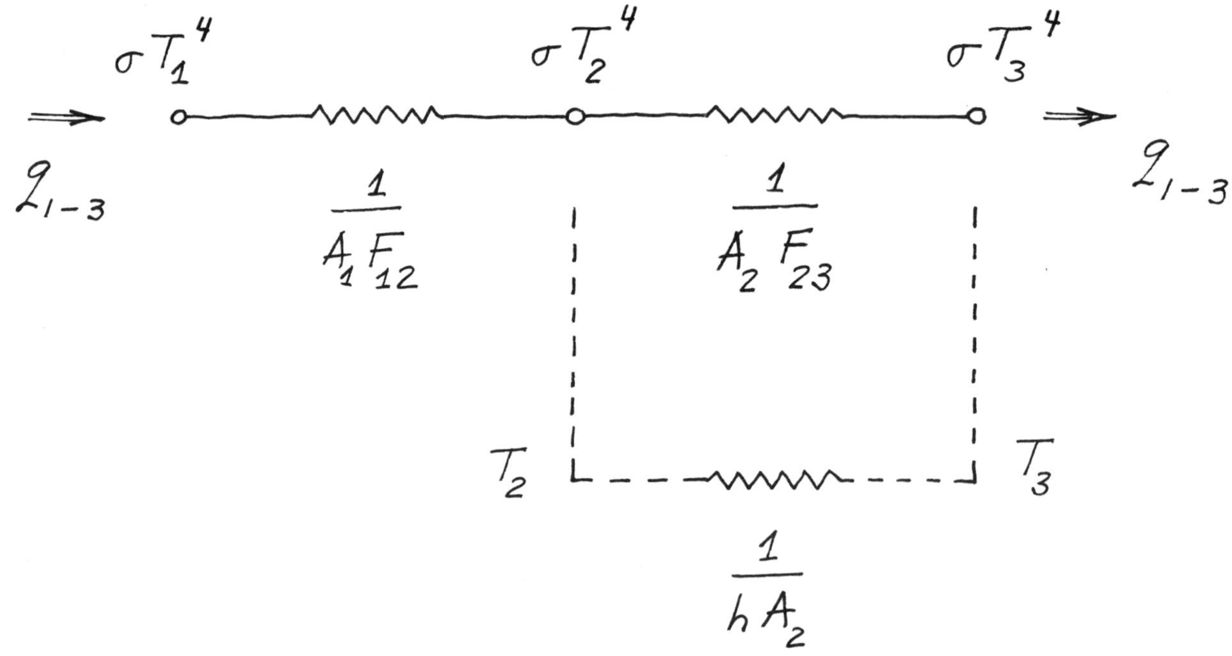

Note that on the outside of the barbecue enclosure there is a parallel path for radiation and convection. The net rate of heat loss from the coal is

$$q_{1\text{-}3} = \underbrace{\frac{\sigma T_1^4 - \sigma T_2^4}{\dfrac{1}{A_1 F_{12}}}}_{\text{total heat transfer}} = \underbrace{\frac{\sigma T_2^4 - \sigma T_3^4}{\dfrac{1}{A_2 F_{23}}}}_{\text{external radiation}} + \underbrace{\frac{T_2 - T_3}{\dfrac{1}{h A_2}}}_{\text{external convection}}$$

The second of these equations reduces sequentially to

$$\frac{A_1}{A_2} T_1^4 + T_3^4 + \frac{h}{\sigma} T_3 = \left(1 + \frac{A_1}{A_2}\right) T_2^4 + \frac{h}{\sigma} T_2$$

$$\frac{0.063}{1.131} (800 + 273)^4 + (25 + 273)^4 + \frac{20}{5.67 \times 10^{-8}} (25 + 273) =$$

$$= \left(1 + \frac{0.063}{1.131}\right)\left(\frac{T_2}{K}\right)^4 + \frac{20}{5.67 \times 10^{-8}} \cdot \frac{T_2}{K}$$

$$1.769 \times 10^{11} = \left(\frac{T_2}{K}\right)^4 + 3.34 \times 10^8 \frac{T_2}{K}$$

The solution to the last equation can be obtained numerically by trial and error,

$$T_2 \cong 428.6 \text{ K} \cong 156°\text{C}$$

b) The relative importance of radiation and convection on the outside of the barbecue enclosure is indicated by the ratio of the two terms on the right side of the $q_{1\text{-}3}$ expression listed above:

$$\frac{\text{radiation}}{\text{convection}} = \frac{\sigma A_2 \left(T_2^4 - T_3^4\right)}{h A_2 (T_2 - T_3)}$$

$$= \frac{5.67 \times 10^{-8} \left(428.6^4 - 298^4\right)}{20\ (428.6 - 298)}$$

$$= 0.56$$

This means that radiation accounts for only 36 percent of the total heat transfer rate from the outer surface of the enclosure:

$$\frac{\text{radiation heat transfer}}{\text{total heat transfer}} = \frac{0.56}{1 + 0.56} = 0.36$$

$$\frac{\text{convection heat transfer}}{\text{total heat transfer}} = \frac{1}{1 + 0.56} = 0.64$$

Problem 10.15. a) The symmetry of the enclosure requires that

$$F_{12} = F_{13} = \frac{1}{2} \qquad (1)$$

Next, the reciprocity property $A_1F_{12} = A_2F_{21}$ yields

$$F_{21} = \frac{A_1}{A_2} F_{12} = \frac{1}{20} \qquad (2)$$

Finally, from the energy conservation statement for the radiation emitted by A_2 (the "enclosure" rule),

$$F_{21} + F_{22} + F_{23} = 1, \quad (3)$$

in which $F_{22} = 0$, we deduce the remaining view factor:

$$F_{23} = 1 - F_{21} = \frac{19}{20} \quad (4)$$

b) The total thermal resistance between the nodes $E_{b,1}$ and $E_{b,2}$ can be estimated by following the steps shown in the figure.

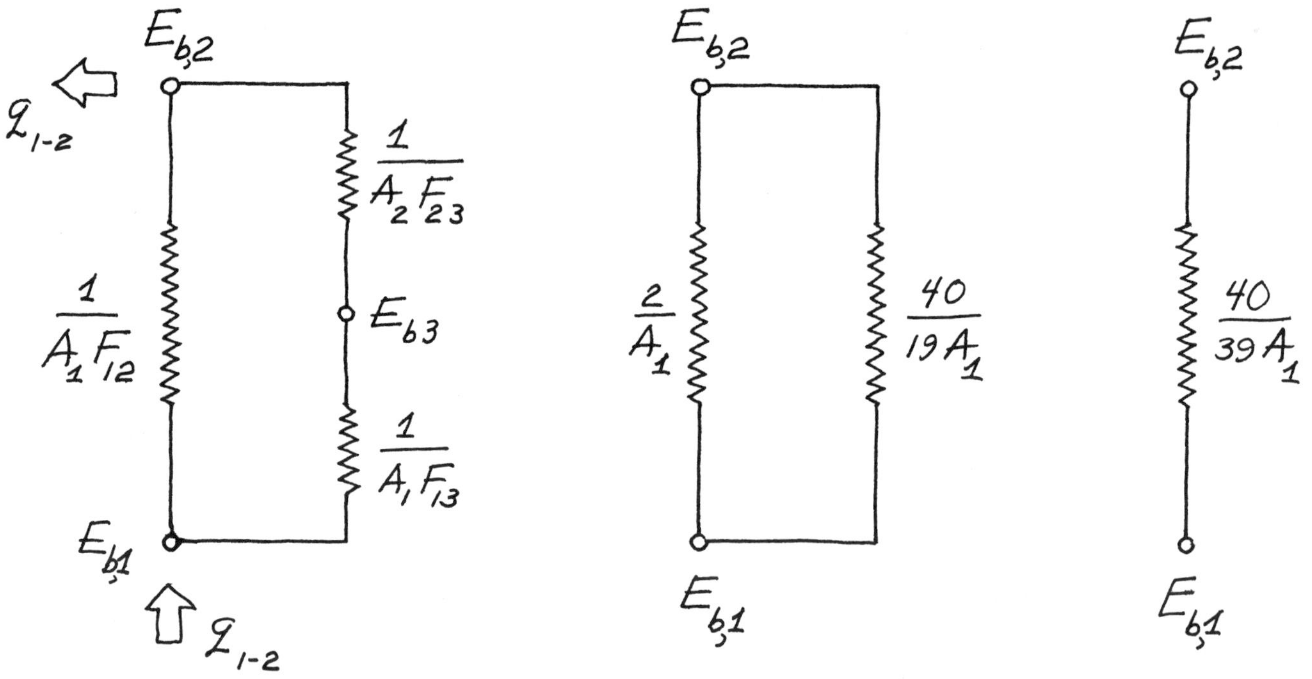

The net heat current from A_1 to A_2 is therefore

$$q_{1-2} = \frac{E_{b,1} - E_{b,2}}{\dfrac{40}{39 A_1}} = \frac{39}{40} \sigma A_1 (T_1^4 - T_2^4) \quad (5)$$

c) Let q_3 represent the heat current that passes through the node labeled $E_{b,3}$ on the first of the resistance networks. This current can be written in two ways

$$q_3 = \frac{E_{b,1} - E_{b,3}}{\dfrac{1}{A_1 F_{13}}} = \frac{E_{b,3} - E_{b,2}}{\dfrac{1}{A_2 F_{23}}} \quad (6)$$

Recalling that $A_2 = 10A_1$, $F_{13} = 1/2$ and $F_{23} = 19/20$, the second of the equations written above becomes

$$\frac{\sigma(T_1^4 - T_3^4)}{\dfrac{2}{A_1}} = \frac{\sigma(T_3^4 - T_2^4)}{\dfrac{20}{(10)A_1(19)}} \qquad (7)$$

The T_3 formula that comes out of this is

$$T_3 = \left(\frac{1}{20} T_1^4 + \frac{19}{20} T_2^4\right)^{1/4} \qquad (8)$$

Problem 10.16. The geometry is represented by the following parameters:

$$A_1 = \pi R_1^2 = \pi\,(0.1)^2\ m^2 = 0.031\ m^2$$

$$A_2 = \pi R_2^2 = \pi\,(0.05)^2\ m^2 = 0.00785\ m^2$$

$$A_3 = A_2 = 0.00785\ m^2$$

$$A_4 = 4\pi R_4^2 = 4\pi\,(0.3)^2\ m^2 = 1.131\ m^2$$

For the parallel-discs configuration, the fifth entry in Table 10.2 lists

$$F_{12} = \frac{1}{2}\left\{X - \left[X^2 - 4\left(\frac{x_2}{x_1}\right)^2\right]^{1/2}\right\}$$

in which

$$x_1 = \frac{R_1}{H} = \frac{0.1}{0.15} = 0.667$$

$$x_2 = \frac{R_2}{H} = \frac{0.05}{0.15} = 0.333$$

$$X = 1 + \frac{1 + x_2^2}{x_1^2} = 1 + \frac{1.111}{0.444} = 3.5$$

therefore

$$F_{12} = \frac{1}{2}\left\{3.5 - \left[3.5^2 - 4\left(\frac{1}{2}\right)^2\right]^{1/2}\right\} = 0.073$$

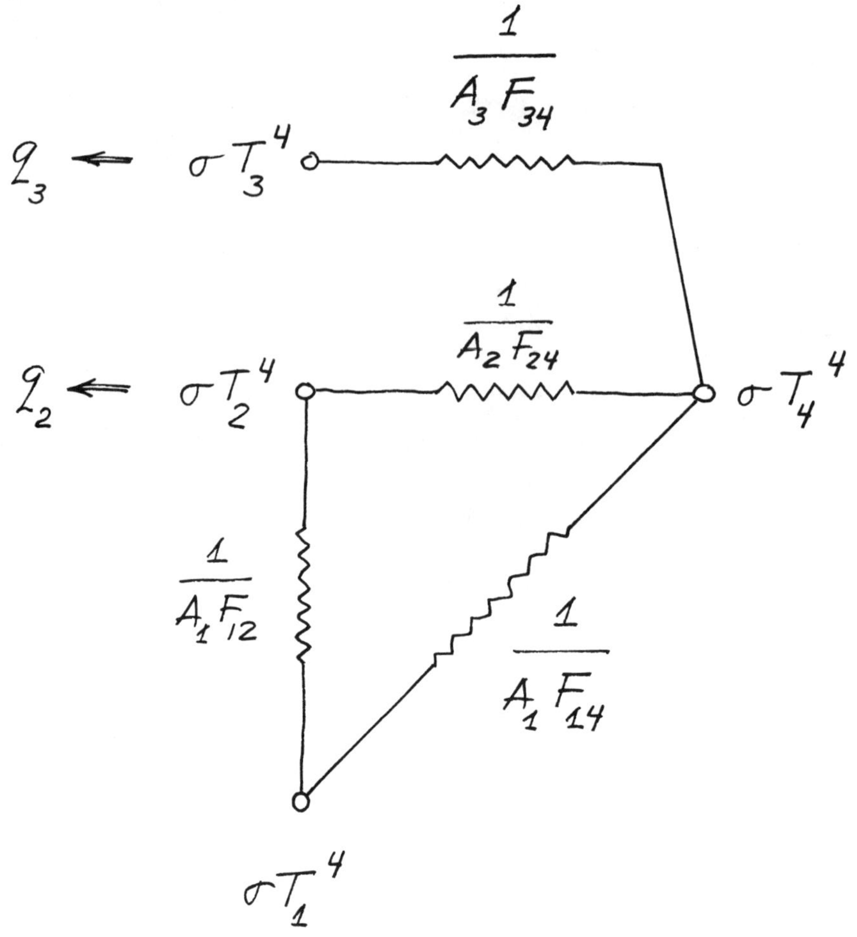

The other view factors are:

$F_{14} = 1 - F_{12} = 0.927$

$F_{34} = 1$

$F_{24} = 1 - F_{21}$, where $A_2F_{21} = A_1F_{12}$

$$F_{21} = \frac{A_1}{A_2} F_{12} = 0.292$$

therefore

$$F_{24} = 1 - 0.292 = 0.708$$

The total heat transfer rate <u>into</u> the bottom surface of the steak (A_2) is

$$q_2 = A_1F_{12}\,\sigma\left(T_1^4 - T_2^4\right) + A_2F_{24}\sigma\left(T_4^4 - T_2^4\right)$$
$$= \left[0.031 \times 0.073 \times 5.67 \times 10^{-8}\left(1073^4 - 298^4\right) + \right.$$
$$\left. + \; 0.00785 \times 0.708 \times 5.67 \times 10^{-8}\left(423^4 - 298^4\right)\right] \text{W}$$
$$= (169.1 + 7.6)\ \text{W} = 176.7\ \text{W}$$

The A_2-averaged heat flux is

$$q_2'' = \frac{q_2}{A_2} = \frac{176.7\ \text{W}}{0.00785\ \text{m}^2} = 2.25 \times 10^4\ \text{W/m}^2$$

The attached radiation network shows that the corresponding quantities for the top of the steak (A_3) are

$$q_3 = A_3F_{34}\,\sigma\left(T_4^4 - T_3^4\right)$$
$$= 0.00785\ \text{m}^2\ 5.67 \times 10^{-8}\ \frac{\text{W}}{\text{m}^2\ \text{K}^4}\left(423^4 - 298^4\right)\text{K}^4$$
$$= 10.74\ \text{W}$$
$$q_3'' = \frac{q_3}{A_3} = \frac{10.74\ \text{W}}{0.00785\ \text{m}^2} = 1.37 \times 10^3\ \text{W/m}^2$$

In conclusion, the top heat flux is much smaller than (only 6 percent of) the bottom heat flux, which is why it is a good idea to turn the steak over, again and again.

<u>Problem 10.17</u>. a) The space between the parallel plates is surrounded by a 3-surface enclosure, with the per-unit-length areas

$$A_1 = 1\ \text{cm}, \qquad A_2 = 4\ \text{cm}, \qquad A_3 = 1\ \text{cm}$$

The radiation heat transfer network is shown in the figure. The view factors are calculated as follows. We recognize first that the A_2 surface is the sum of the left and right faces,

$$A_2 = A_{left} + A_{right}$$

The view factor from A_1 to A_{right} is given by the second entry in Table 10.2, in which

$$x = \frac{2\ \text{cm}}{1\ \text{cm}} = 2$$

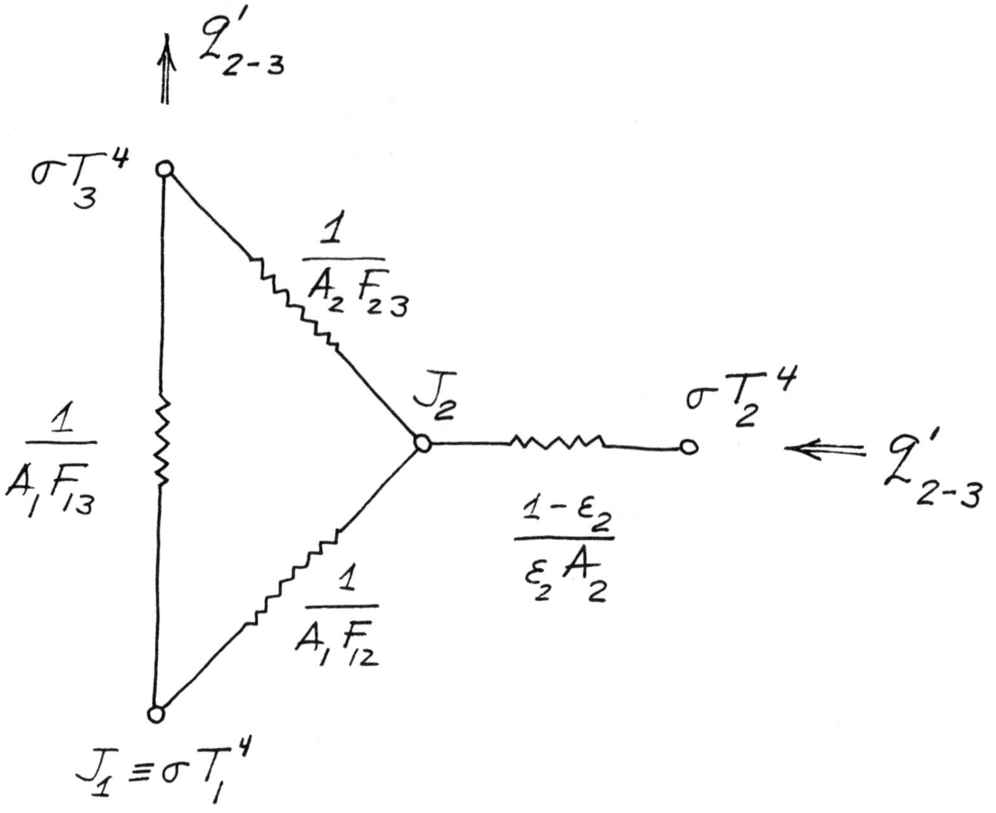

hence

$$F_{1\text{-right}} = \frac{1}{2}\left[1 + x - \left(1 + x^2\right)^{1/2}\right] = 0.382$$

Symmetry requires that $F_{1\text{-right}} = F_{1\text{-left}}$, therefore, based on the additivity property,

$$F_{12} = F_{1\text{-left}} + F_{1\text{-right}} = 2 \times 0.382$$

$$= 0.764$$

The enclosure relation provides the value of F_{13}:

$$F_{12} + F_{13} = 1$$

$$F_{13} = 1 - F_{12} = 0.236$$

Finally, we rely again on symmetry,

$$F_{23} = F_{21}$$

and on the reciprocity property,

$$A_1F_{12} = A_2F_{21}$$

in order to calculate F_{23}:

$$F_{23} = F_{21} = \frac{A_1}{A_2} F_{12} = \frac{1 \text{ cm}}{4 \text{ cm}} 0.764$$

$$= 0.191$$

The thermal resistances that participate in the network are

$$\frac{1}{A_1F_{13}} = \frac{1}{1 \text{ cm } 0.236} = 4.237/\text{cm}$$

$$\frac{1}{A_2F_{23}} = \frac{1}{4 \text{ cm } 0.191} = 1.309/\text{cm}$$

$$\frac{1}{A_1F_{12}} = \frac{1}{1 \text{ cm } 0.764} = 1.309/\text{cm}$$

$$\frac{1-\varepsilon_2}{\varepsilon_2 A_2} = \frac{1-0.7}{0.7 \times 4 \text{ cm}} = 0.107/\text{cm}$$

These can be combined step by step (graphically) until the total thermal resistance between σT_2^4 and σT_3^4 emerges as 1.059/cm + 0.107/cm = 1.166/cm. The net heat transfer rate is therefore

$$q'_{2-3} = \frac{\sigma T_2^4 - \sigma T_3^4}{1.166 \text{ cm}^{-1}}$$

$$= \frac{5.67 \times 10^{-8} \text{W}}{\text{m}^2 \text{ K}^4 \ 1.166 \text{ cm}^{-1}} \left[(150 + 273.15)^4 - (25 + 273.15)^4\right] \text{K}^4$$

$$= 11.75 \frac{\text{W}}{\text{m}}$$

b) Let q_1' be the heat current that passes (i.e. is conserved) through the node σT_1^4:

$$q'_1 = \frac{\sigma T_1^4 - \sigma T_3^4}{4.237/\text{cm}} = \frac{J_2 - \sigma T_3^4}{5.546/\text{cm}}$$

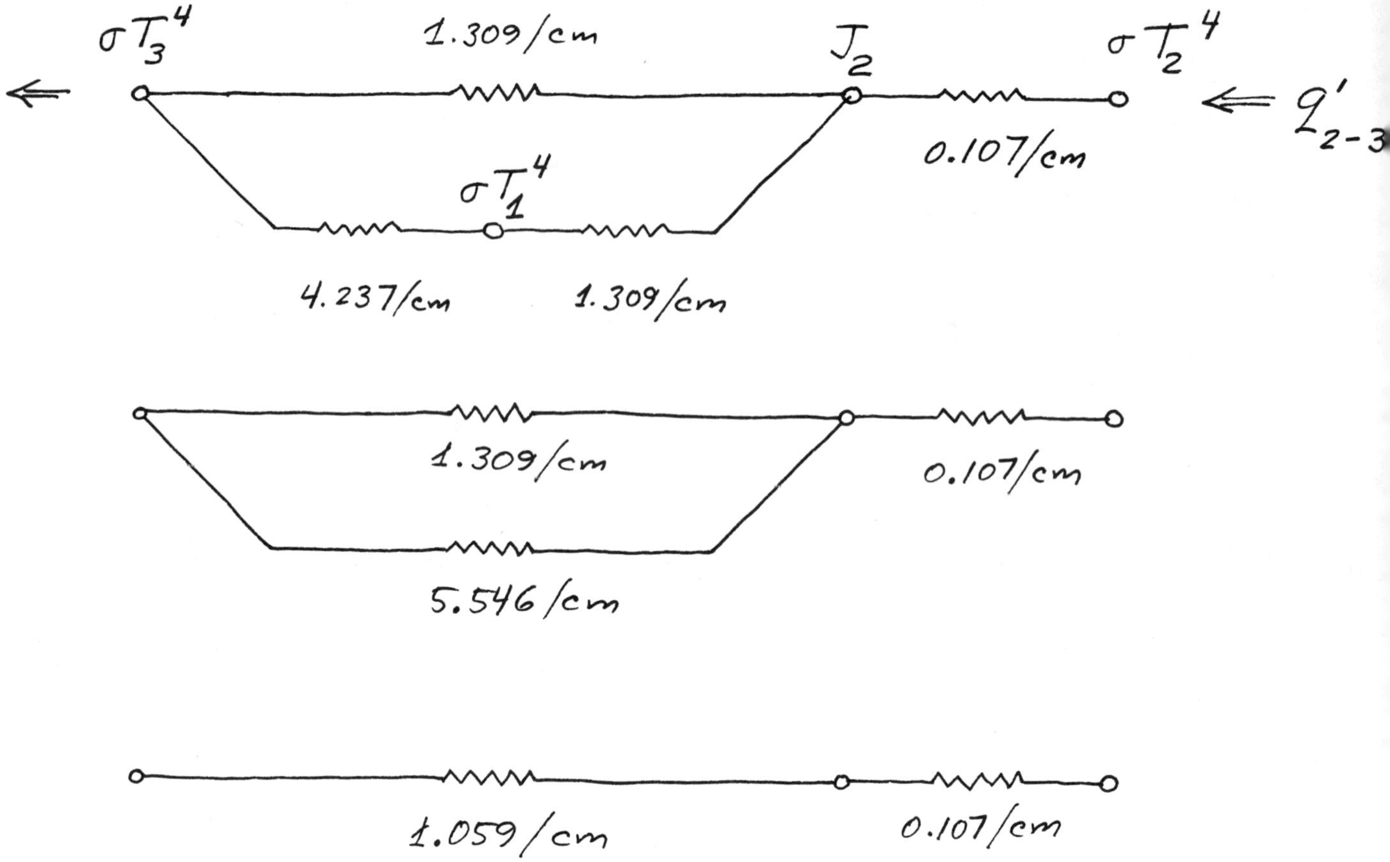

This provides one relation between the unknown $T_1{}^4$ and J_2/σ:

$$\frac{T_1^4 - T_3^4}{J_2/\sigma - T_3^4} = 0.764 \qquad \text{(i)}$$

The second relation follows from the internal resistance of surface A_2,

$$q'_{2\text{-}3} = \frac{\sigma T_2^4 - J_2}{0.107/\text{cm}} = 11.75\,\frac{\text{W}}{\text{m}}$$

which yields

$$\frac{J_2}{\sigma} = 2.98 \times 10^{10}\ \text{K}^4 \qquad \text{(ii)}$$

By comlbining (i) and (ii) we obtain

$$T_1^4 = 2.466 \times 10^{10}\ \text{K}^4$$

or

$$T_1 = 396.3\ \text{K} \cong 123°\text{C}$$

Problem 10.18. According to eq. (10.85), the local net heat flux from A_1 to A_2, at the location x, is

$$\frac{q_{1-2}}{A_1} = q''_{1-2} = \frac{\sigma}{a}\left(T_1^4 - T_2^4\right)$$

where a is shorthand for

$$a = \frac{1}{\varepsilon_1} + \frac{A_1}{A_2}\left(\frac{1}{\varepsilon_2} - 1\right)$$

$$= \frac{1}{0.6} + \left(\frac{15}{16}\right)^2\left(\frac{1}{0.6} - 1\right) = 2.25$$

Note that the local net heat flux q''_{1-2} is based on the surface A_1; consequently, the net heat transfer rate from A_1 to A_2 is

$$q_{1-2} = \int_0^L q''_{1-2}\, \pi D_1\, dx$$

$$= \pi D_1 \frac{\sigma}{a} \int_0^L \left\{T_1{}^4 - \left[T_2(0) + \left(\frac{dT_2}{dx}\right)x\right]^4\right\} dx$$

$$= \underbrace{\pi D_1 L}_{A_1} \quad \frac{\sigma}{a} T_1{}^4 \quad \underbrace{\int_0^1 \left\{1 - \left[b + c\xi\right]^4\right\} d\xi}_{0.544}$$

because, under the integral sign,

$$b = \frac{T_2(0)}{T_1} = \frac{(20 + 273.15)\text{ K}}{(220 + 273.15)\text{ K}} = 0.594$$

$$c = \left(\frac{dT_2}{dx}\right)\frac{L}{T_1} = 200\, \frac{\text{K}}{\text{km}}\, \frac{1\text{ km}}{493.15\text{ K}} = 0.406$$

Continuing the q_{1-2} calculation, we obtain

$$q_{1-2} = \pi D_1 L \frac{\sigma}{a} T_1{}^4 \times 0.544$$

$$= \pi\ 15\text{ cm}\ 10^3\text{m}\ 5.67 \times 10^{-8}\, \frac{\text{W}}{\text{m}^2\text{ K}^4}\, \frac{1}{2.25}\, (493.15)^4\text{ K}^4 \times 0.544$$

$$\cong 3.82 \times 10^5\text{W}$$

Problem 10.19. In accordance with eq. (10.86), the heat transfer from the inner surface of the cardboard box to the much smaller surface (A) of the ice slab is

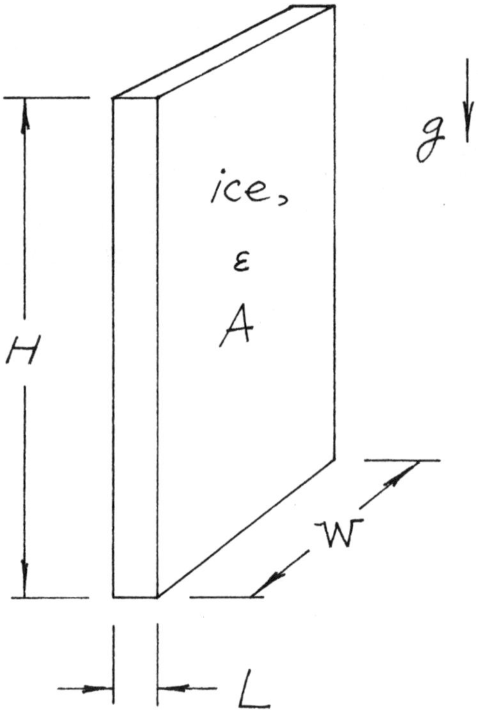

$$q = \sigma A \varepsilon \left(T_\infty^{\,4} - T_W^{\,4}\right)$$

where $\varepsilon \cong 0.92$ is the total hemispherical emissivity of smooth ice, Table 10.4. The area of the ice slab is

$$A = 2\,(HW + LW + LH)$$

$$= 1040.45 \text{ cm}^2$$

which corresponds to the total heat transfer rate

$$q = 5.67 \times 10^{-8} \frac{\text{W}}{\text{m}^2\,\text{K}^4}\, 1040.45 \text{ cm}^2\, 0.92\, (291.15^4 - 273.15^4)\ \text{K}^4$$

$$= 8.79 \text{ W}$$

To this heat input corresponds the following volumetric flowrate of meltwater:

$$\frac{\dot{m}}{\rho} = \frac{q}{\rho h_{sf}} = \frac{8.79 \text{ W}}{1 \frac{\text{g}}{\text{cm}^3}\, 333.4 \frac{\text{J}}{\text{g}}} = 0.0264 \frac{\text{cm}^3}{\text{s}}$$

$$= 94.9 \text{ cm}^3/\text{h}$$

This flowrate can be compared with the natural convection melting rate, which is expressed by eq. (d) of Project 7.1,

$$\frac{\dot{m}}{\rho} = \frac{2W}{\rho} \frac{k}{h_{sf}} (T_\infty - T_W)\, \overline{Nu}_H \qquad \text{(d)}$$

Evaluating the air properties at 10°C, we have

$$k = 0.58 \frac{W}{m\,K} \qquad \frac{g\beta}{\alpha\nu} = \frac{125}{cm^3\,K}$$

$$Ra_H = \frac{125}{cm^3\,K}(38.3)^3 cm^3\ 18°C = 1.264 \times 10^8$$

and, according to eq. (7.62'),

$$\overline{Nu_H} = 0.68 + 0.515\,Ra_H^{1/4} = 55.29$$

In conclusion, the melting rate (d) caused by natural convection over the two large vertical sides is

$$\frac{\dot{m}}{\rho} = \frac{2 \times 25.5\ cm}{1 \frac{g}{cm^3}}\ 0.58 \frac{W}{m\,K} \frac{g}{333.4\ J}\ 18°C\ 55.29$$

$$= 0.88 \frac{cm^3}{s} = 3179 \frac{cm^3}{h}$$

The melting rate due to radiation alone (94.9 cm^3/h) represents only 3 percent of the melting rate due to laminar natural convection.

Problem 10.20. The side of the ash hopper (T_1), the pit wall (T_2) and the ambient (T_3) form the two-dimensional rectangular enclosure illustrated in the problem statement. The pit wall is a reradiating (adiabatic) surface, and the ambient can be treated as black. We know the following:

$T_1 = 300°C$ $\qquad T_2 = ?$ $\qquad T_3 = 40°C$

$L_1 = 5m$ $\qquad L_2 = 7.5m$ $\qquad L_3 = 2.5m$

$\varepsilon_1 = 0.6$

Unknown is T_2. With reference to the network shown in the attached figure, we write the following radiation-current expressions:

$$q_1' = \frac{\varepsilon_1 L_1}{1 - \varepsilon_1}\left(\sigma T_1^4 - J_1\right) \tag{1}$$

$$q_1' = L_1 F_{12}\left(J_1 - \sigma T_2^4\right) + L_1 F_{13}\left(J_1 - \sigma T_3^4\right) \tag{2}$$

$$q_2' = L_1 F_{12}\left(J_1 - \sigma T_2^4\right) \tag{3}$$

$$q_2' = L_3 F_{32}\left(\sigma T_2^4 - \sigma T_3^4\right) \tag{4}$$

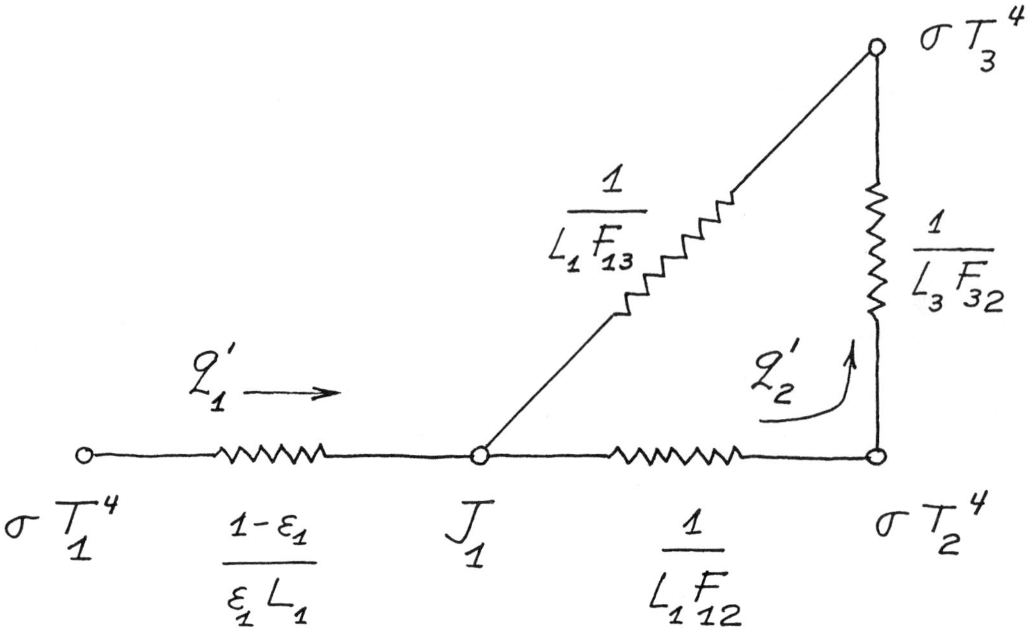

By eliminating q_2' between the last two equations we obtain

$$J_1 = \sigma T_2^4 + \frac{L_3}{L_1}\frac{F_{32}}{F_{12}}\sigma\left(T_2^4 - T_3^4\right) \tag{5}$$

for which the view factor can be calculated sequentially, by starting with the second formula listed in Table 10.2:

$$F_{13} = \frac{1}{2}\left[1 + x - \left(1 + x2\right)^{1/2}\right] \qquad \left(\text{note: } x = \frac{2.5}{5} = 0.5\right)$$

$$= 0.191$$

$$F_{12} = 1 - F_{13} = 0.809$$

$$L_1 F_{13} = L_3 F_{31}$$

$$F_{31} = F_{13}\frac{L_1}{L_3} = 0.191\,\frac{5}{2.5} = 0.382$$

$$F_{32} = 1 - F_{31} = 0.618$$

Equation (5) becomes:

$$\frac{J_1}{\sigma} = 1.382\ T_2^4 - 0.382\ T_3^4 \qquad (6)$$

A second relationship between T_2^4 and J_1/σ is obtained by eliminating q_1' between eqs. (1) and (2):

$$\frac{J_1}{\sigma} = 0.6\ T_1^4 + 0.323\ T_2^4 + 0.0764\ T_3^4 \qquad (7)$$

Finally, by eliminating J_1/σ between eqs. (6) and (7) we arrive at

$$T_2^4 = 0.567\ T_1^4 + 0.433\ T_3^4$$

which yields numerically

$$T_2 = [0.567\ (300 + 273.15)^4 + 0.433\ (40 + 273.15)^4]^{1/4} K$$

$$= 506\ K = 232°C$$

Problem 10.21. The heat transfer interactions proceed according to the network shown in the attached sketch. The hopper surface temperature T_1 is fixed, therefore the heat transfer by natural convection from T_1 to the T_3-temperature air that fills the core of the hopper-pit space has no influence on T_2. The latter depends on the radiative properties of the (T_1,T_2,T_3) enclosure, and on the natural convection heat transfer from T_2 to T_3, through the heat transfer conductance h_2L_2. The temperature T_2 "floats" because the pit wall is adiabatic.

The analysis consists of writing the following expressions for the heat currents (W/m) shown on the figure:

$$q_1' = \frac{\varepsilon_1 L_1}{1 - \varepsilon_1}\left(\sigma T_1^4 - J_1\right) \qquad (1)$$

$$q_1' = L_1 F_{12}\left(J_1 - \sigma T_2^4\right) + L_1 F_{13}\left(J_1 - \sigma T_3^4\right) \qquad (2)$$

$$q_2' = L_1 F_{12}\left(J_1 - \sigma T_2^4\right) \qquad (3)$$

$$q_2' = L_3 F_{32}\, \sigma \left(T_2^4 - T_3^4\right) + h_2 L_2 (T_2 - T_3) \qquad (4)$$

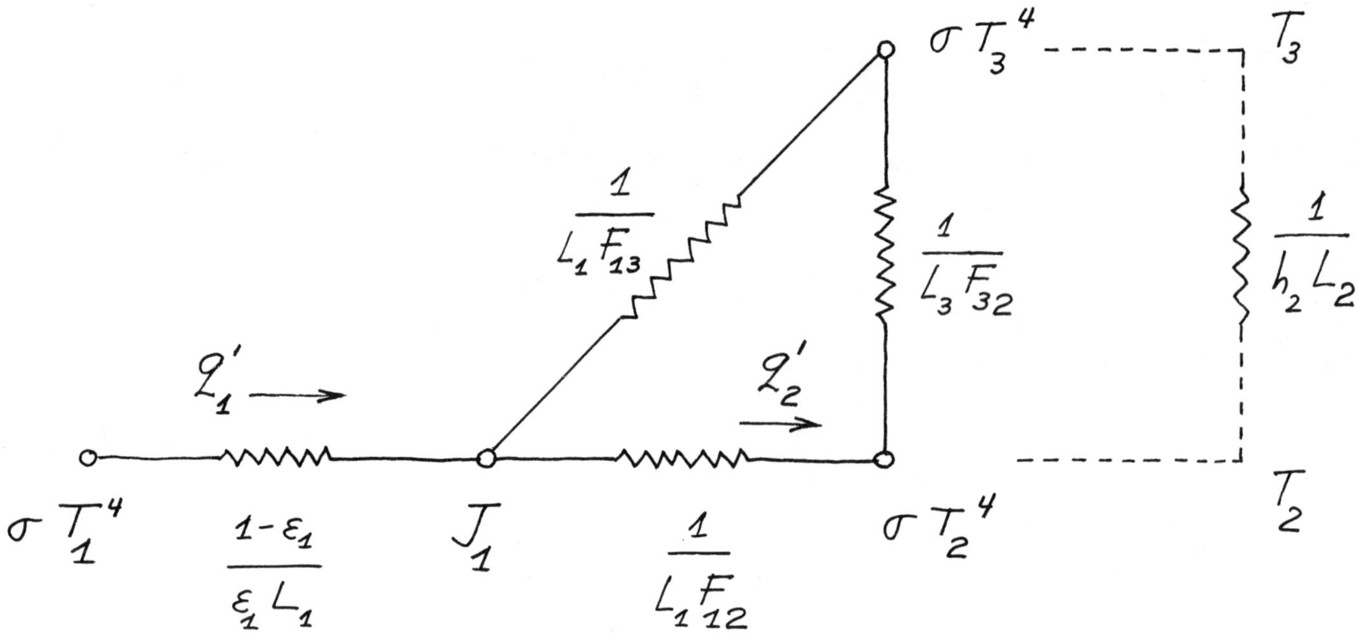

Next, we eliminate q_1' between eqs. (1) and (2), for which the necessary view factors are:

$$F_{13} = \frac{1}{2}\left[1 + x - (1 + x^2)^{1/2}\right] \qquad \text{(note: } x = \frac{2.5}{5} = 0.5\text{, Table 10.2, second entry)}$$

$$= 0.191$$

$$F_{12} = 1 - F_{13} = 0.809$$

$$L_1 F_{13} = L_3 F_{31}$$

$$F_{31} = F_{13} \frac{L_1}{L_3} = 0.191 \frac{5}{2.5} = 0.382$$

$$F_{32} = 1 - F_{31} = 0.618$$

The result of combining eqs. (1) and (2) is

$$\frac{J_1}{\sigma} = 0.6\, T_1^4 + 0.324\, T_2^4 + 0.0764\, T_3^4 \qquad (5)$$

A second equation between J_1/σ and T_2 follows from eqs. (3) and (4), after eliminating q_2':

$$T_2^4 - T_3^4 + C_1 (T_2 - T_3) - C_2 \left(\frac{J_1}{\sigma} - T_2^4 \right) = 0 \qquad (6)$$

with

$$C_1 = \frac{h_2 L_2}{F_{32} L_3 \sigma} \qquad (7)$$

$$C_2 = \frac{L_1 F_{12}}{L_3 F_{32}} = \frac{5}{2.5} \frac{0.809}{0.618} = 2.62 \qquad (8)$$

An order of magnitude estimate for C_1 can be obtained by calculating h_2 as the height averaged heat transfer coefficient for natural convection over the 5m-tall wall of the pit. Assuming that the "core" air that descends slowly into the 2.5m-wide space is at the ambient temperature T_∞ = 40°C, and that the eventual pit wall temperature is lower than in the preceding problem (i.e. lower than when natural convection cooling is neglected), say, $T_2 \cong$ 160°C, the temperature difference across the natural convection boundary layer is

$$\Delta T \cong (160 - 40)°\text{C} = 120°\text{C}$$

and the film temperature is

$$T_{film} = \frac{T_\infty + T_2}{2} = \frac{40 + 160}{2} °\text{C} = 100°\text{C}$$

With this information, we calculate in order

$$Ra = \left(\frac{g\beta}{\alpha\nu} \right)_{\substack{\text{air} \\ 100°\text{C}}} H^3 \Delta T = \frac{34.8}{\text{cm}^3\text{K}} (500 \text{ cm})^3 \ 120 \text{ K}$$

$$= 5.22 \times 10^{11} \qquad \text{(turbulent flow)}$$

$$\overline{Nu} = \left(0.825 + \frac{0.387 \, Ra^{1/6}}{1.181} \right)^2 \qquad \text{(see eq. 7.70 with Pr = 0.72)}$$

$$= 913.2$$

$$h_2 = \overline{Nu} \frac{k \text{ (air at 100°C)}}{H}$$

$$= 913.2 \frac{0.032 \text{ W}}{\text{m K}} \frac{1}{5\text{m}} = 5.84 \frac{\text{W}}{\text{m}^2 \text{ K}}$$

$$C_1 = \frac{h_2 L_2}{F_{32} L_3 \sigma} = \frac{5.84 \text{ W}}{\text{m}^2 \text{ K}} \frac{(5 + 2.5) \text{ m}}{2.5\text{m}} \frac{1}{0.618} \frac{\text{m}^2 \text{ K}^4}{5.67 \times 10^{-8} \text{W}}$$

$$= 5 \times 10^8 \text{ K}^3$$

Note that in the calculation of C_1 we have assumed that h_2 (vertical natural convection boundary layer) applies over the entire wall of the pit, including the bottom surface. This approximation is justified, on the background of the more drastic assumption that the entire pit wall is at one temperature, T_2.

In summary, the calculation of the equilibrium T_2 reduces to eliminating J_1/σ between eqs. (5) and (6), and solving the resulting equation numerically. The result is

$$T_2 \cong 450 \text{ K} \cong 176°\text{C},$$

showing that the effect of natural convection lowers the pit wall temperature by some 56°C (from T_2 = 232°C in the preceding problem).

The calculated pit wall temperature, T_2 = 176°C, is a bit higher than the value assumed in the calculation of ΔT, T_{film} and Ra (namely T_2 = 160°C). We could, in principle, repeat the steps ΔT, T_{film}, Ra,...T_2, by starting with T_2 = 176°C as a better guess for the wall temperature. This refinement, however, is not justified because of the approximate character of the enclosure model, and the assumption that h_2 has the same value over the entire pit surface.

Problem 10.22. The resistance network that represents the enclosure model described in the problem statement is shown in the attached figure. The radiosities J_i and J_e refer to the internal and external surfaces of the hopper wall. The total heat transfer rate from the ash-pile surface (T_1) to the ambient (T_3) is

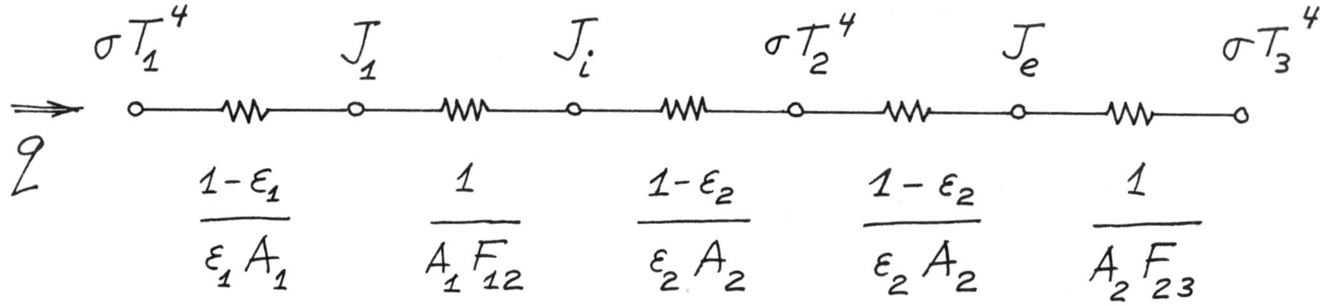

$$q = \frac{\sigma T_1^4 - \sigma T_3^4}{\frac{1-\varepsilon_1}{\varepsilon_1 A_1} + \frac{1}{A_1 F_{12}} + 2\frac{1-\varepsilon_2}{\varepsilon_2 A_2} + \frac{1}{A_2 F_{23}}}$$

$$= \frac{\sigma A_1 \left(T_1^4 - T_3^4\right)}{\frac{1-0.9}{0.9} + 1 + 2\frac{1-0.6}{0.6}\frac{11\ m2}{27\ m2} + \frac{11\ m2}{27\ m2}}$$

$$= 0.485\ \sigma A_1 (T_1{}^4 - T_3{}^4) \qquad (1)$$

in which we noted that $F_{12} = F_{23} = 1$. In the steady state, the heat transfer rate q is sustained by the flow of hot ash through the inner chamber. The new ash enters (i.e. lands on top of the ash being cooled) with the initial temperature T_{in} = 900°C, and exits (it is buried) with the temperature T_1:

$$q = (\dot{m}c)_{ash} (T_{in} - T_1) \qquad (2)$$

By eliminating q between eqs. (1) and (2) we obtain

$$T_1{}^4 - T_3{}^4 = (T_{in} - T_1)\ C_1 \qquad (3)$$

in which

$$C_1 = \frac{\dot{m}c}{0.485\ \sigma A_1}$$

$$= \frac{550\ kg}{3600\ s}\ 1\ \frac{10^3\ J}{kg\ K}\ \frac{1}{0.485}\ \frac{m2\ K^4}{5.67 \times 10^{-8} W}\ \frac{1}{11\ m2}$$

$$= 5.05 \times 10^8\ K^3$$

Equation (3) can be solved for T_1 iteratively, by first rearranging the equation as

$$T_{1,new} = \left[T_3^4 + C_1 \left(T_{in} - T_{1,old}\right)\right]^{1/4}$$

and substituting T_{in} = 1173.15 K and T_3 = 313.15 K. The solution converges rapidly ($T_{1,new} = T_{1,old}$), and the result is

$$T_1 = 704.5\ K = 431.3°C$$

The total heat transfer rate is, cf. eq. (2),

$$q = \dot{m}c\,(T_{in} - T_1)$$

$$= \frac{550\ \text{kg}}{3600\ \text{s}}\ 1\ \frac{10^3 \text{J}}{\text{kg K}}\ (900°\text{C} - 431.3°\text{C})$$

$$= 71600\ \text{W}$$

Finally, in order to calculate T_2 we recognize an alternative way of expressing the total heat transfer rate,

$$q = \frac{\sigma T_1^4 - \sigma T_2^4}{\frac{1 - \varepsilon_1}{\varepsilon_1 A_1} + \frac{1}{A_1} + \frac{1 - \varepsilon_2}{\varepsilon_2 A_2}}$$

$$= \frac{\sigma A_1 \left(T_1^4 - T_2^4\right)}{\frac{0.1}{0.9} + 1 + \frac{0.4}{0.6}\frac{11}{27}}$$

$$= 0.723\ \sigma A_1 \left(T_1^4 - T_2^4\right) \qquad (4)$$

By eliminating q between eqs. (1) and (4) we arrive at

$$\frac{T_1^4 - T_2^4}{T_1^4 - T_3^4} = 0.671$$

$$T_2^4 = 0.329\ T_1^4 + 0.671\ T_3^4$$

$$T_2 = \left[0.329\ (704.5)^4 + 0.671\ (40 + 273.15)^4\right]^{1/4}\ \text{K}$$

$$= 543.9\ \text{K} = 271°\text{C}$$

Problem 10.23. The heat current absorbed by the earth is

$$q_{in} = \alpha\, I A_C$$

where $A_C = \pi R^2$ is the earth's cross-sectional area. The heat current that flows from the earth to the surroundings is, cf. eq. (10.86)

$$q_{out} = \sigma A \varepsilon \left(T^4 - T_\infty^4\right)$$

$$\cong \sigma\, \varepsilon\, T^4$$

where we recognized that $T_\infty^4 << T^4$. The area A is $4\pi R^2$. In the steady state, $q_{in} = q_{out}$ leads to an equation for the average temperature of the earth's surface:

$$T = \left(\frac{\alpha I A_c}{\sigma A \varepsilon}\right)^{1/4} = \left(\frac{\alpha I}{4\varepsilon\sigma}\right)^{1/4}$$

$$= \left(\frac{0.7 \times 1360 \frac{W}{m^2}}{4 \times 0.95 \times 5.67 \times 10^{-8} \frac{W}{m^2\,K^4}}\right)^{1/4}$$

$$= 257.9\ K \cong -15°C$$

The actual average temperature is 30°C higher than this estimate because the atmosphere partially impedes (intercepts) the heat current q_{out} calculated above. This phenomenon is better known as the atmospheric "greenhouse" effect.

Problem 10.24. We evaluate, in order, the emissivity of CO_2 (the lone participating gas in the mixture), the absorptivity of CO_2 with respect to radiation arriving from the wall, and the net heat transfer rate from the gas to the wall.

a) Emissivity:

$P_c = 0.3 \times 2$ atm $= 0.6$ atm

$L_e = 0.6D = 0.6$m $= 1.97$ ft (Table 10.5)

$P_c L_e = 1.18$ atm·ft

Figure 10.26 yields $\varepsilon_c \cong 0.15$ for $T_g = 1273$ K and $P_c\, L_e = 1.18$ atm·ft

Figure 10.27 yields $C_c \cong 1.09$ for $P = 2$ atm and $P_c\, L_e = 1.18$ atm·ft

$$\varepsilon_g = 0.15 \times 1.09 = 0.164 \qquad (10.106)$$

b) Absorptivity:

$$P_c\, L_e \frac{T_s}{T_g} = 1.18 \text{ atm·ft} \frac{473}{1273} = 0.44 \text{ atm·ft}$$

Figure 10.26 yields $\varepsilon_c = 0.107$ for $T_g \rightarrow T_s = 473$ K and

$P_c\, L_e \rightarrow 0.44$ atm·ft

Figure 10.27 yields $C_c = 1.14$ for $P = 2$ atm and $P_c L_e \to 0.44$ atm·ft

$$\alpha_c = 0.107 \times 1.14 \left(\frac{1273}{473}\right)^{0.65} = 0.232 \qquad (10.108)$$

c) Net heat transfer rate:

$A_s = \pi D^2 = \pi (1m)^2 = 3.14\ m^2$

$$q_{g\to s} = \varepsilon_g \sigma T_g^4 A_s$$

$$= 0.164 \times 5.67 \times 10^{-8} \frac{W}{m^2 K^4} (1273\ K)^4\ 3.14\ m^2$$

$$= 76.7\ kW$$

$$q_{s\to g} = \alpha_g \sigma T_s^4 A_s$$

$$= 0.232 \times 5.67 \times 10^{-8} \frac{W}{m^2 K^4} (473\ K)^4\ 3.14\ m^2$$

$$= 2.1\ kW$$

$$\underset{(net)}{q_{g\text{-}s}} = q_{g\to s} - q_{s\to g}$$

$$= (76.7 - 2.1)\ kW = 74.6\ kW$$

Problem 10.25. Water vapor is the only radiating gas in the mixture. We calculate in order the gas emissivity, gas absorptivity, and net gas-wall heat transfer:

$L_e = 0.95\ D = 1.25$ ft (Table 10.5)

$P_w = 0.794\ P = 0.794$ atm

$P_w L_e = 0.99$ atm·ft

The abscissa of Fig. 10.28 does not extend all the way to $T_g = 3000$ K. We extend linearly the $P_w L_e \cong 1$ atm·ft curve to $T_g = 3000$ K, and find

$$\varepsilon_w \cong 0.057$$

On the abscissa of Fig. 10.29 we use

$$\frac{1}{2}(P + P_w) = \frac{1}{2}(1 + 0.794)\ \text{atm} \cong 0.9\ \text{atm}$$

and after locating the $P_w L_e \cong 1$ atm·ft line we find

$$C_w \cong 1.34$$

The emissivity of the gas is

$$\varepsilon_g \cong 0.057 \times 1.34 = 0.076$$

In the calculation of α_g, we are guided by eq. (10.111):

$$P_w L_e \frac{T_s}{T_g} = 0.99\ \text{atm·ft}\ \frac{600\ \text{K}}{3000\ \text{K}} \cong 0.2\ \text{atm·ft}$$

$\varepsilon_w \cong 0.135$ (Fig. 10.28, at 600 K and 0.2 atm·ft)

$C_w \cong 1.46$ (Fig. 10.28, at 0.9 atm on the abscissa, and 0.2 atm·ft on the curve)

$$\alpha_g = 0.135 \times 1.46 \left(\frac{3000}{600}\right)^{0.45} \qquad (10.111)$$

$$= 0.407$$

The net heat transfer rate into the cylinder wall (per unit axial length) is calculated in three steps:

$$q'_{g\to s} = \varepsilon_g\, \sigma\, T_g^4\, \pi D$$

$$= 0.076 \times 5.67 \times 10^{-8}\ \frac{\text{W}}{\text{m}^2\ \text{K}^4}\ (3000\ \text{K})^4\ \pi\ 0.4\text{m}$$

$$= 438.6\ \frac{\text{kW}}{\text{m}}$$

$$q'_{s\to g} = \alpha_g\, \sigma\, T_s^4\, \pi D$$

$$= 0.407 \times 5.67 \times 10^{-8}\ \frac{\text{W}}{\text{m}^2\ \text{K}^4}\ (600\ \text{K})^4\ \pi\ 0.4\text{m}$$

$$= 3.8\ \frac{\text{kW}}{\text{m}}$$

$$\dot{q}'_{g\text{-}s} = \dot{q}'_{g\rightarrow s} - \dot{q}'_{s\rightarrow g} = 434.8 \frac{kW}{m}$$
(net)

Problem 10.26. There are two radiating gases in the mixture, carbon dioxide and water vapor. Their mole fractions and partial pressures are:

$$x_c = \frac{1}{1 + 2 + 7.52} = 0.095, \qquad P_c = 0.095 \text{ atm}$$

$$x_w = \frac{2}{1 + 2 + 7.52} = 0.19, \qquad P_w = 0.19 \text{ atm}$$

The mixture emissivity ε_g can be calculated based on eq. (10.112), in the following sequence:

CO_2 alone:

$L_e = 0.95\ D = 0.19\text{m} = 0.623\ \text{ft}$

$P_c\ L_e = 0.06\ \text{atm·ft}$

$\varepsilon_c \cong 0.0212$ (Fig. 10.26, $T_g = 2320$ K, $P_c\ L_e = 0.06$ atm·ft)

$C_c = 1$ (Fig. 10.27, $P = 1$ atm)

$\varepsilon_c = 0.0212 \times 1 = 0.0212$

H_2O alone:

$P_w\ L_e = 0.19\ \text{atm}\ 0.623\ \text{ft} = 0.118\ \text{atm·ft}$

$\varepsilon_w \cong 0.1$ (Fig. 10.28, $T_g = 2320$ K, $P_w\ L_e \cong 0.12$ atm·ft)

$$\frac{1}{2}(P + P_w) = \frac{1}{2}(1 + 0.19)\ \text{atm} \cong 0.6\ \text{atm}$$

$C_w \cong 1.14$ (Fig. 10.29)

$\varepsilon_w = 0.1 \times 1.14 = 0.114$

CO_2 and H_2O together:

$L_e\ (P_c + P_w) = 0.623\ \text{ft}\ (0.095 + 0.19)\ \text{atm} \cong 0.18\ \text{ft·atm}$

$$\frac{P_w}{P_c + P_w} = \frac{0.19}{0.095 + 0.19} = 0.67$$

$T_g = 2320 \text{ K} = 2047°\text{C}$

$\Delta\varepsilon \cong 0.001$ (Fig. 10.30, the third graph)

In conclusion, we learn that the gas emission is dominated by the contribution due to water vapor:

$$\varepsilon_g = \varepsilon_c + \varepsilon_w - \Delta\varepsilon$$

$$\cong 0.0212 + 0.114 - 0.001 = 0.134$$

If we neglect the radiation emitted by the wall (and absorbed by the gas), the net heat transfer rate into the wall is

$$q'_{\substack{g\text{-}s \\ net}} = q'_{g\to s} - q'_{s\to g}$$

$$\cong q'_{g\to s} = \varepsilon_g \, \sigma \, T_g^4 \, \pi \, D$$

$$= 0.134 \times 5.67 \times 10^{-8} \frac{W}{m^2 \, K^4} (2320 \text{ K})^4 \, \pi \, 0.2\text{m}$$

$$= 138 \frac{\text{kW}}{\text{m}}$$

Problem 10.27. There are two radiating gases in the mixture, carbon dioxide and water vapor. Their mole fractions and partial pressures are

$$x_c = \frac{1}{1 + 2 + 6 + 30.08} = \frac{1}{39.08} = 0.026, \qquad P_c = 0.026 \text{ atm}$$

$$x_w = \frac{2}{39.08} = 0.051 \qquad P_w = 0.051 \text{ atm}$$

The mixture emissivity ε_g is furnished by eq. (10.112), in three steps:

CO_2 alone:

$$L_e = 1.8s = 1.8 \times 0.5\text{m} = 0.9\text{m} = 2.953 \text{ ft}$$

$P_c L_e = 0.026 \text{ atm } 2.953 \text{ ft} = 0.077 \text{ atm·ft}$

$\varepsilon_c \cong 0.07$ (Fig. 10.26, $T_g = 700$ K, $P_c L_e \cong 0.08$ atm·ft)

$C_c = 1$ (Fig. 10.27, P = 1 atm)

$\varepsilon_c = 0.07 \times 1 = 0.07$

H_2O alone:

$P_w L_e = 0.051 \text{ atm } 2.953 \text{ ft} = 0.15 \text{ atm·ft}$

$\varepsilon_w \cong 0.105$ (Fig. 10.28, $T_g = 700$ K, $P_w L_e = 0.15$ atm·ft)

$\frac{1}{2}(P + P_w) = \frac{1}{2}(1 + 0.051) \text{ atm} \cong 0.53 \text{ atm}$

$C_w \cong 1.04$ (Fig. 10.29)

$\varepsilon_w = 0.105 \times 1.04 = 0.109$

$CO_2 + H_2O$ together:

$L_e (P_c + P_w) = 2.953 \text{ ft } (0.026 + 0.051) \text{ atm} \cong 0.23 \text{ ft·atm}$

$$\frac{P_w}{P_c + P_w} = \frac{0.051}{0.026 + 0.051} = 0.67$$

$T_g = 700 \text{ K} = 427°\text{C}$

$\Delta\varepsilon = 0.001$, or less (Fig. 10.30, the second graph)

$$\varepsilon_g = \varepsilon_c + \varepsilon_w - \Delta\varepsilon \tag{10.112}$$

$$= 0.07 + 0.109 - 0.001 = 0.178$$

The absorptivity of CO_2 alone is furnished by eq. (10.108):

$$P_c L_e \frac{T_s}{T_g} = 0.077 \text{ atm·ft} \frac{460 \text{ K}}{700 \text{ K}} = 0.051 \text{ atm·ft}$$

$\varepsilon_c = 0.068$ (Fig. 10.26, 460 K, 0.05 atm·ft)

$C_c = 1$ (Fig. 10.27, P = 1 atm)

$$\alpha_c = 0.068 \times 1 \times \left(\frac{700\ K}{460\ K}\right)^{0.65} = 0.089$$

The absorptivity of H_2O alone is provided by eq. (10.111):

$$P_w L_e \frac{T_s}{T_g} = 0.15\ \text{atm·ft}\ \frac{460\ K}{700\ K} \cong 0.1\ \text{atm·ft}$$

$\varepsilon_w \cong 0.135$ (Fig. 10.28, 460 K, 0.1 atm·ft)

$C_w \cong 1.03$ (Fig. 10.29)

$$\alpha_w = 0.135 \times 1.03 \times \left(\frac{700\ K}{460\ K}\right)^{0.45} = 0.168$$

The absorptivity of the gas mixture is

$$\alpha_g = \alpha_c + \alpha_w = 0.257$$

The net heat transfer rate received by the two parallel walls is calculated in the following three steps:

$$A = 20\ m^2 + 20\ m^2 = 40\ m^2$$

$$q_{g\to s} = \varepsilon_g\, \sigma\, T_g^4\, A$$

$$= 0.178 \times 5.67 \times 10^{-8} \frac{W}{m^2\ K^4} (700\ K)^4\ 40\ m^2$$

$$= 96.93\ kW$$

$$q_{s\to g} = \alpha_g\, \sigma\, T_s^4\, A$$

$$= 0.257 \times 5.67 \times 10^{-8} \frac{W}{m^2\ K^4} (460\ K)^4\ 40\ m^2$$

$$= 26.10\ kW$$

$$\underset{(net)}{q_{g\text{-}s}} = q_{g\to s} - q_{s\to g} = 70.8\ kW$$

Problem 10.28. The solution is listed in general terms in eqs. (10.116) - (10.117). This problem is made simpler by

$$A_1 = A_2 = A$$

$$F_{12} = 1, \quad F_{1g} = 1, \quad F_{2g} = 1$$

$$\varepsilon_1 = \varepsilon_2 = \varepsilon_g = 0.5$$

$$\alpha_g = \varepsilon_g = 0.5$$

$$\tau_g = 1 - \alpha_g = 0.5$$

By carefully substituting in eqs. (10.116)-(10.117) we obtain

$$R_\Delta = \frac{4}{3\,A}$$

$$q_{1-2}(\text{gas}) = \frac{3}{10}\,\sigma A\left(T_1^4 - T_2^4\right)$$

When the space between the two plates is occupied by vacuum, the resistance R_Δ is replaced by $1/F_{12}A$, where $F_{12} = 1$. The appropriate network can be seen in Fig. 10.21. The heat transfer rate turns out to be

$$q_{1-2}\ (\text{vacuum}) = \frac{1}{3}\,\sigma A\left(T_1^4 - T_2^4\right)$$

In conclusion, the presence of the gray gas induces a 10 percent drop in the heat transfer rate across vacuum,

$$\frac{q_{1-2}\ (\text{gas})}{q_{1-2}\ (\text{vacuum})} = \frac{3/10}{1/3} = 0.9$$

Problem 10.29. In order to estimate the emissivity of water vapor, we must assume that its temperature is known. A reasonable guess is

$$T_g = \frac{1}{2}\left(T_1 + T_2\right) = 850\ \text{K}$$

and the calculation consists of the following steps

$$L_e = 1.8\ s = 1.8 \times 10\ \text{cm} = 0.591\ \text{ft} \qquad \text{(Table 10.5)}$$

$$P_w = P = 1\ \text{atm}$$

$$P_w\,L_e = 0.591\ \text{atm·ft}$$

$\varepsilon_w \cong 0.195$ (Fig. 10.28, at 850 K and $\cong$ 0.6 atm·ft)

$$\frac{1}{2}(P + P_w) = \frac{1}{2}(1 + 1) \text{ atm} = 1 \text{ atm}$$

$C_w \cong 1.45$ (Fig. 10.29, at 1 atm and $\cong$ 0.6 atm·ft)

$$\varepsilon_g = 0.195 \times 1.45 = 0.283$$

According to the gray gas model, we assume that

$$\alpha_g = \varepsilon_g = 0.283$$

which means also that

$$\tau_g = 1 - 0.283 = 0.717$$

The goodness of the gray-gas model $\alpha_g = 0.283$ can be evaluated by calculating α_g in accordance with the method of subsection 10.5.2, specifically, eq. (10.111). Let α_{g1} be the absorptivity of steam (850 K, 1 atm) with respect to the radiation that arrives from $T_1 = 1000$ K. We calculate in order:

$$P_w L_e \frac{T_1}{T_g} = 0.591 \text{ atm·ft} \frac{100 \text{ K}}{850 \text{ K}} =$$
$$= 0.695 \text{ atm·ft}$$

$\varepsilon_w = 0.19$ (Fig. 10.28, at 1000 K and $\cong$ 0.7 atm·ft)

$C_w \cong 1.44$ (Fig. 10.29, at 1 atm and $\cong$ 0.7 atm·ft)

$$\alpha_{g1} = 0.19 \times 1.44 \left(\frac{850 \text{ K}}{1000 \text{ K}}\right)^{0.45} = 0.254$$

Similarly, for the absorptivity (α_{g2}) with respect to radiation from $T_2 = 700$ K, we have

$$P_w L_e \frac{T_2}{T_g} = 0.591 \text{ atm·ft} \frac{700 \text{ K}}{850 \text{ K}} = 0.487 \text{ atm·ft}$$

$\varepsilon_w = 0.2$ (Fig. 10.28, at 700 K and $\cong$ 0.5 atm·ft)

$C_w = 1.46$ (Fig. 10.29, at 1 atm and $\cong$ 0.5 atm·ft)

$$\alpha_{g2} = 0.2 \times 1.46 \left(\frac{850 \text{ K}}{700 \text{ K}}\right)^{0.45} = 0.319$$

In conclusion, these calculations show that the true absorptivities α_{g1} and α_{g2} are 10 percent lower and, respectively, 13 percent higher than the α_g value of 0.283 assumed based on the gray gas model.

Problem 10.30. a) Gray gas means that $\alpha_g = \varepsilon_g = 0.3$, and $\tau_g = 1 - 0.3 = 0.7$. The radiation network is shown in Fig. 10.31, however, the present configuration is simpler:

$$F_{12} = 1, \quad F_{1g} = 1, \quad F_{2g} = 1$$

$$A_1 = A_2 = A$$

$$R_\Delta = \frac{\frac{1}{0.7\,A}\left(\frac{1}{0.3\,A} + \frac{1}{0.3\,A}\right)}{\frac{1}{0.7\,A} + \frac{1}{0.3\,A} + \frac{1}{0.3\,A}} = \frac{1.18}{A}$$

$$q_{1\text{-}2} = \frac{\sigma\left(T_1^4 - T_2^4\right)}{\frac{1 - 0.8}{0.8\,A} + \frac{1.18}{A} + \frac{1 - 0.7}{0.7\,A}} = \frac{\sigma A}{1.86}\left(T_1^4 - T_2^4\right)$$

$$\frac{q_{1\text{-}2}}{A} = \frac{1}{1.86}\, 5.67 \times 10^{-8}\, \frac{W}{m^2\,K^4}\left(1000^4 - 700^4\right) K^4$$

$$= 23.2\, \frac{kW}{m^2}$$

b) In order to determine T_g, we note that symmetry requires that

$$\sigma\, T_g^4 = \frac{1}{2}\left(J_1 + J_2\right) \qquad (1)$$

This can be seen on the network of Fig. 10.31, because the two resistances connected to the node $\sigma\, T_g^4$ are equal. In conclusion, we must determine J_1 and J_2 before we can calculate T_g.

The net heat current $q_{1\text{-}2}$ can be expressed in (at least) two additional ways,

$$q_{1\text{-}2} = \frac{\varepsilon_1\, A}{1 - \varepsilon_1}\left(\sigma\, T_1^4 - J_1\right)$$

$$= 4\, A\left(\sigma\, T_1^4 - J_1\right)$$

$$\frac{J_1}{\sigma} = T_1^4 - \frac{q_{1-2}/A}{4\sigma}$$

$$= 10^{12}\ K^4 - \frac{23.2 \times 10^3\ W/m^2}{4 \times 5.67 \times 10^{-8}\ W/m^2\ K^4}$$

$$= 0.898 \times 10^{12}\ K^4$$

and

$$q_{1-2} = \frac{\varepsilon_2 A}{1 - \varepsilon_2}\left(J_2 - \sigma T_2^4\right)$$

$$= 2.33\ A\left(J_2 - \sigma T_2^4\right)$$

$$\frac{J_2}{\sigma} = T_2^4 + \frac{q_{1-2}/A}{2.33\ \sigma}$$

$$= (700\ K)^4 + \frac{23.2 \times 10^3\ W/m^2}{2.33 \times 5.67 \times 10^{-8}\ W/m^2\ K^4}$$

$$= 0.417 \times 10^{12}\ K^4$$

Now we can use eq. (1),

$$T_g^4 = \frac{1}{2}\left(\frac{J_1}{\sigma} + \frac{J_2}{\sigma}\right) = 0.657 \times 10^{12}\ K^4$$

$$T_g = 900\ K$$

In conclusion, the temperature of the gas as a gray medium is closer to the higher of the two side-wall temperatures.

Problem 10.31. The configuration is similar to that of Fig. 10.31, except that the gas node σT_g^4 is not floating (it receives heat transfer from the chemical reaction). Floating now is the node of the second surface, because that is insulated on its back side. Consequently σT_2^4 falls on top of the J_2 node as there is not heat transfer between the two. The appropriate radiation network is attached. The net heat transfer rate from σT_g^4 to σT_1^4 is clearly

$$q_{g-1} = \frac{\sigma\left(T_g^4 - T_1^4\right)}{\frac{1 - \varepsilon_1}{\varepsilon_1 A} + R_\Delta}$$

where

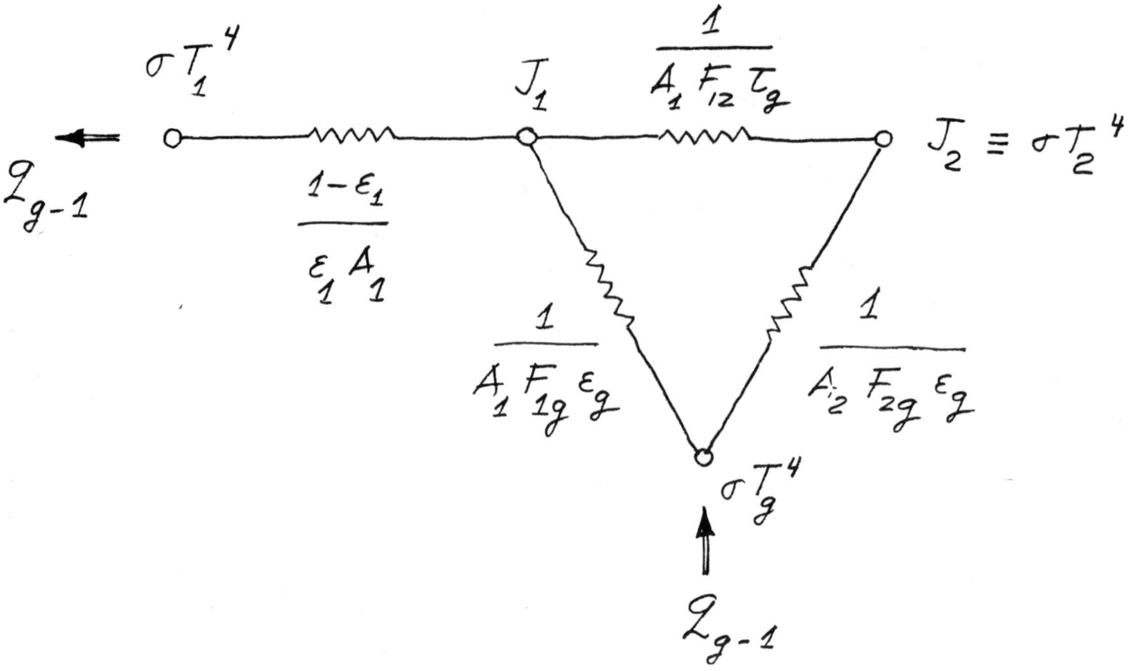

$$R_\Delta = \frac{(A_1 F_{1g} \varepsilon_g)^{-1} \left[(A_1 F_{12} \tau_g)^{-1} + (A_2 F_{2g} \varepsilon_g)^{-1} \right]}{(A_1 F_{1g} \varepsilon_g)^{-1} + (A_1 F_{12} \tau_g)^{-1} + (A_2 F_{2g} \varepsilon_g)^{-1}}$$

Problem 10.32. The configuration is similar to that of Fig. 10.31, except that the σT_2^4 node is now floating, while σT_g^4 is not. The heat transfer rate q_{g-1} proceeds from σT_g^4 to σT_1^4. The node σT_2^4 falls right on top of node J_2:

a) The present configuration is even simpler, because

$$F_{12} = 1, \qquad F_{1g} = 1, \qquad F_{2g} = 1$$

$$\varepsilon_1 = 0.5, \qquad \varepsilon_g = 0.5, \qquad \tau_g = 1 - \alpha_g = 0.5$$

The net heat transfer rate is therefore

$$q_{g-1} = \frac{\sigma \left(T_g^4 - T_1^4 \right)}{\dfrac{1 - \varepsilon_1}{\varepsilon_1 A} + R_\Delta} \tag{1}$$

where

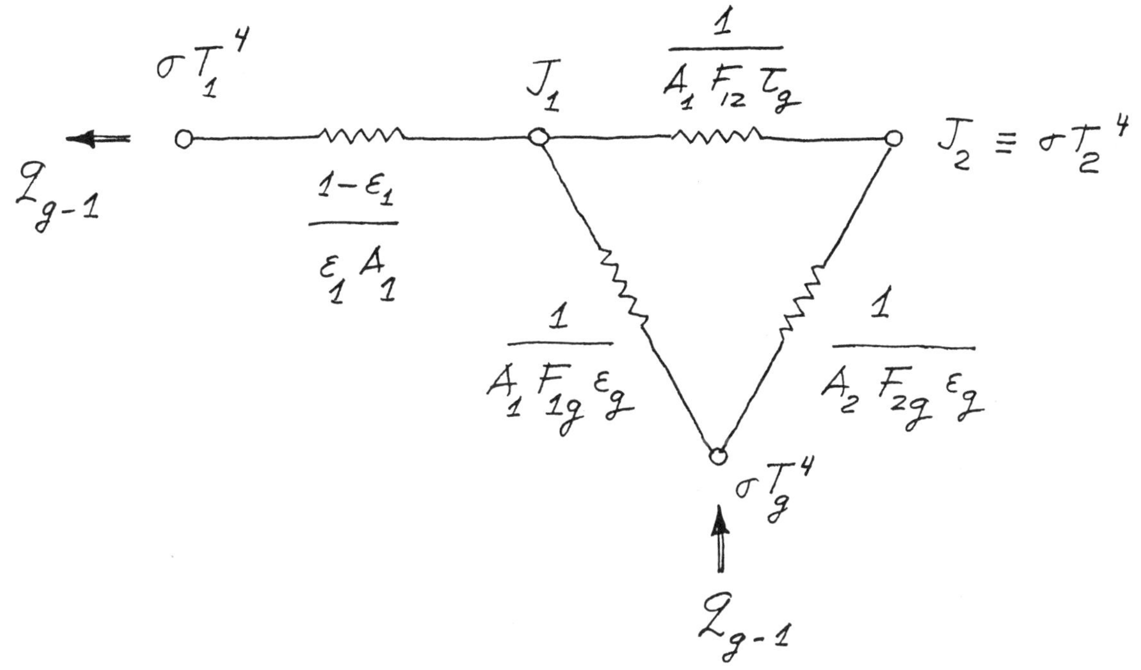

$$R_\Delta = \frac{1}{A} \frac{\frac{1}{\varepsilon_g}\left(\frac{1}{\tau_g} + \frac{1}{\varepsilon_g}\right)}{\frac{1}{\varepsilon_g} + \frac{1}{\tau_g} + \frac{1}{\varepsilon_g}} = \frac{1}{A} \frac{2\,(2+2)}{2+2+2} = \frac{4/3}{A}$$

Equation (1) yields

$$q_{g-1} = \frac{\sigma\left(T_g^4 - T_1^4\right)}{\frac{1-0.5}{0.5\,A} + \frac{4/3}{A}} = \frac{3}{7}\sigma A\left(T_g^4 - T_1^4\right) \qquad (2)$$

b) In the configuration with the vacuum gap we have, cf. eq. (10.84),

$$q_{g-1} = \frac{\sigma A\left(T_g^4 - T_1^4\right)}{\frac{1}{0.5} + \frac{1}{0.5} - 1} = \frac{1}{3}\sigma A\left(T_g^4 - T_1^4\right) \qquad (3)$$

We divide eqs. (2) and (3) side by side,

$$\frac{q_{g-1}\ (\text{gas})}{q_{g-1}\ (\text{vacuum})} = \frac{3}{7}\,3 = 1.29$$

and conclude that the heat transfer rate is 29 percent higher when the heat source fills the entire space between the two plates [part (a)], than when the heat source is flattened into a sheet [part (b)].

Project 10.1. a) This project is being proposed here as a bridge between this heat transfer course (just completed) and the thermodynamics course that preceded it. The steady-state operation of the reversible heat engine sandwiched between the collector (source) and radiator (sink) is ruled by the first law,

$$\dot{W} = q_H - q_L \tag{1}$$

and the second law

$$\frac{q_H}{T_H} = \frac{q_L}{T_L} \tag{2}$$

Combined, eq. (1) and (2) produce the formula for $\dot{W}$: The maximization of $\dot{W}$ is the objective of this project,

$$\dot{W} = q_H \left(1 - \frac{T_L}{T_H}\right) \tag{3}$$

When the solar radiation that reaches A_H is concentrated "ideally", A_S is the only surface that can be seen from A_H, therefore $F_{HS} = 1$ i.e. there is no heat transfer between A_H and the T_∞-cold background,

$$q_H = \underset{\text{net}}{q_{S-H}} = \sigma A_H \left(T_S^4 - T_H^4\right) \tag{4}$$

The radiator A_L can exchange heat only with the T_∞ background, therefore $F_{L\infty} = 1$, and

$$q_L = \underset{\text{net}}{q_{L-\infty}} = \sigma A_L \left(T_L^4 - T_\infty^4\right) \tag{5}$$

Since T_∞ is of the order of a few degrees Kelvin, $T_\infty{}^4$ can be neglected next to $T_L{}^4$ in eq. (5):

$$q_L \cong \sigma A_L T_L^4 \tag{6}$$

By eliminating q_H and q_L between eqs. (4), (6) and the second law (2), we obtain a relationship between T_L and T_H,

$$\xi - 1 = \frac{A_L}{A_H}\left(\frac{T_L}{T_H}\right)^3 \tag{7}$$

in which ξ is shorthand for

$$\xi = \left(\frac{T_S}{T_H}\right)^4 \tag{8}$$

And, if we take into account the total area constraint,

$$A_H + A_L = A \text{ (fixed)} \tag{9}$$

or

$$A_H = \alpha A \tag{10}$$

$$A_L = (1 - \alpha)\ A \tag{11}$$

the relationship between T_L and T_H becomes

$$\frac{T_L}{T_H} = \left[\frac{\alpha}{1 - \alpha}(\xi - 1)\right]^{1/3} \tag{12}$$

In the same notation, the heat input to the engine is written as

$$q_H = \sigma A T_S^4\ \alpha\left(1 - \frac{1}{\xi}\right) \tag{13}$$

The instantaneous power output, $\dot{W}$, eq. (3), can now be nondimensionalized as follows:

$$\underbrace{\left(\frac{\dot{W}}{\sigma A\ T_S^4}\right)}_{\widetilde{W}} = \alpha\left(1 - \frac{1}{\xi}\right)\left\{1 - \left[\frac{\alpha}{1 - \alpha}(\xi - 1)\right]^{1/3}\right\} \tag{14}$$

The dimensionless power output $\widetilde{W}$ emerges a function of two parameters, ξ (or T_H, when T_S is fixed), and α (the way in which the area inventory is split between A_H and A_L). The function $\widetilde{W}$ has a maximum with respect to both ξ and α, namely (after the numerical maximization of $\widetilde{W}$):

$$\xi_{opt} = 1.538 \tag{15}$$

$$\alpha_{opt} = 0.35 \tag{16}$$

$$\widetilde{W}_{max} = 0.0414 \tag{17}$$

According to this design, the collector (A_H) uses only about 1/3 of the total surface area available, which means that A_L is about twice the size of A_H.

In terms of absolute temperatures, if the solar surface is black at $T_S = 5762$ K, then $\xi_{opt} = 1.538$ means that $T_H = 5174$ K, and after using eq. (12), $T_L = 3423$ K. These temperatures are well above the limits that can be withstood by construction materials. In an actual design, these temperatures will be considerably lower not only because of the material constraints but also because of the nonideality of the A_H collector (i.e. the leakage of radiation from T_H to the T_∞-cold background). This and additional effects (e.g. the fact that A_H and A_L are not black) can be analyzed in appropriately stated reformulations of this basic thermal design problem.

b) For example, when A_H, the sun and the background form a three-surface enclosure, the q_H expression (4) is replaced by:

$$q_H = A_H F_{HS} \sigma (T_S^4 - T_H^4) \quad \underbrace{A_H \; F_{H\infty}}_{1 - F_{HS}} \quad \sigma \underbrace{(T_H^4 - T_\infty^4)}_{\text{neglect}}$$

$$= \sigma A_H (F_{HS} T_S^4 - T_H^4) \tag{18}$$

This shows that the new "constant" $F_{HS} T_S^4$ replaces the constant T_S^4 of part (a). The results of the $\dot{W}$ maximization procedure can be written down while looking at eqs. (15)-(17):

$$\frac{F_{HS} T_S^4}{T_H^4} = 1.538 \tag{19}$$

$$\frac{A_H}{A} = 0.35 \tag{20}$$

$$\dot{W} = 0.0414 \, \sigma A \, F_{HS} T_S^4 \tag{21}$$

In the case of T_S = 5762 K, the numerical values of the collector and radiator are

$$T_H = 5174 \, F_{HS}^{1/4} \text{ K} \qquad T_L = 3423 \, F_{HS}^{1/4} \text{ K} \tag{21}$$

In conclusion, the position occupied by the power plant on the temperature scale drops according to the factor $F_{HS}{}^{1/4}$. For example, if $F_{HS} = 10^{-4}$ the optimum power plant design should be sandwiched between T_H = 517 K and T_L = 342 K.

Project 10.2. The radiation and convection paths are indicated in the attached figure. It is assumed that the forced convection (h_2, h_3, h_4) is effected by an air stream whose temperature remains approximately equal to the ambient (T_4) during its flow through the system. This is why the forced convection resistances associated with h_2, h_3 and h_4 are connected to the temperature node T_4. Note the difference between convection nodes (type T) and radiation nodes (type σT^4): the convection resistances cannot be drawn directly on the network formed by the radiation resistances.

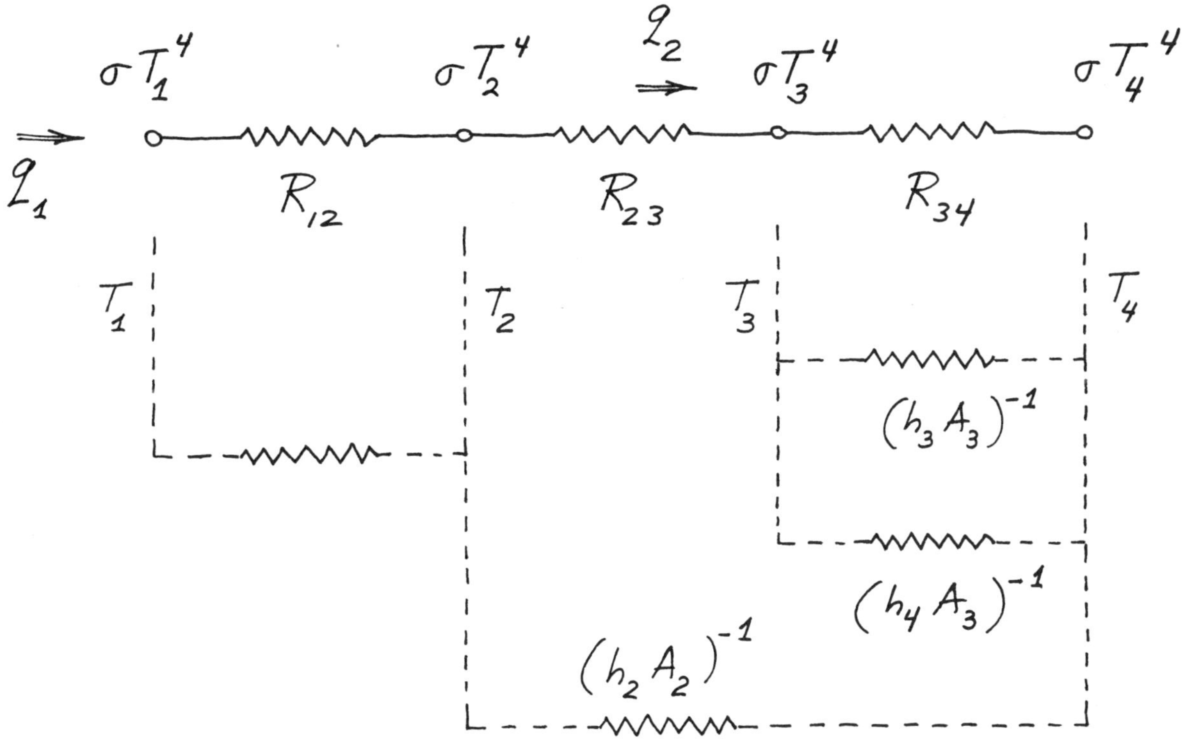

Associated with this drawing are the following quantities:

$$R_{12} = \frac{1 - \varepsilon_1}{\varepsilon_1 A_1} + \frac{1}{A_1 F_{12}} + \frac{1 - \varepsilon_2}{\varepsilon_2 A_2}$$

$$= \frac{1 - 0.9}{0.9 \times 11 \text{ m}^2} + \frac{1}{11 \text{ m}^2} + \frac{1 - 0.6}{0.6 \times 27 \text{ m}^2}$$

$$= \frac{1}{7.94 \text{ m}^2}$$

$$R_{23} = \frac{1 - \varepsilon_2}{\varepsilon_2 A_2} + \frac{1}{A_2 F_{23}} + \frac{1 - \varepsilon_3}{\varepsilon_3 A_3}$$

$$= \frac{1 - 0.6}{0.6 \times 27 \text{ m}^2} + \frac{1}{27 \text{ m}^2} + \frac{1 - 0.6}{0.6 \times 52 \text{ m}^2} =$$

$$= \frac{1}{13.4 \text{ m}^2}$$

$$R_{34} = \frac{1 - \varepsilon_3}{\varepsilon_3 A_3} + \frac{1}{A_3 F_{34}} =$$

$$= \frac{1 - 0.6}{0.6 \times 52\ m^2} + \frac{1}{52\ m^2} = \frac{1}{31.2\ m^2}$$

The convection resistances $(h_3 A_3)^{-1}$ and $(h_4 A_3)^{-1}$ are connected in parallel: their equivalent is

$$\frac{1}{UA_3} = \frac{(h_3 A_3)^{-1} (h_4 A_3)^{-1}}{(h_3 A_3)^{-1} + (h_4 A_3)^{-1}} = \frac{1}{2h_{FC} A_3}$$

where h_3 and h_4 have been assumed equal to the same forced-convection coefficient h_{FC}.

The analysis consists of writing the following equations for q_1 and q_2:

$$q_1 = h_1 A_2 (T_1 - T_2) + \frac{1}{R_{12}} \sigma\left(T_1^4 - T_2^4\right)$$

$$q_1 = h_2 A_2 (T_2 - T_4) + \frac{1}{R_{23}} \sigma\left(T_2^4 - T_3^4\right)$$

$$q_1 = \dot{m}c (T_{in} - T_1)$$

$$q_2 = UA (T_3 - T_4) + \frac{1}{R_{34}} \sigma\left(T_3^4 - T_4^4\right)$$

$$q_2 = \frac{1}{R_{23}} \sigma\left(T_2^4 - T_3^4\right)$$

By eliminating q_1 and q_2 we obtain a system for only T_1, T_2, and T_3,

$$\left.\begin{aligned} C_1 (T_1 - T_2) + C_4 \left(T_1^4 - T_2^4\right) - C_3 (T_{in} - T_1) = 0 \\ C_2 (T_2 - T_4) + C_5 \left(T_2^4 - T_3^4\right) - C_3 (T_{in} - T_1) = 0 \\ C_6 (T_3 - T_4) + C_7 \left(T_3^4 - T_4^4\right) - C_5 \left(T_2^4 - T_3^4\right) = 0 \end{aligned}\right\} \quad (S)$$

in which T_{in} = 1173 K, and the coefficients assume the following values:

$$C_1 = h_1 A_2 = 5 \frac{W}{m^2\ K} 27\ m^2 = 135 \frac{W}{K}$$

$$C_2 = h_2 A_2 = 5 \frac{W}{m^2\ K} 27\ m^2 = 135 \frac{W}{K}$$

$$C_3 = \dot{m}c = \frac{550 \text{ kg}}{3600 \text{ s}} \; 1 \; \frac{10^3 \text{J}}{\text{kg K}} = 152.8 \frac{\text{W}}{\text{K}}$$

$$C_4 = \frac{\sigma}{R_{12}} = 5.67 \times 10^{-8} \frac{\text{W}}{\text{m}^2 \text{ K}^4} \; 7.94 \text{ m}^2 = 45 \times 10^{-8} \frac{\text{W}}{\text{K}^4}$$

$$C_5 = \frac{\sigma}{R_{23}} = 5.67 \times 10^{-8} \frac{\text{W}}{\text{m}^2 \text{ K}^4} \; 13.4 \text{ m}^2 = 76 \times 10^{-8} \frac{\text{W}}{\text{K}^4}$$

$$C_6 = UA_3 = 2 \times 5 \frac{\text{W}}{\text{m}^2 \text{ K}} \; 52 \text{ m}^2 = 520 \frac{\text{W}}{\text{K}}$$

$$C_7 = \frac{\sigma}{R_{34}} = 5.67 \times 10^{-8} \frac{\text{W}}{\text{m}^2 \text{ K}^4} \; 31.2 \text{ m}^2 = 176.9 \frac{\text{W}}{\text{K}^4}$$

Note the use of $h_2 = h_3 = h_4 = h_{FC} = 5\text{W/m}^2\text{K}$ in the evaluation of C_2 and C_6.

The numerical solution to the system (S) (three equations, for T_1, T_2, T_3) is listed in the following table. The same calculation was repeated for two alternative values of the forced-convection heat transfer coefficient. The table shows that the shield temperature is indeed sensitive to the h_{FC} value, and that an h_{FC} of order $10\text{W/m}^2\text{K}$ or more is needed in order to maintain the shield temperature at a safe level (below 70°C).

h_{FC} $\left(\frac{\text{W}}{\text{m}^2\text{K}}\right)$	T_1 (K)	T_2 (K)	T_3 (K)
2.5	691	565	408
5	675	532	371
10	654	488	339

This project could be extended by proposing to the student to choose an air flowrate and a geometry for the space between T_2 and T_3, in order to achieve the needed h_{FC} value revealed by the present analysis.

Project 10.3. a) The natural convection heat current is decoupled from (i.e. flows in parallel with) the radiation heat current because the end temperatures for both currents are specified (T_1,T_2).

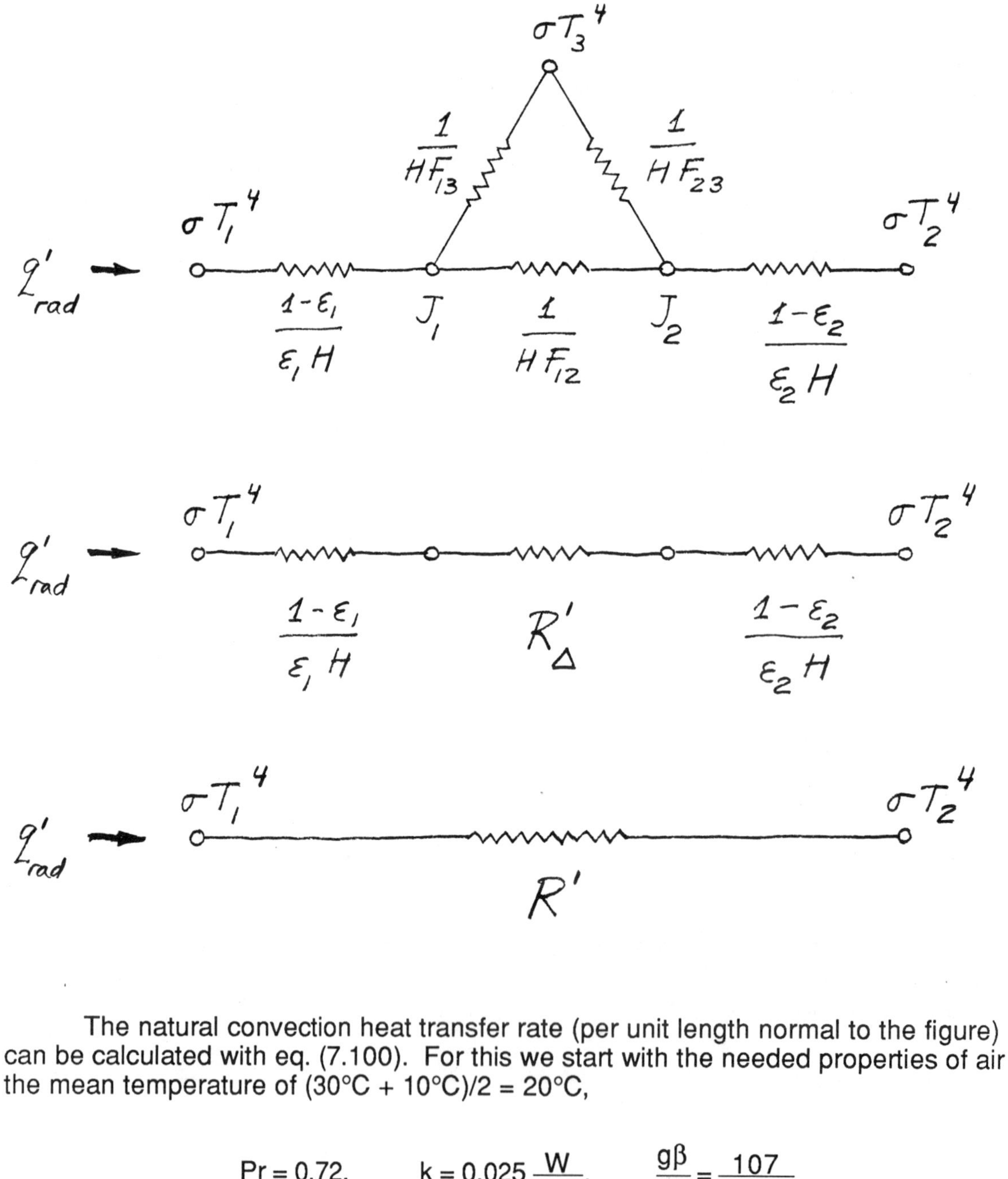

The natural convection heat transfer rate (per unit length normal to the figure) can be calculated with eq. (7.100). For this we start with the needed properties of air at the mean temperature of (30°C + 10°C)/2 = 20°C,

$$Pr = 0.72, \qquad k = 0.025 \frac{W}{m \cdot K}, \qquad \frac{g\beta}{\alpha\nu} = \frac{107}{cm^3\,K}$$

$$Ra_H = \frac{g\beta}{\alpha\nu} H^3 (T_1 - T_2)$$

$$= \frac{107}{cm^3\,K} (300\text{ cm})^3 (30 - 10)\text{ K} = 5.78 \times 10^{10}$$

(turbulent)

$$\overline{Nu}_H = 0.18 \frac{0.72}{0.2 + 0.72} Ra_H^{0.29} \qquad (7.100)$$

$$= 0.141\, Ra_H^{0.29} = 186.3$$

$$\overline{Nu}_H = \frac{\overline{q''}\, H}{k (T_1 - T_2)}$$

$$q'_{conv} = \overline{q''} H = k (T_1 - T_2)\, \overline{Nu}_H$$

$$= 0.025 \frac{W}{m \cdot K} (30 - 10)\text{ K } 186.3$$

$$= 93.1 \frac{W}{m}$$

The enclosure is made up of three surfaces, the third of which (top + bottom walls) is reradiating. The radiative heat transfer network is attached: it is the same as in Example 10.6. We calculate, in order,

$F_{12} \cong 0.41$, from Fig. 10.11 with $\frac{Y}{H} = 1$ and $\frac{X}{H} \geq 30$

$$F_{13} = 1 - F_{12} = 0.59$$

$F_{23} = F_{13} = 0.59$, because of symmetry

The following quantities are expressed per unit length in the direction perpendicular to the plane of the figure. We use in the reverse order the results of Example 10.6:

$$R'_s = \frac{1}{H\,F_{13}} + \frac{1}{H\,F_{23}} = \frac{1}{3m}\left(\frac{1}{0.59} + \frac{1}{0.59}\right) = \frac{1.13}{m} \qquad (c)$$

$$R'_\Delta = \frac{R'_s \dfrac{1}{H\,F_{12}}}{R'_s + \dfrac{1}{H\,F_{12}}} = \frac{\dfrac{1.13}{m}\,\dfrac{1}{3m\;0.41}}{\dfrac{1.13}{m} + \dfrac{1}{3m\;0.41}} = \frac{0.473}{m} \qquad (b)$$

$$R' = \frac{1-0.9}{0.9 \times 3m} + \frac{0.473}{m} + \frac{1-0.9}{0.9 \times 3m} = \frac{0.547}{m}$$

$$q'_{rad} = \frac{\sigma\left(T_1^4 - T_2^4\right)}{R'}$$

$$= \frac{m}{0.547} 5.67 \times 10^{-8} \frac{W}{m^2\,K^4} \left[(273+30)^4 - (273+10)^4\right] K^4$$

$$= 208.8 \frac{W}{m}$$

The total heat transfer rate is the sum of the contributions made by convection and radiation,

$$q' = q'_{conv} + q'_{rad} \cong 302 \frac{W}{m}$$

This shows that radiation accounts for almost 70 percent of the total heat transfer rate.

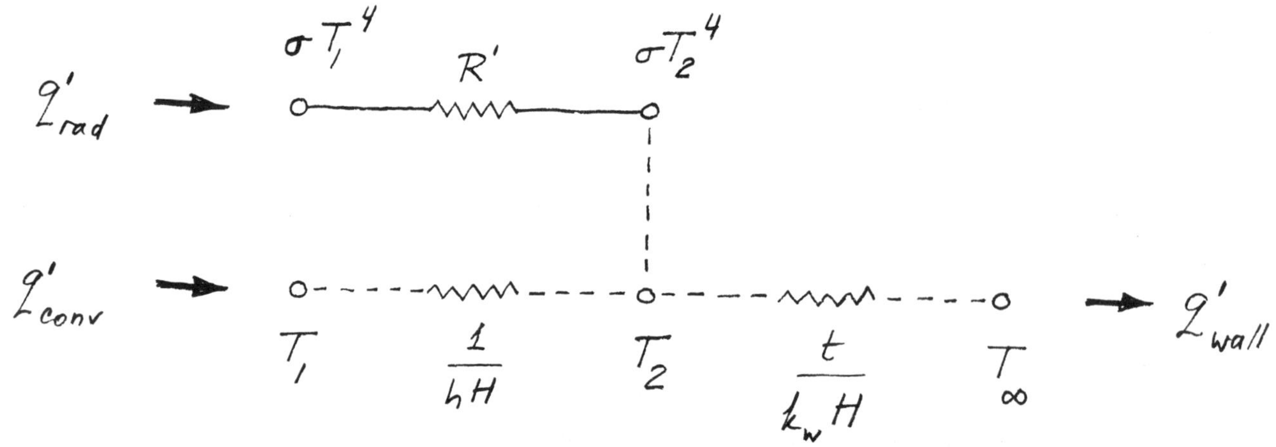

b) When the brick wall is present, the inner wall temperature T_2 floats to some unknown level between $T_1 = 30°C$ and $T_\infty = 10°C$. The heat transfer processes (convection + radiation) inside the cavity are of the same kind (i.e. in parallel) as in part (a). This time, however, q'_{conv} and q'_{rad} are driven from T_1 (given) to T_2 (unknown). The calculations outlined in part (a) (for $T_2 = 10°C$) can be rewritten for any T_2 in this manner:

$$q'_{conv} = 93.1\ \frac{W}{m}\frac{(T_1 - T_2)^{1+0.29}}{(T_1 - T_\infty)^{1+0.29}}$$

$$= 93.1\ \frac{W}{m}\frac{(1-\tau)^{1.29}}{\left(1 - \frac{273+10}{273+30}\right)^{1.29}}$$

$$= 3102.3\ \frac{W}{m}(1-\tau)^{1.29}$$

where

$$\tau = \frac{T_2}{T_1}, \qquad (T_1,\ T_2 \text{ in degrees Kelvin})$$

and

$$q'_{rad} = 208.8\ \frac{W}{m}\frac{T_1^4 - T_2^4}{T_1^4 - T_\infty^4}$$

$$= 208.8\ \frac{W}{m}\frac{1-\tau^4}{1-\left(\frac{283}{303}\right)^4}$$

$$= 873.6\ \frac{W}{m}\left(1-\tau^4\right)$$

The heat current sum $q'_{conv} + q'_{rad}$ flows through the brick wall of fixed conductance,

$$q'_{conv} + q'_{rad} = q'_{wall} \qquad (1)$$

where

$$q'_{wall} = k_w \frac{H}{t}(T_2 - T_\infty)$$

$$= k_w \frac{H}{t} T_1 \left(\tau - \frac{T_\infty}{T_1}\right)$$

$$\cong 0.45 \frac{W}{m \cdot K}\ \frac{3m}{0.3m}\ 303\ K\left(\tau - \frac{283}{303}\right)$$

$$= 1363.5\ \frac{W}{m}(\tau - 0.934)$$

Substituting the three q' expressions into eq. (1) we obtain an equation for τ,

$$873.6\ (1-\tau^4) + 3102.3\ (1-\tau)^{1.29} = 1363.5\ (\tau - 0.934)$$

This can be solved by trial and error, and the result is

$$\tau = 0.9842$$

or

$$T_2 = 298 \text{ K} \cong 25°\text{C}$$

In conclusion, 75 percent of the overall temperature difference takes place across the brick wall. The total heat transfer rate is

$$q'_{wall} = 1363.5 \frac{\text{W}}{\text{m}}(0.9842 - 0.934)$$
$$= 68.45 \frac{\text{W}}{\text{m}}$$

The brick wall reduces the heat transfer rate to 23 percent of the heat transfer rate calculated in part (a).

Project 10.4. Consider one "elementary" unit of the fibrous insulation, namely the space between the midplanes of shield i and i + 1,

$$\frac{D}{2} + s + \frac{D}{2} = s + D = \frac{L}{n} \qquad (1)$$

On the right side of eq. (1), n is the (large) number of shields,

$$n = \frac{L}{s + D} = \frac{T_0 - T_L}{T_i - T_{i+1}} \qquad (2)$$

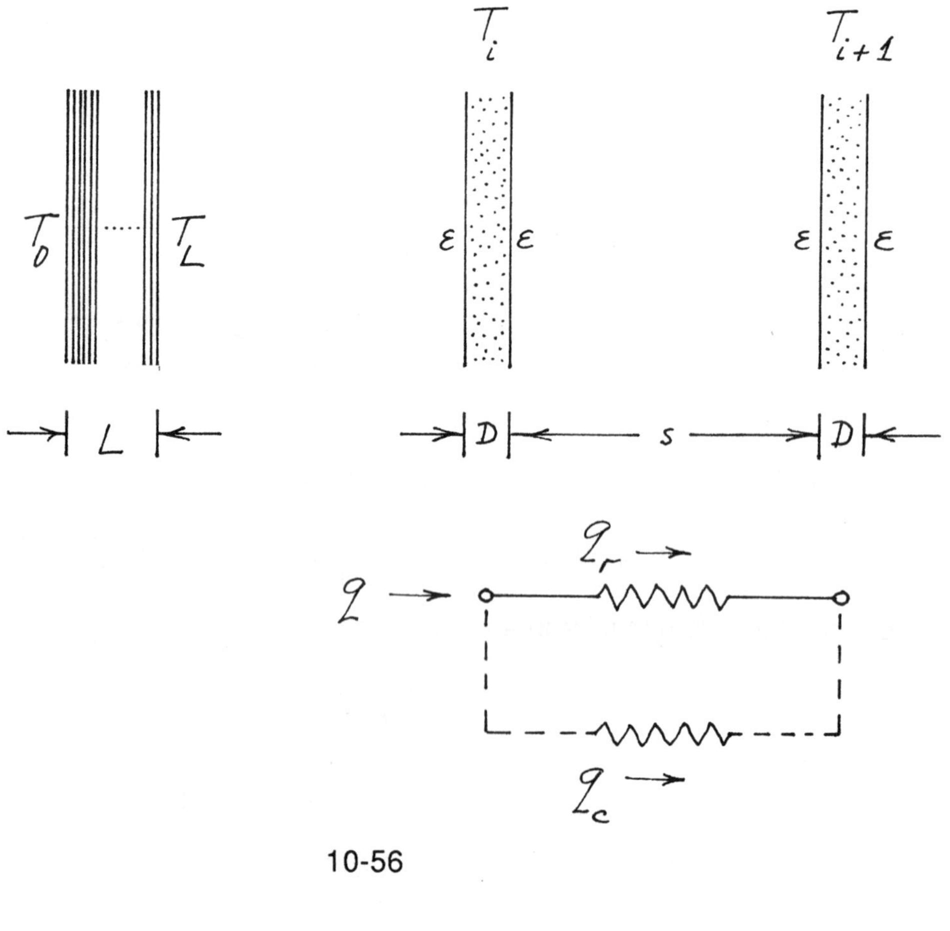

According to the network shown in the sketch, the heat transfer rate is due to the parallel flow of radiation through transparent air (q_r) and conduction through air (q_c),

$$q = q_r + q_c \tag{3}$$

where

$$q_r = \frac{\sigma A \left(T_i^4 - T_{i+1}^4\right)}{\frac{2}{\varepsilon} - 1} \cong \frac{\sigma A}{\frac{2}{\varepsilon} - 1} 4\, T_i^3 \left(T_i - T_{i+1}\right) \tag{4}$$

$$q_c = k \frac{A}{s} \left(T_i - T_{i+1}\right) \tag{5}$$

In eq. (5), k is the thermal conductivity of air. In developing the right side of eq. (4), we have assumed that the fiber-to-fiber temperature difference is much smaller than the local absolute temperature,

$$\frac{T_i - T_{i+1}}{T_i} << 1 \tag{6}$$

By combining eqs. (1)-(5), it is possible to write the total heat transfer rate q as a function of the spacing s,

$$q = k \frac{A}{L} \left(T_0 - T_L\right) \left(\frac{D}{s} + B\right) \left(\frac{s}{D} + 1\right) \tag{7}$$

where B is the dimensionless group

$$B = \frac{\sigma\, 4\, T_i^3\, D}{k\left(\frac{2}{\varepsilon} - 1\right)} \tag{8}$$

Equation (7) shows that $q \sim s^{-1}$ when s is small, and $q \sim s$ when s is large, i.e. that q has a minimum with respect to s. We find that optimal spacing by solving $\partial q / \partial s = 0$,

$$s_{opt} = D\, B^{-1/2} \tag{9}$$

$$q_{min} = k \frac{A}{L} \left(T_0 - T_L\right) \left(1 + B^{1/2}\right)^2 \tag{10}$$

In conclusion, eq. (9) shows that the optimal spacing increases as $D^{1/2}$. The porosity of the fibrous structure is

$$\varphi = \frac{s}{s + D} \tag{11}$$

therefore

$$\varphi_{opt} = \frac{1}{1 + B^{1/2}} \tag{12}$$

Numerical example:

$\varepsilon \cong 1$ $\quad\quad$ $\sigma = 5.67 \times 10^{-8} \dfrac{W}{m^2 K^4}$

$T_i \cong 300$ K $\quad\quad$ $k = 0.025 \dfrac{W}{m \cdot K}$

Equations (9) and (12) yield the following values:

D	10 μm	100 μm
s_{opt}	202 μm	639 μm
φ_{opt}	0.95	0.86

Worth noting here is that 100-100μm represents the D range of animal hair, and that the observed (natural) porosity of animal fur is in the range 0.95-0.99. This porosity is comparable with the φ_{opt} values calculated above.

Project 10.5. i) The total heat transfer rate from T_i to T_∞ can be expressed while looking at the cylindrical shell,

$$q = \frac{2\pi k L}{\ln \frac{r_o}{r_i}} (T_i - T_o) \tag{1}$$

or by looking at the net radiation outside the shell, cf. eq. (10.86),

$$q = \sigma\, 2\pi\, r_o\, L\, \varepsilon \left(T_o^4 - T_\infty^4\right) \tag{2}$$

We eliminate q between eqs. (1) and (2), and obtain an implicit relationship between the outer temperature T_o and the outer radius r_o,

$$\frac{1 - \theta}{\ln \rho} = B\, \rho\, \theta^4 \tag{3}$$

where

$$\theta = \frac{T_o}{T_i} < 1 \tag{4}$$

$$\rho = \frac{r_o}{r_i} > 1 \tag{5}$$

$$B = \frac{\varepsilon \sigma r_i}{k} T_i^3 \tag{6}$$

Equation (3) is based on the additional assumption that

$$\left(\frac{T_\infty}{T_i}\right)^4 << \theta^4 \tag{7}$$

Equation (3) can be solved by trial-and-error to determine the outer temperature, (θ) as a function of the outer radius (ρ) for a fixed set of properties (B), in other words,

$$\theta = \theta(\rho, B) \tag{8}$$

This result can then be substituted in eq. (1) to calculate the total heat transfer rate,

$$\frac{q}{2\pi k L} = \frac{1 - \theta(\rho, B)}{\ln \rho} \tag{9}$$

Numerically one finds that when B is lower than approximately 0.2, the heat transfer rate reaches a maximum at the critical outer radius $\rho_c = r_{o,c}/r_i$ listed in the table. When $B \gtrsim 0.2$, there is no maximum: the heat transfer rate decreases monotonically as ρ increases, i.e. as more insulation is added.

B	ρ_c
0.2	2.06
0.1	9.59
0.05	32.9

ii) Numerical example:

$k = 0.02 \frac{W}{m \cdot K}$ $\qquad \varepsilon = 0.2$

$r_i = 10$ cm $\qquad T_i = 500$ K

Since B is greater than roughly 0.2,

$$B = \frac{\varepsilon \sigma r_i}{k} T_i^3 = 7.1,$$

the addition of a layer of insulation will lead to a decrease in the heat transfer rate q.

Chapter 11

MASS TRANSFER PRINCIPLES

Problem 11.1. Starting with the mole fraction definition (11.14), we write in order

$$x_i = \frac{N_i}{N} = \frac{m_i}{M_i}\frac{M}{m}$$

$$= \frac{m_i}{M_i}\frac{M}{m}\frac{V}{V}$$

in which $\rho_i = m_i/V$, and $\rho = m/V$. In conclusion, the right hand side reduces to

$$x_i = \rho_i \frac{M}{\rho M_i}$$

Problem 11.2. a) In order to calculate the concentration of hydrogen on the solid side of each surface we need at estimate for the solubility of hydrogen in neoprene. From Table 11.6 we learn that

$$S \cong 2.1 \times 10^{-3} \frac{\text{kmol}}{\text{m}^3\cdot\text{bar}}$$

therefore, we can invoke now eq. (11.46) to calculate:

$$C_0 = SP_0 = 2.1 \times 10^{-3} \frac{\text{kmol}}{\text{m}^3\cdot\text{bar}} \, 2 \text{ bar}$$

$$= 0.0042 \text{ kmol/m}^3$$

$$C_L = SP_L = 2.1 \times 10^{-3} \frac{\text{kmol}}{\text{m}^3\cdot\text{bar}} \, 1 \text{ bar}$$

$$= 0.0021 \text{ kmol/m}^3$$

On the gas side of each surface, the hydrogen may be modeled as an ideal gas at T = 300 K,

$$PV = N\overline{R}T \qquad \text{(a)}$$

where $\overline{R}$ = 8.314 kJ/kmol·K is the universal ideal-gas constant, V is the volume of one gas sample, and N is the number of moles of hydrogen in that sample. The ratio

$$C = \frac{N}{V} \tag{b}$$

represents the molar concentration of hydrogen in the sample, therefore, combining eqs. (a) and (b) we obtain

$$C = \frac{P}{\overline{R}T} \tag{c}$$

Applying this result to the gas side of each surface leads to

$$C_{0,gas} = \frac{P_0}{\overline{R}T} = \frac{2\ \text{bar}}{300\ \text{K}} \frac{\text{kmol·K}}{8.314\ \text{kJ}}$$

$$= 0.08\ \text{kmol/m}^3$$

$$C_{L,gas} = \frac{P_L}{\overline{R}T} = \frac{1\ \text{bar}}{300\ \text{K}} \frac{\text{kmol·K}}{8.314\ \text{kJ}}$$

$$= 0.04\ \text{kmol/m}^3$$

b) We can now sketch the concentration distribution through and in the vicinity of the interface, while keeping in mind the linear profile indicated by eq. (11.29). Note the C discontinuity across each neoprene-hydrogen interface.

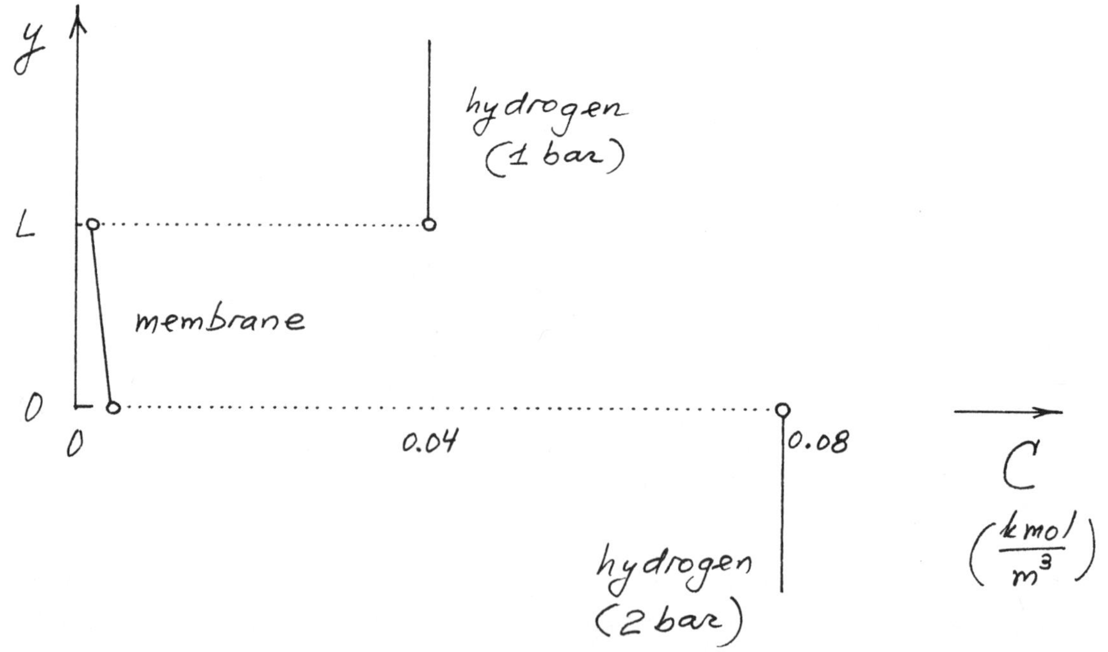

c) Table 11.4 shows that $D \cong 4.31 \times 10^{-6} cm^2/s$. The molar flux through the neoprene membrane is therefore

$$\hat{j}_{H_2} = D\frac{C_0 - C_L}{L} = 4.31 \times 10^{-6}\frac{cm^2}{s}\frac{(0.0042 - 0.0021)\ kmol/m^3}{0.1\ mm}$$

$$= 9.05 \times 10^{-9}\frac{kmol}{m^2 s}$$

The corresponding mass flux is (note that $M_{H_2} = 2.018$ kg/kmol):

$$j_{H_2} = M_{H_2}\hat{j}_{H_2} = 2.018\frac{kg}{kmol}\,9.05 \times 10^{-9}\frac{kmol}{m^2 s}$$

$$= 1.83 \times 10^{-8}\frac{kg}{m^2 s}$$

Problem 11.3. The needed property values are supplied by Tables 11.4 and 11.6:

$$D \cong 4.5 \times 10^{-11}\frac{cm^2}{s} \qquad S = 3.4 \times 10^{-4}\frac{kmol}{m^3 bar}$$

The helium concentrations on the solid side of each surface are

$$C_0 = SP_0 = 3.4 \times 10^{-4}\frac{kmol}{m^3 bar}\,2\ bar$$

$$= 6.8 \times 10^{-4}\frac{kmol}{m^3}$$

$C_L = SP_L = 0$, because $P_L = 0$ (the tube is surrounded by vacuum)

The molar flowrate of helium (per unit length of tube) is given by eq. (11.33), in which $r_L = 2$ mm and $r_0 = 1.5$ mm:

$$\dot{N}'_{He} = \frac{2\pi D}{\ln (r_L / r_0)} C_0$$

$$= \frac{2\pi}{\ln (2/1.5)} 4.5 \times 10^{-11} \frac{cm^2}{s} 6.8 \times 10^{-4} \frac{kmol}{m^3}$$

$$= 6.7 \times 10^{-17} \frac{kmol}{s \cdot m}$$

Problem 11.4. With reference to the left side of Fig. 11.5 we write

$$\rho_L = 10^{-2} \frac{g}{cm^3}, \qquad \rho_0 = 0$$

and j_{H_2} for the mass flux of hydrogen through the wall of thickness L = 1mm:

$$j_{H_2} = D \frac{\rho_L - \rho_0}{L}$$

$$= 10^{-5} \frac{cm^2}{s} \frac{10^{-2}\ g/cm^3}{0.1\ cm} = 10^{-6} \frac{g}{cm^2\ s}$$

$$= 10^{-5} \frac{kg}{m^2 \cdot s}$$

The corresponding number of cubic meters at 1 atm that leak through the wall (per $m^2 \cdot s$) is obtained by dividing j_{H_2} by the density of hydrogen at standard conditions (1 atm, 25°C, Appendix D):

$$\frac{j_{H_2}}{\rho_{H_2}} = \frac{10^{-5} \frac{kg}{m^2 \cdot s}}{0.0823 \frac{kg}{m^3}} = 1.2 \times 10^{-4} \frac{m^3 \text{ (at 1 atm, 25°C)}}{m^2 \cdot s}$$

$$= 121.5 \frac{cm^3 \text{ (at 1 atm, 25°C)}}{m^2 \cdot s}$$

Observation: Appendix D lists the density of n-Hydrogen at 1 atm and 300 K. Instead of interpolating linearly in the table to find ρ at 25°C (298.15 K), it is advisable to use the ideal-gas equation of state $P = \rho RT$, which for P = constant (1 atm) reduces to:

$$\rho\,(1\text{ atm}, 298.15\text{ K}) = \frac{300\text{ K}}{298.15\text{ K}}\,\rho\,(1\text{ atm}, 300\text{ K})$$

$$= \frac{300}{298.15}\,0.0818\,\frac{\text{kg}}{\text{m}^3} = 0.0823\,\frac{\text{kg}}{\text{m}^3}$$

Problem 11.5. According to eq. (11.37), the mass transfer rate through the spherical shell is

$$\dot{m} = \frac{4\pi D}{r_0^{-1} - r_L^{-1}}(\rho_0 - \rho_L) \qquad (1)$$

where

$$\rho_0 = M_{H_2O}\,\frac{\rho}{M}\,x_0 \qquad (2)$$

$$\rho_L = M_{H_2O}\,\frac{\rho}{M}\,x_L \qquad (3)$$

and the ratio (ρ/M) refers to the mixture of dry air and water vapor. The value of ρ/M is roughly the same as for dry air (1 atm, 20°C)

$$\frac{\rho}{M} = \frac{1.205\text{ kg/m}^3}{28.97\text{ kg/kmol}} = 0.0416\,\frac{\text{kmol}}{\text{m}^3}$$

On the inner surface, the air is saturated with water vapor,

$$x_0 = \frac{P_{sat}\,(20°\text{C})}{1\text{ atm}} = \frac{0.02339\text{ bar}}{1.01325\text{ bar}}$$

$$= 0.0231$$

On the outside, the air has 10 percent relative humidity,

$$P_L = 0.1\,P_{sat}\,(20°\text{C}) = 0.00234\text{ bar}$$

$$x_L = \frac{P_L}{1\text{ atm}} = 0.0023$$

The corresponding water vapor densities are given by eqs. (2) and (3),

$$\rho_0 = 18.015\,\frac{\text{kg}}{\text{kmol}}\,0.0416\,\frac{\text{kmol}}{\text{m}^3}\,0.0231 = 0.0173\,\frac{\text{kg}}{\text{m}^3}$$

$$\rho_L = 18.015\,\frac{\text{kg}}{\text{kmol}}\,0.0416\,\frac{\text{kmol}}{\text{m}^3}\,0.0023 = 0.0017\,\frac{\text{kg}}{\text{m}^3}$$

and the water mass flowrate is given by eq. (1):

$$\dot{m} = \frac{4\pi \times 2.6 \times 10^{-5} \frac{m^2}{s}}{\left(\frac{1}{4} - \frac{1}{5}\right)\frac{1}{mm}} (0.0173 - 0.0017) \frac{kg}{m^3}$$

$$\sim 10^{-4} \frac{g}{s}$$

Problem 11.6. a) The properties of interest are listed in Tables 11.4 and 11.6:

$$D \cong 4 \times 10^{-10} \frac{cm^2}{s} \qquad S = 4 \times 10^{-4} \frac{kmol}{m^3 \cdot bar}$$

The helium concentrations on the glass side of each spherical surface are

$$C_0 = SP_0 = 4 \times 10^{-4} \frac{kmol}{m^3 \cdot bar} 1 \text{ bar}$$

$$= 4 \times 10^{-4} \frac{kmol}{m^3}$$

$C_L = SP_L = 0$, because $P_L = 0$ (the helium presence is negligible in the surrounding air)

Equation (11.36) shows how to calculate the instantaneous molar flowrate of the helium that diffuses through the spherical shell:

$$\dot{N} = \frac{4\pi D}{r_0^{-1} - r_L^{-1}} C_0 = \frac{4\pi}{\left(\frac{1}{1} - \frac{1}{2.5}\right)\frac{1}{mm}} 4 \times 10^{-10} \frac{cm^2}{s} 4 \times 10^{-4} \frac{kmol}{m^3}$$

$$= 3.35 \times 10^{-19} \frac{kmol}{s}$$

b) Let N represent the instantaneous number of moles of helium trapped inside the spherical shell. Treating this system as an ideal gas,

$$N = \frac{PV}{\bar{R}T} \qquad \text{(i)}$$

where the group $V/\bar{R}T$ is a constant, we note that N decreases to N/2 when the pressure drops from P to P/2. To trace the history of the amount of trapped helium is analogous to plotting the decay of the helium pressure P versus time.

An equation for P(t) is obtained by combining eq. (i) with the mass (moles, actually) conservation statement

$$\dot{N} = -\frac{dN}{dt} \qquad \text{(ii)}$$

in which $\dot{N}$ is the instantaneous flowrate of the escaping helium, cf. eq. (11.36),

$$\dot{N} = \frac{4\pi D}{r_0^{-1} - r_L^{-1}} SP \qquad \text{(iii)}$$

The result is the differential equation

$$\frac{dP}{P} = -a\,dt \qquad \text{(iv)}$$

in which "a" is shorthand for the constant group

$$a = \frac{4\pi DS}{r_0^{-1} - r_L^{-1}} \frac{\overline{R}T}{V} = 2 \times 10^{-9}\ s^{-1}$$

In the numerical evaluation of this group, we used $\overline{R}$ = 8.314 kJ/kmol·K, T = 300 K, and $V = 4\pi(1\ mm)^3/3$. Integrated from the initial condition

$$P = 1 \text{ bar, at } t = 0$$

eq. (iv) yields a formula for the pressure history, namely

$$\ln\left(\frac{P}{1\ bar}\right) = -a\,t$$

The time interval needed for P to decrease to 0.5 bar is

$$t = \frac{\ln(0.5)}{-a} = 3.5 \times 10^8 \text{ seconds}$$

$$= 4011 \text{ days} \cong 11 \text{ years}$$

Problem 11.7. Since the bottom surface is impermeable to air, the water layer can be viewed as the L-thin half of the plate sketched in Fig. 11.8. Let C_0 represent the concentration of air in water at the surface exposed to the atmosphere, that is where the water is saturated with air. The water volume has absorbed 50 percent of all the air that it is capable of absorbing, when

$$\overline{C} = \frac{1}{2} C_0$$

Substituting this $\overline{C}$ value and $C_{in} = 0$ in the ordinate parameter of Fig. 11.8 we obtain

$$\frac{C_0 - \overline{C}}{C_0 - C_{in}} = \frac{1}{2}$$

The top curve of Fig. 11.8 or, more precisely, eq. (11.53), shows that the time that corresponds to this event is

$$\frac{D}{L^2} t = 0.196$$

In the case of L = 1 cm and $D = 2.5 \times 10^{-9} m^2/s$ (the diffusivity of air in water, Table 11.3), the actual time interval is

$$t = 0.196 \frac{L^2}{D} = 0.196 \frac{1 \text{ cm}^2}{2.5 \times 10^{-9} m^2/s}$$

$$= 7840 \text{ s} = 2.18 \text{ hours}$$

Problem 11.8. a) Rewriting the left side of eq. (11.51) in terms of mass concentrations, we obtain in order

$$\frac{50 - 120}{16 - 120} = \text{erf}\left[\frac{y}{2(D \cdot t)^{1/2}}\right]$$

$$0.673 = \text{erf}\left[\frac{y}{2(D \cdot t)^{1/2}}\right]$$

$$0.694 = \frac{y}{2(D \cdot t)^{1/2}}$$

The diffusion time corresponding to the depth y = 0.5 mm is therefore

$$t = \left(\frac{y}{2 \times 0.694}\right)^2 \frac{1}{D}$$

$$= \frac{0.25 \times 10^{-6} m^2}{1.927} \frac{s}{6 \times 10^{-10} m^2} = 216 \text{ s}$$

$$= 3.6 \text{ minutes}$$

b) When the average concentration of carbon reaches 100 kg/m^3, the ordinate parameter of Fig. 11.8 reaches the value

$$\frac{120 - 100}{120 - 16} = 0.192$$

The lowest curve of Fig. 11.8 shows that the dimensionless time that corresponds to this value is

$$\frac{D}{r_0^2} t \cong 0.117$$

A more precise estimate can be obtained by using eq. (11.55),

$$\frac{D}{r_0^2} t = 0.1166$$

or, after substituting $r_0 = 0.5$ cm and $D = 6 \times 10^{-10} m^2/s$,

$$t = 0.1166 \frac{0.25 \times 10^{-4} m^2}{6 \times 10^{-10} m^2/s}$$

$$= 4859 \text{ s} = 81 \text{ minutes}$$

Problem 11.9. The required time is the time of mass diffusion across the wall of thickness L,

$$t \sim \frac{L^2}{D}$$

The diffusivity is, approximately (Table 11.4),

$$D \sim 4 \times 10^{-10} \frac{cm^2}{s}$$

therefore the penetration time is

$$t \sim \frac{(0.2\ \text{cm})^2}{4 \times 10^{-10}\ \frac{\text{cm}^2}{\text{s}}} = 10^8\ \text{s} \cong 3\ \text{years}$$

In the case of a polyethylene wall, D is of the order of 3×10^{-6} cm^2/s, therefore

$$t \sim \frac{(0.2\ \text{cm})^2}{3 \times 10^{-6}\ \frac{\text{cm}^2}{\text{s}}} = 1.3 \times 10^4\ \text{s} \cong 3.7\ \text{hours}$$

Problem 11.10. When the volume averaged concentration of carbon reaches 110 kg/m^3, the ordinate parameter in Fig. 11.8 reaches the value

$$\frac{120 - 110}{120 - 16} = 0.0962$$

This value is off the graph; instead, we use eq. (11.54) and write, in order,

$$0.0962 = \frac{4}{(2.405)^2} \exp\left[- (2.405)^2 \frac{D}{r_o^2} t\right]$$

$$\ln (0.139) = - (2.405)^2 \frac{D}{r_o^2} t$$

$$\frac{D \cdot t}{r_o^2} = 0.341$$

$$t = 0.341 \frac{(0.01\ \text{m})^2}{6 \times 10^{-10}\ \text{m}^2/\text{s}} = 5.7 \times 10^4\ \text{s}$$

$$= 15.8\text{h}$$

Problem 11.11. The diffusivity of atomic hydrogen in nickel at 400°C can be estimated based on the information given in Table 11.4:

$$D = 0.0045\ \frac{cm^2}{s}\ \exp\left[-\frac{36\ 000\ \frac{kJ}{kmol}}{8.314\ \frac{kJ}{kmol \cdot K}\ (400 + 273.15)\ K}\right]$$

$$= 7.24 \times 10^{-6}\ \frac{cm^2}{s}$$

For the degassing time $D \cdot t/L^2$ we cannot use Fig. 11.8 because the abscissa is not long enough (note that the value on the ordinate would be 0.1). Instead, we use eq. (11.53), which is particularly accurate at long times:

$$\frac{0 - 0.1}{0 - 1} = \frac{8}{\pi^2} \exp\left(-\frac{\pi^2}{4}\frac{D}{L^2}t\right)$$

$$-2.093 = -\frac{\pi^2}{4}\frac{D}{L^2}t$$

$$0.848 = \frac{D}{L^2}t$$

$$t = 0.848\ \frac{L^2}{D}$$

$$= 0.848\ \frac{(0.1\ cm)^2}{7.24 \times 10^{-6}\ \frac{cm^2}{s}} = 1172 s = 19.5\ \text{minutes}$$

Problem 11.12. The lower part of Table 11.4 shows that the diffusivity of carbon in austenite is

$$D = D_0 \exp\left(-\frac{Q}{\overline{R}T}\right)$$

$$= 0.2\ \frac{cm^2}{s}\ \exp\left[-\frac{138\ 200\ kJ/kmol}{8.314\ \frac{kJ}{kmol \cdot K}\ (900 + 273.15)\ K}\right]$$

$$= 1.4 \times 10^{-7}\ \frac{cm^2}{s}$$

The required time is delivered by eq. (11.51), in which $C_0 = 0$ and $C = \frac{1}{2} C_{in}$:

$$\frac{\frac{1}{2}C_{in} - 0}{C_{in} - 0} = \text{erf}\left[\frac{y}{2\,(D\cdot t)^{1/2}}\right]$$

$$\frac{y}{2\,(D\cdot t)^{1/2}} = 0.477$$

$$t = \frac{y^2}{4D\,(0.477)^2}$$

$$= \frac{(0.1\ \text{cm})^2}{4 \times 1.4 \times 10^{-7}\ \frac{\text{cm}^2}{\text{s}}\ (0.477)^2} = 78\,483\text{s} = 21.8\text{h}$$

Problem 11.13. The distance of penetration by diffusion is

$$\delta_1 \sim (D_1 \cdot t)^{1/2}$$

for which the diffusivity is (Table 11.4):

$$D_1 = 0.02\ \frac{\text{cm}^2}{\text{s}}\ \exp\left[-\frac{84\,200\ \text{kJ/kmol}}{8.314\ \frac{\text{kJ}}{\text{kmol}\cdot\text{K}}\ (900 + 273.15)\ \text{K}}\right]$$

$$= 3.56 \times 10^{-6}\ \frac{\text{cm}^2}{\text{s}}$$

To double the penetration depth for the same time t we must have

$$\frac{\delta_2}{\delta_1} = 2 = \left(\frac{D_2}{D_1}\right)^{1/2}$$

therefore

$$D_2 = 4\,D_1 = 1.425 \times 10^{-5}\ \frac{\text{cm}^2}{\text{s}}$$

$$1.425 \times 10^{-5}\ \frac{\text{cm}^2}{\text{s}} = 0.02\ \frac{\text{cm}^2}{\text{s}}\ \exp\left(-\frac{Q}{\bar{R}T_2}\right)$$

$$\frac{Q}{\bar{R}T_2} = 7.25$$

$$T_2 = \frac{84\,200\ \text{kJ/kmol}}{8.314\ \frac{\text{kJ}}{\text{kmol}\cdot\text{K}}\ 7.25} = 1398\ \text{K} = 1124°\text{C}$$

Problem 11.14. The relevant properties of atmospheric air at 25°C are

$$\rho_a = 1.185 \frac{kg}{m^3} \qquad \nu_a = 1.55 \times 10^{-5} \frac{m^2}{s}$$

The Reynolds number based on pool length shows that the boundary layer flow is turbulent,

$$Re_L = \frac{U_\infty L}{\nu_a} = 30 \frac{10^3 m}{3600 s} \frac{3m}{1.55 \times 10^{-5}\ m^2/s}$$

$$= 1.61 \times 10^6 \qquad \text{(turbulent)}$$

Tables 11.2 and 11.7 show that the properties of water vapor in air are

$$D = 2.88 \times 10^{-5} \frac{m^2}{s} \qquad Sc = 0.6$$

therefore, starting with eq. (11.81) we calculate

$$\overline{Sh}_L = 0.037\ Sc^{1/3} \left(Re_L^{4/5} - 23\ 550\right)$$

$$= 0.037\ (0.6)^{1/3} \left[\left(1.61 \times 10^6\right)^{4/5} - 23\ 550\right]$$

$$= 2147$$

$$\overline{h}_m = \overline{Sh}_L \frac{D}{L} = 2147 \frac{2.88 \times 10^{-5}\ m^2/s}{3m}$$

$$= 0.0206 \frac{m}{s}$$

The instantaneous water flowrate removed by the wind is

$$\dot{m} = \overline{h}_m A \left(\rho_w - \rho_\infty\right)$$

in which $A = (3m)^2 = 9m^2$. For calculating the water density at the pool surface (ρ_w) we note that the saturation pressure of water vapor at 25°C is

$$P_{sat}\ (25°C) = 3169 \frac{N}{m^2}$$

and that the pressure of the air-water vapor mixture is 1 atm = $1.0133 \times 10^5\ N/m^2$. The mole fraction of water vapor at the surface is

$$x_w = \frac{P_{sat}\,(25°C)}{1\ \text{atm}} = \frac{3169}{1.0133 \times 10^5}$$

$$= 0.0313$$

The water vapor density (ρ_w) that corresponds to this mole fraction can be calculated by using eq. (11.24) and, later, eq. (11.23):

$$\rho_w = M_{H_2O}\, C_w$$

$$= M_{H_2O} \frac{\rho_a}{M_a} x_w = 18.02 \frac{\rho_a}{28.97}\, 0.0313$$

$$= 0.0195\, \rho_a$$

Turning our attention to the calculation of ρ_∞, we recall that the relative humidity is defined as the ratio

$$\phi = \frac{P_v}{P_{sat}(T)}$$

In our case, $\phi_\infty = 0.3$ and $P_{sat}(25°C) = 3169\ N/m^2$, therefore the partial pressure of water vapor outside the concentration boundary layer is

$$P_{v,\infty} = 0.3 \times 3169 \frac{N}{m^2} = 951 \frac{N}{m^2}$$

Beyond this point, the calculation of ρ_∞ follows the steps used earlier in the calculation of ρ_w:

$$x_{,\infty} = \frac{951}{1.0133 \times 10^5} = 0.0094$$

$$\rho_\infty = \frac{M_{H_2O}}{M_a} x_\infty\, \rho_a = \frac{18.02}{28.97}\, 0.0094\, \rho_a$$

$$= 0.0058\, \rho_a$$

Now we have all the necessary information for calculating the water mass transfer rate:

$$\dot{m} = \bar{h}_m A (\rho_w - \rho_\infty)$$

$$= 0.0206 \frac{m}{s} 9m^2 (0.0195 - 0.0058)\ 1.185 \frac{kg}{m^3}$$

$$= 0.003 \frac{kg}{s}$$

Let Δt be the time interval in which the pool water level drops by $\Delta z = 2mm$,

$$\dot{m}\, \Delta t = \rho_{water} A\, \Delta z$$

The density of water at 25°C is 997 kg/m^3, therefore

$$\Delta t = 997 \frac{kg}{m^3} \frac{9m^2\ 0.002\ m}{0.003\ kg/s}$$

$$= 5982\ s = 1.66h$$

Problem 11.15. The penetration time for pure mass diffusion would be

$$t \sim \frac{x^2}{D} = \frac{(5m)^2}{10^{-5}\ m^2/s} \sim 2.5 \times 10^6 s$$

$$\sim 29 \text{ days}$$

This time interval is six orders of magnitude greater than the observed one (10s), therefore the mass transfer mechanism is not that of diffusion.

In the case of convection, if the air flow (draft) proceeds parallel to the length $x = 5m$, the velocity scale of this flow is

$$U \sim \frac{x}{t} \sim \frac{5m}{10s} = 0.5 \frac{m}{s}$$

Problem 11.16. In the following analysis ρ represents the density of water vapor (kg/m^3) in the "humid air" mixture. The flowrate of water vapor through the duct inlet is $UA_c\rho_{in}$, while the corresponding flowrate through the outlet is $UA_c\rho_{out}$. The rate at which the air stream removes water from the duct surface is therefore

$$\dot{m} = UA_c\rho_{out} - UA_c\rho_{in}$$

$$= UA_c(\rho_{out} - \rho_{in}) \quad (1)$$

This result shows that we must first determine ρ_{out} if we are to calculate the water removal rate $\dot{m}$. By following the steps of the heat transfer analysis of section 6.4, we account first for the conservation of species (water vapor) in a duct element of length dx:

$$j_w \, p \, dx = UA_c \, d\rho \tag{2}$$

The wall mass flux j_w is proportional to the local difference $(\rho_w - \rho)$, where ρ is the local bulk density of water vapor in the stream,

$$j_w = h_m (\rho_w - \rho) \tag{3}$$

By eliminating j_w between eqs. (2) and (3) we obtain

$$\frac{d\rho}{\rho_w - \rho} = \frac{h_m p}{UA_c} dx \tag{4}$$

Integrating once

$$-\ln(\rho_w - \rho) = \frac{h_m p}{UA_c} x + C \tag{5}$$

and invoking the inlet condition ($\rho = \rho_{in}$ at $x = 0$),

$$-\ln(\rho_w - \rho_{in}) = 0 + C \tag{6}$$

leads to the distribution of water vapor along the stream, $\rho(x)$:

$$\frac{\rho_w - \rho}{\rho_w - \rho_{in}} = \exp\left(-\frac{h_m \, px}{UA_c}\right) \tag{7}$$

In particular, $\rho = \rho_{out}$ at $x = L$, and eq. (7) becomes

$$\frac{\rho_w - \rho_{out}}{\rho_w - \rho_{in}} = \exp\left(-\frac{h_m}{U}\frac{A}{A_c}\right) \tag{8}$$

where $A = pL$ is the total mass transfer area. In conclusion, eq. (8) contains the ρ_{out} answer that is needed for continuing with eq. (1):

$$\dot{m} = UA_c(\rho_w - \rho_{in})\left[1 - \exp\left(-\frac{h_m}{U}\frac{A}{A_c}\right)\right] \tag{9}$$

Problem 11.17. a) First, we determine the characteristics of the channel flow:

$$D_h = 2d = 2 \text{ cm} \qquad \text{(hydraulic diameter)}$$

$$Re_{D_h} = \frac{UD_h}{\nu_a} = 0.1 \frac{m}{s} \frac{0.02 \text{ m}}{1.55 \times 10^{-5} \text{ m}^2/\text{s}}$$

$$= 129 \qquad \text{(laminar flow)}$$

$$X \cong 0.05\, D_h\, Re_{D_h} = 0.05 \times 0.02\text{m} \times 129$$

$$\cong 0.13\text{m} \qquad \text{(flow entrance length)}$$

The entrance length is much shorter than the length of the channel ($X << L$), therefore over most of the length L the flow is fully developed and laminar.

b) The heat-transfer analog of the present problem is the parallel-plate channel with isothermal walls (the penultimate entry in Table 6.1):

$$Nu_{D_h} = 7.54 = Sh_{D_h} = \frac{h_m D_h}{D}$$

Noting that for water vapor in atmospheric air at 25°C the mass diffusivity is $D = 2.88 \times 10^{-5} \text{ m}^2/\text{s}$, we can finally calculate the mass transfer coefficient:

$$h_m = \frac{D}{D_h} Sh_{D_h} = \frac{2.88 \times 10^{-5} \text{ m}^2/\text{s}}{0.02\text{m}} 7.54$$

$$= 0.0109 \frac{m}{s}$$

c) For the total rate of water removal from the channel walls we turn to the formula derived in the preceding problem,

$$\dot{m} = UA_c(\rho_w - \rho_{in})\left[1 - \exp\left(-\frac{h_m}{U}\frac{A}{A_c}\right)\right]$$

where

$$\rho_{in} = 0 \frac{kg}{m^3} \qquad \text{(dry air at inlet)}$$

$$A = 2\,LW \qquad \text{(total mass transfer area)}$$

W = wall width, perpendicular to the plane of the figure

$$A_c = Wd \qquad \text{(flow cross-section)}$$

The water vapor density in the humid air mixture right at the wall (ρ_w) can be calculated in the following steps:

$$P_{sat}(25°C) = 3169 \frac{N}{m^2} \qquad \text{(pressure of saturated water vapor at 25°C)}$$

$$x_w = \frac{P_{sat}(25°C)}{1 \text{ atm}} = \frac{3169 \text{ N/m}^2}{1.0133 \times 10^5 \text{ N/m}^2}$$

$$= 0.0313 \qquad \text{(mole fraction of water vapor at the wall)}$$

$$\rho_w = M_{H_2O}\, c_w$$

$$= M_{H_2O} \frac{\rho_a}{M_a} x_w = 18.02 \frac{\rho_a}{28.97} 0.0313$$

$$= 0.0195\, \rho_a = 0.0195 \times 1.185 \frac{kg}{m^3}$$

$$= 0.0231 \frac{kg}{m^3}$$

The $\dot{m}$ expression becomes

$$\dot{m} = 0.1 \frac{m}{s} W\, 0.01m\, 0.0231 \frac{kg}{m^3}\left[1 - \exp\left(- \frac{0.0109 \text{ m/s}}{0.1 \text{ m/s}} \frac{2W\, 1.5m}{W\, 0.01m}\right)\right]$$

$$= W\, 2.31 \times 10^{-5} \frac{\text{kg/s}}{\text{m}}$$

d) The total water that is initially present on the two walls is

$$m = 2WL\,\delta\,\rho_{\text{liquid water at 25°C}}$$

$$= 2W\ 1.5\text{m}\ 10^{-5}\text{m}\ 997\ \frac{\text{kg}}{\text{m}^3}$$

$$= W\ 0.03\ \frac{\text{kg}}{\text{m}}$$

The time needed for removing this quantity of water is

$$t = \frac{m}{\dot{m}} = \frac{W\ 0.03\ \text{kg/m}}{W\ 2.31 \times 10^{-5}\ (\text{kg/m})/\text{s}}$$

$$= 1300\ \text{s} \cong 22\ \text{minutes}$$

<u>Problem 11.18</u>. The total rate of water removal is

$$\dot{m} = UA_c(\rho_w - \rho_{in})\left[1 - \exp\left(-\frac{h_m}{U}\frac{A}{A_c}\right)\right] \qquad (1)$$

in which $A_c = sW$, $A = LW$ (note: only one side is wet), and

$$\frac{h_m A}{U A_c} = \frac{h_m LW}{U sW} = \frac{0.0175\ \frac{\text{m}}{\text{s}}}{0.5\ \frac{\text{m}}{\text{s}}}\ \frac{1.5\text{m}}{0.4 \times 10^{-2}\text{m}}$$

$$= 13.13$$

W = the width of the channel, perpendicular to the flow

$$\rho_w = \left(43360\ \frac{\text{cm}^3}{\text{g}}\right)^{-1} = 0.0231\ \frac{\text{kg}}{\text{m}^3},$$ (the mass density of saturated water vapor at 25°C, from saturated steam tables)

$$\rho_{in} = 0.2 \times 0.0231\ \frac{\text{kg}}{\text{m}^3} = 0.0046\ \frac{\text{kg}}{\text{m}^3},$$ (the mass density of water vapor in humid air at 20 percent relative humidity)

We obtain:

$$\frac{\dot{m}}{W} = 0.5\,\frac{m}{s}\,0.004m\,(0.0231 - 0.0046)\,\frac{kg}{m^3}\,[1 - \exp(-13.13)] \cong 2 \times 10^{-6}$$

$$= 3.7 \times 10^{-5}\,\frac{kg/s}{m} \qquad (2)$$

The humidity at the outlet can be calculated by invoking the overall conservation of water as a species,

$$\dot{m} = UA_c\left(\rho_{out} - \rho_{in}\right) \qquad (3)$$

By eliminating $\dot{m}$ between eqs. (1) and (3), and noting that the quantity in square brackets in eq. (3) is essentially equal to 1, we conclude that

$$\rho_{out} = \rho_w = 0.0231\,\frac{kg}{m^3}$$

The outflowing stream is saturated with water vapor.

Problem 11.19. The relevant properties of air, and water vapor in air (at 20°C and 1 atm) are

$$\nu = 0.15\,\frac{cm^2}{s}$$

$$D = 2.6 \times 10^{-5}\,\frac{m^2}{s}$$

$$Sc = 0.6$$

The average mass transfer coefficient is delivered by eq. (11.90), in which

$$Re_{D_o} = \frac{U_\infty D_o}{\nu} = 100\,\frac{cm}{s}\,1\,cm\,\frac{s}{0.15\,cm^2}$$

$$= 667$$

$$\overline{Sh}_{D_o} = 2 + \left(0.4\,Re_{D_o}^{1/2} + 0.06\,Re_{D_o}^{2/3}\right)Sc^{0.4}$$

$$= 14.15$$

$$\frac{\bar{h}_m D_o}{D} = \overline{Sh}_{D_o}$$

$$\bar{h}_m = 14.15\,\frac{2.6 \times 10^{-5}\,\frac{m^2}{s}}{0.01m} = 0.037\,\frac{m}{s}$$

The air boundary layer around the nut, and the porous shell form two mass-transfer resistances in series (in the radial direction), respectively, $1/\bar{h}$ and L/D, where L = 1mm (the thickness of the shell). The ratio of the two represents a "mass-transfer Biot number",

$$\frac{L/D}{1/\bar{h}} = \frac{\bar{h}L}{D} = \frac{0.037 \frac{m}{s}\, 0.001m}{2.6 \times 10^{-5} \frac{m^2}{s}} = 1.4$$

which shows that the two resistances are of the same order of magnitude.

Problem 11.20. The pertinent data for humid air at 25°C and 1 atm are

$\nu \cong 0.15 \frac{cm^2}{s}$	(Appendix D, dry air)
$D = 2.6 \times 10^{-5} \frac{m^2}{s}$	(Table 11.2)
$Sc = 0.6$	(Table 11.7)
$\rho_w = 0.0231 \frac{kg}{m^3}$	(saturated steam at 25°C)
$\rho_\infty = 0.2\, \rho_w = 0.0046 \frac{kg}{m^3}$	(water vapor in air, at 20 percent relative humidity)

We rely on eq. (11.90), and calculate in order

$$Re_{D_o} = \frac{U_\infty D_o}{\nu} = 31.3 \frac{m}{s}\, 0.074m \frac{s}{0.15 \times 10^{-4}\, m^2}$$

$= 1.54 \times 10^5$, which is only slightly outside the range of eq. (11.90)

$$\overline{Sh}_{D_o} = 2 + \left(0.4\, Re_{D_o}^{1/2} + 0.06\, Re_{D_o}^{2/3}\right) Sc^{0.4}$$

$$= 2 + 329.8\, (0.6)^{0.4} = 270.9$$

$$\bar{h}_m = \overline{Sh}_{D_o} \frac{D}{D_o} = 270.9 \frac{2.6 \times 10^{-5} \frac{m^2}{s}}{0.074m}$$

$$= 0.0952 \frac{m}{s}$$

$$\bar{j}_w = \bar{h}_m (\rho_w - \rho_\infty)$$

$$= 0.0952 \frac{m}{s} (0.0231 - 0.0046) \frac{kg}{m^3} = 1.76 \frac{g}{m^2 s}$$

$$A = \pi D_o^2 = \pi (0.074 m)^2 = 0.0172 m^2$$

$$\dot{m} = \bar{j}_w A = 1.76 \frac{g}{m^2 s} 0.0172 m^2 =$$

$$= 0.0303 \frac{g}{s}$$

$$t = \frac{L}{U_\infty} = \frac{19.4 m}{31.3 m/s} = 0.62 s \quad \text{(time of travel)}$$

$$m = \dot{m} t = 0.0303 \frac{g}{s} 0.62 s$$

$$= 0.019 g \quad \text{(mass of evaporated water)}$$

The evaporated mass 0.019g has an equivalent volume of 0.019 cm^3 (note: the density of water around room temperature $\cong 1 g/cm^3$). This volume is negligible when compared with the original amount (2 cm^3) smeared all over the ball.

Problem 11.21. a) When the air is dry, the "mixture" has the same density as the partial mass concentration of dry air:

$$\rho_1 = \rho_a + \rho_v$$

$$= (1.185 + 0) \frac{kg}{m^3} = 1.185 \frac{kg}{m^3}$$

When the humid air (25°C, 1 atm) is saturated with water vapor, we have the following data:

$$x_v = 0.0313 \quad \text{(see Example 11.3)}$$

$$\rho_v = \left(43360 \frac{cm^3}{g}\right)^{-1} = 0.0231 \frac{kg}{m^3} \quad \text{(see properties of saturated steam)}$$

The mass concentration of dry air in this second mixture is related to ρ_v by

$$\rho_a = \frac{1 - x_v}{x_v} \frac{M_a}{M_v} \rho_v \qquad (1)$$

This relation can be derived by noting the ideal gas equations of state, individually, for dry air and water vapor at the mixture T and P:

$$P_a V = m_a R_a T = m_a \frac{\bar{R}}{M_a} T$$

$$P_v V = m_v R_v T = m_v \frac{\bar{R}}{M_v} T$$

Dividing side by side

$$\frac{x_a}{x_v} = \frac{\rho_a}{\rho_v} \frac{M_v}{M_a}$$

and noting that $x_a = 1 - x_v$, we arrive at eq. (1). Numerically, eq. (1) yields:

$$\rho_a = \frac{1 - 0.0313}{0.0313} \frac{28.97}{18.02} 0.0231 \frac{kg}{m^3} = 1.1493 \frac{kg}{m^3}$$

The density of the saturated humid air is

$$\rho_2 = \rho_a + \rho_v$$

$$= 1.1493 \frac{kg}{m^3} + 0.0231 \frac{kg}{m^3} = 1.1724 \frac{kg}{m^3}$$

The group $\rho\beta_c$, eq. (11.99), can be calculated by assuming that ρ varies linearly as ρ_v increases:

$$\rho\beta_c = -\left(\frac{\partial \rho}{\partial \rho_v}\right)_{T,P} \cong -\frac{\rho_2 - \rho_1}{\rho_v - 0}$$

$$= -\frac{1.1724 - 1.185}{0.0231 - 0} = 0.54$$

This $\rho\beta_c$ estimate is only 11 percent below the value 0.61 listed in Table 11.7.

When the air is dry, the mixture M is the same as for dry air,

$$M_1 = M_a = 28.97 \frac{kg}{kmol}$$

When the humid air is saturated with water vapor, the M value of the mixture is

$$M_2 = x_a M_a + x_v M_v$$

$$= (1 - 0.0313)\ 28.97 \frac{kg}{kmol} + 0.0313 \times 18.02 \frac{kg}{kmol}$$

$$= 28.63 \frac{kg}{kmol}$$

The ρ/M ratios of the dry and saturated mixtures are

$$\left(\frac{\rho}{M}\right)_{dry} = \frac{\rho_1}{M_1} = \frac{1.185\ kg/m^3}{28.97\ kg/kmol} = 0.0409 \frac{kmol}{m^3}$$

$$\left(\frac{\rho}{M}\right)_{saturated} = \frac{\rho_2}{M_2} = \frac{1.1724\ kg/m^3}{28.63\ kg/kmol} = 0.0410 \frac{kmol}{m^3}$$

These last two lines show that the ratio ρ/M is practically constant as the relative humidity increases from 0 to 100 percent.

Problem 11.22. In order to calculate the mass-transfer Rayleigh number

$$Ra_{m,H} = \frac{g H^3}{\nu D} \beta_c (\rho_w - \rho_\infty)$$

we evaluate in order

$D = 2.6 \times 10^{-5} \frac{m^2}{s}$	(Table 11.2)
$\rho\beta_c = 0.61$	(Table 11.7)
$Sc = 0.6$	(Table 11.7)
$\nu = 0.155 \frac{cm^2}{s}$	(Dry air, Appendix D)
$\rho_w = \left(43360 \frac{cm^3}{h}\right)^{-1} = 0.0231 \frac{kg}{m^3}$	(Saturated steam tables, 25·C)

The calculation of the mass concentration in the atmosphere (40 percent relative humidity) involves these steps:

$$\phi = \frac{P_{v,\infty}}{P_{sat}\ (25°C)} = 0.4$$

$$P_{v,\infty} = 0.4 \times 3169 \frac{N}{m^2} = 1268 \frac{N}{m^2}$$

$$\rho_\infty = \frac{P_{v,\infty}}{P_{v,w}} \rho_w = \frac{1268 \text{ N/m}^2}{3169 \text{ N/m}^2} 0.0231 \frac{kg}{m^3}$$

$$= 0.00924 \frac{kg}{m^3}$$

The density of humid air increases by roughly one percent as the relative humidity drops from 100 all the way to zero. Therefore it is reasonable to approximate the mixture density ρ (which appears in $\rho\beta_c$) in terms of the density of dry air at 25°C and 1 atm:

$$\rho \cong \rho_a = 1.185 \frac{kg}{m^3}$$

The dimensionless group $\beta_c(\rho_w - \rho_\infty)$ that appears in the $Ra_{m,H}$ definition has the value

$$\beta_c \left(\rho_w - \rho_\infty\right) = \rho\beta_c \frac{\rho_w - \rho_\infty}{\rho}$$

$$= 0.61 \frac{0.0231 - 0.00924}{1.185} = 0.00713$$

The Rayleigh number becomes

$$Ra_{m,H} = \frac{9.81 \frac{m}{s^2} (1m)^3}{2.6 \times 10^{-5} \frac{m^2}{s} 0.155 \times 10^{-4} \frac{m^2}{s}} 0.00713 = 2.43 \times 10^{10}$$

and the correlation (11.103) yields

$$\overline{Sh}_H = \left(0.825 + 0.32 \, Ra_{m,H}^{1/6}\right)^2 = 325.9$$

$$\overline{h}_m = \overline{Sh}_H \frac{D}{H}$$

$$= 325.9 \frac{2.6 \times 10^{-5} \text{ m}^2/\text{s}}{1m} = 0.00847 \frac{m}{s}$$

The evaporation rate produced by the 1m^2 vertical area is

$$\dot{m} = \overline{h}_m A (\rho_w - \rho_\infty)$$

$$= 0.00847 \frac{m}{s} 1 m^2 (0.0231 - 0.00924) \frac{kg}{m^3}$$

$$= 1.17 \times 10^{-4} \frac{kg}{s} = 0.42 \text{ kg/h}$$

The flow of humid air is upward, because the density of more humid air is smaller than the density of less humid air.

Problem 11.23. In laminar boundary layer flow driven by heat transfer along a vertical wall we have the following horizontal (entrainment) velocity scales:

$$u \sim \frac{\alpha}{H} Ra_H^{1/4}, \quad (Pr \gtrsim 1) \qquad (7.25a)$$

$$u \sim \frac{\alpha}{H} (Ra_H \, Pr)^{1/4}, \quad (Pr \lesssim 1) \qquad (7.35a)$$

The analogous scales for a laminar vertical boundary layer driven by mass transfer are

$$u \sim \frac{D}{H} Ra_{m,H}^{1/4}, \quad (Sc \gtrsim 1)$$

$$u \sim \frac{D}{H} (Ra_{m,H} \, Sc)^{1/4}, \quad (Sc \lesssim 1)$$

in other words, by

$$u \sim \frac{D}{H} Ra_{m,H}^{1/4} \, Sc^n$$

where

$$n = \begin{cases} 0 & \text{for} \quad Sc \gtrsim 1 \\ \frac{1}{4} & \text{for} \quad Sc \lesssim 1 \end{cases}$$

The horizontal mass flux of mixture fluid associated with this entrainment is

$$\rho u \sim \rho \frac{D}{H} Ra_{m,H}^{1/4} \, Sc^n \qquad (1)$$

The mass flux through the wall can be estimated in the same way, by changing the notation in

$$\overline{Nu}_H \sim Ra_H^{1/4}, \quad (Pr \gtrsim 1) \qquad (7.29)$$

$$\overline{Nu}_H \sim (Ra_H\, Pr)^{1/4}, \qquad (Pr \lesssim 1) \tag{7.37}$$

We obtain

$$\overline{Sh}_H \sim Ra_{m,H}^{1/4}\; Sc^n$$

in which

$$\overline{Sh}_H = \frac{\bar{h}_m H}{D} = \frac{\bar{j}_w H}{(\rho_{i,w} - \rho_{i,\infty}) D}$$

The mass flux through the wall is therefore

$$\bar{j}_w \sim \frac{D}{H}(\rho_{i,w} - \rho_{i,\infty})\, Ra_{m,y}^{1/4}\; Sc^n \tag{2}$$

The vertical surface may be modelled as impermeable (zero through-flow) when

$$|\bar{j}_w| << \rho u$$

or, after using eqs. (1) and (2),

$$\frac{D}{H}|\rho_{i,w} - \rho_{i,\infty}|\, Ra_{m,y}^{1/4}\; Sc^n << \rho \frac{D}{H} Ra_{m,H}^{1/4}\; Sc^n$$

$$|\rho_{i,w} - \rho_{i,\infty}| << \rho$$

In conclusion, the impermeable-surface assumption is valid when the species of interest is present in small quantities (traces) in the mixture.
